THE PRIMARY STRUCTURE
OF TRANSFER RNA

THE PRIMARY STRUCTURE OF TRANSFER RNA

T. V. VENKSTERN
Institute of Molecular Biology
Academy of Sciences of the USSR
Moscow, USSR

Translation Editor
JAMES T. MADISON
Research Chemist
United States Plant, Soil, and Nutrition Laboratory
United States Department of Agriculture
Ithaca, New York

Translated from Russian by
BASIL HAIGH

PLENUM PRESS • NEW YORK–LONDON

Tat'yana Vladimirovna Venkstern was born in Moscow in 1917. She was graduated from the Biological Faculty of Moscow University in 1941 and has the degree of Doctor of Biological Sciences. She occupies the post of Senior Research Assistant at the Institute of Molecular Biology, Academy of Sciences of the USSR. She is one of the discoverers of the primary structure of valine tRNA from bakers' yeast. At present she is studying enzymic methylation of tRNA.

The original Russian text, published by Nauka Press in Moscow in 1970, has been corrected by the author for the present edition. This translation is published under an agreement with Mezhdunarodnaya Kniga, the Soviet book export agency.

Первичная структура транспортных рибонуклеиновых кислот
Т. В. Венкстерн

PERVICHNAYA STRUKTURA TRANSPORTNYKH RIBONUKLEINOVYKH KISLOT
T. V. Venkstern

Library of Congress Catalog Card Number 79-186259

ISBN-13: 978-1-4684-1973-3   e-ISBN-13: 978-1-4684-1971-9
DOI: 10.1007/978-1-4684-1971-9

© 1973 Plenum Press, New York
Softcover reprint of the hardcover 1st edition 1973

A Division of Plenum Publishing Corporation
227 West 17th Street, New York, N. Y. 10011

United Kingdom edition published by Plenum Press, London
A Division of Plenum Publishing Company, Ltd.
Davis House (4th Floor), 8 Scrubs Lane, Harlesden, London, NW10 6SE, England

PREFACE TO THE ENGLISH EDITION

The study of the primary structure of nucleic acids, i.e., the determination of the nucleotide sequence in ribosomal, transfer, and messenger types of RNA and in DNA, is an essential preliminary to the attempt to correlate the structures of these compounds with their functions. This is one of the most urgent problems in molecular biology, for until it is solved it is impossible to understand fully the mechanisms of fundamental living processes. Research has naturally tended to concentrate on the nucleic acids of smallest molecular weight, namely the transfer RNAs, and tremendous progress has been made in recent years in the unraveling of their structure.

In 1958–1959, when structural investigation was still in its infancy, it was even doubted whether the structure of any transfer RNA could be determined. No methods of isolating homogeneous preparations were available, and it was uncertain whether the very slight physical and chemical differences between tRNAs of different specificity would be sufficient to allow successful fractionation. When the first enriched preparations were obtained, there arose the problem of whether it was possible to determine the nucleotide sequence having regard to the limited number of heterogeneous structural elements present in tRNA. These doubts were resolved in 1965, when the structure of the first nucleic acid, alanine-specific transfer RNA, was unraveled in Holley's laboratory. Only six years have elapsed since that time, but the number of transfer RNAs with an established nucleotide sequence is now more than twenty. The structure of tRNAs specific to several different amino acids is now known. In some cases the primary structure of a tRNA of identical amino-acid specificity, but obtained from different sources, has been determined. The structure of some isoacceptor tRNAs has been established. Sufficient material is thus now available to allow preliminary conclusions to be drawn regarding the architecture of molecules of this class of nucleic acid and its link with function, the structural basis of the taxonomic differences and multiplicity of the transfer RNAs.

Work on the amber-supperessor tRNA^{Tyr} and the tRNA^{Tyr} of other mutants of *Escherichia coli* has shown how closely structural research is interwoven with genetic problems. The study of the formylmethionine tRNAs helps to elucidate the mechanism of initiation of protein synthesis.

It thus follows that the determination of tRNA structure is, at least in principle, a problem already solved. With the improvement in methods of isolation of the tRNAs, with an increase in the resolving power of methods used to fractionate the various fragments of the molecules, and with the development of new methods of analysis using radioisotopes, and designed for work with the minimum quantity of material, there is every reason to suppose that in the very near future new structural formulae will be discovered, so that wider and fuller generalizations will be possible. Nevertheless it is perfectly obvious that the determination of primary structure is an extremely delicate problem, calling for considerable skill and requiring many of the latest methods of investigation. It is also clear that, despite the existence of a general plan of analysis, which has fully justified itself in the execution of this research, the study of each new tRNA brings its own specific difficulties and circumstances. There is therefore an urgent need at the present time for a general review of experience acquired during the structural investigation of the transfer RNAs. Another reason why this is necessary is because the study of the structure of macromolecular nucleic acids is at present based mainly on the methods developed for use with the transfer RNAs.

The problem of determining the primary structure of any nucleic acid can be subdivided into the following stages: isolating the nuclecic acid as a homogeneous preparation, establishing the number and type of nucleotide components of which it is composed, and identifying the order in which these components are arranged in the molecule. All these aspects of the problem are examined in this monograph. Considerable attention is paid to the comparative analysis of the primary structure of tRNA, which, although it has proven less informative than was at first considered, is nevertheless an essential stage which bridges the gap between purely structural and structural–functional investigations. Matters of functional topography are dealt with only very briefly, because this is an independent field, undergoing extremely rapid development at the present time as a result of advances made in structural research, and requiring special treatment. A separate chapter surveys the latest facts discovered in relation to the minor nucleotides, paying particular attention to their possible function and to experimental approaches to the study of this problem.

More than three years have elapsed since the publication of the Russian edition of this monograph. During this period tremendous progress has been made in the study of the primary structure of nucleic acids. I have taken the opportunity of adding new material to the book and of bringing it up to date where this was particularly necessary. However, I must confess that it was a practical impossibility to cover the whole of the necessary literature, considering the short time which I had at my disposal. It is equally certain that by the time of appearance of the English edition of the book fresh advances will have been made in the study of the transfer and the high-molecular-weight nucleic acids. I can only hope that the factual material and the research experience which are summarized in this book will prove helpful to those engaged in the study of nucleic-acid structure.

T.V.V.

CONTENTS

ABBREVIATIONS

Nucleic Acids

DNA	Deoxyribonucleic acid.
RNA	Ribonucleic acid.
tRNA, rRNA, mRNA	Transfer, ribosomal, and messenger RNAs, respectively.
$tRNA^{Ala}$, $tRNA^{Val}$, etc.	Alanine-specific tRNA, valine-specific tRNA, etc.
$tRNA_1^{Val}$, $tRNA_2^{Val}$, etc.	First $tRNA^{Val}$, second $tRNA^{Val}$, etc.
$tRNA_f^{Met}$	N-formylmethionyl tRNA.
val-tRNA	tRNA aminoacylated with valine.

Nuclear Bases and Nucleosides

Ade, Gua, Cyt, Ura, Hyp	Adenine, guanine, cytosine, uracil, and hypoxanthine, respectively.
A, G, C, U	Adenosine, guanosine, cytidine, uridine.
N	Nucleoside in general.
Pu	Purine.
Py	Pyrimidine.

Minor Components

1meA	1-Methyladenosine [9(β-D-ribofuranosyl)-1-methyl-6-aminopurine].
2meA	2-Methyladenosine [9(β-D-ribofuranosyl)-2-methyl-6-aminopurine].
6meA	N^6-Methyladenosine [9(β-D-ribofuranosyl)-6-methylaminopurine].

1

6me$_2$A	N^6-Dimethyladenosine [9(β-D-ribofuranosyl)-6-dimethylaminopurine].
1meG	1-Methylguanosine [9(β-D-ribofuranosyl)-1-methyl-2-amino-6-hydroxypurine].
2meG	N^2-Methylguanosine [9(β-D-ribofuranosyl)-2-methylamino-6-hydroxypurine].
2me$_2$G	N^2-Dimethylguanosine [9(β-D-ribofuranosyl)-2-dimethylamino-6-hydroxypurine].
7meG	7-Methylguanosine [9(β-D-ribofuranosyl)-2-amino-6-hydroxy-7-methylpurine].
I	Inosine [9(β-D-ribofuranosyl)-6-hydroxypurine].
1meI	1-Methylinosine [9(β-D-ribofuranosyl)-1-methyl-6-hydroxypurine].
ψ	Pseudouridine [5(β-D-ribofuranosyl)-2,4-dihydroxypyrimidine].
T	Ribothymidine [1(β-D-ribofuranosyl)-5-methyl-2,4-dihydroxypyrimidine].
H$_2$U	5,6-Dihydrouridine [1(β-D-ribofuranosyl)-2,4-dihydroxy-5,6-dihydropyrimidine].
5meC	5-Methylcytidine [1(β-D-ribofuranosyl)-5-methyl-2-hydroxy-4-aminopyrimidine].
3meC	3-Methylcytidine [1(β-D-ribofuranosyl)-3-methyl-2-hydroxy-4-aminopyrimidine].
4acC	N^4-Acetylcytidine [1(β-D-ribofuranosyl)-2-hydroxy-4-acetylamino-pyrimidine].
ipeA	N^6-Isopentenyladenosine [9(β-D-ribofuranosyl)-6-(3-methylbuten-2-ylamino)purine].
mtpA	2-Methylthio-N^6-isopentenyladenosine [9(β-D-ribofuranosyl)-2-methylthio-6-(3-methyl-buten-2-ylamino)purine].
2thioU	2-Thiouridine [9(β-D-ribofuranosyl)-2-thio-4-hydroxypyrimidine].
4thioU	4-Thiouridine [9(β-D-ribofuranosyl)-2-hydroxy-4-thiopyrimidine].
OmeN	Nucleosides methylated at the 2-OH group of the ribose.
thrA	9(β-D-Ribofuranosyl)-N-(purinyl-6-carbamoyl)-threonine.
oacU	Uridine-5-oxyacetic acid.
Y	A minor component from the 3′-side of the anticodon in yeast tRNA[Phe].

A*, U*, and so on   Unidentified nucleosides. Sometimes the num-
                           bers indicating the position of the substituent
                           are omitted.

Mononucleotides and Oligonucleotides

p   Phosphoryl (in the symbol of a nucleoside);
                           placed on the right it denotes 3' phosphoryl
                           (Ap represents 3'-adenylic acid); placed on
                           the left it denotes 5'-phosphoryl (pA repre-
                           sents 5'-adenylic acid); $N > p$ or Np! denotes
                           2',3'-nucleoside cyclic phosphate.
                           In oligonucleotides of known sequence (e.g.,
                           ApCpUpCp) p can be replaced by a hyphen:
                           A-C-U-Cp.
                           In some cases, e.g., in drawings of the clover-
                           leaf structure of the tRNA molecule, the
                           hyphen is omitted.
(A, C, U, C)p, etc.   Oligonucleotides of unknown sequence.
poly(A), etc.   Homopolymers.
poly(A,U), etc.   Heteropolymers.
poly(A)·poly(U), etc.   Associated chains.
P   Oligonucleotides formed by complete hydrolysis
                           of RNA by pyrimidyl RNase.
T   Oligonucleotides formed by complete hydrolysis
                           of RNA by guanyl RNase.
Metamers   Fragments formed by partial hydrolysis of
                           RNA by any RNase and including several
                           oligonucleotides.
Halves   Fragments of the tRNA molecule formed by
                           cleavage of one phosphodiester bond located
                           in the anticodon region.

Enzymes

RNase   Ribonuclease.
Pyrimidyl RNase   Pancreatic pyrimidyl RNase.
Guanyl RNase   Guanyl RNase from *Actinomyces aureoverti-*
                           cillatus Kras et Di Chen, strain 1306, or from
                           takadiastase.
T_1-RNase   Guanyl RNase from takadiastase.

T$_2$-RNase	Nonspecific RNase from takadiastase.
PMEase	Phosphomonoesterase from prostate gland or from *Escherichia coli.*
PDEase	Phosphodiesterase from spleen or snake venom.
tRNA-methylase, rRNA-methylase, etc.	Methylases of corresponding nucleic acids.

Celluloses

DEAE-cellulose	Diethylaminoethylcellulose.
BD-cellulose	Benzoylated DEAE-cellulose.
TEAE-cellulose	Triethylaminoethylcellulose.

Spectrophotometric Parameters

λ	Wavelength, nm.
$E_{260}, E_{280}, E_{max}$	Extinction at 260 nm, at 280 nm, and at the maximum. $E = -\log_{10}T = 2 - \log_{10}(100T)$, where T denotes the transmission of the specimen in a cell with 1-cm optical section against that of the solvent, whose transmission is taken as 1.
$\varepsilon_{260}, \varepsilon_{280}, \varepsilon_{max}$	Molar coefficients of extinction at 260 nm, 280 nm, and at the maximum.

Chapter 1

TRANSFER RNAS AS OBJECTS OF STRUCTURAL RESEARCH

The primary structure of the nucleic acids is the basis on which all other properties of these biopolymers depend. That is why determination of the nucleotide sequence in nucleic-acid molecules is receiving such great attention at the present time. This problem can be solved only if certain conditions are satisfied, for otherwise it would be useless even to make the attempt. The first condition is homogeneity of the test material. In fact, structure can be determined only if it is absolutely certain that the test object is a chemically individual compound. Next, the specimen must be available in adequate quantitites for analysis. Finally, the size of the molecule is important, because the macromolecular nature of most nucleic acids has until recently gravely hindered their study by existing methods.

Even a cursory study of the transfer (soluble, adaptor) RNAs, which were discovered in 1957 (Holley, 1957; Hoagland et al., 1957; Ogata and Nohara, 1957), showed that this class of nucleic acids is endowed with certain properties which offer some hope of success.

Some of the reasons the tRNAs were chosen originally as objects of structural investigation may be enumerated. First was their low molecular weight, only 25,000–30,000, indicating that the polynucleotide chain of tRNA consists of between 70 and 80 monomers. Second, in the case of tRNA there is a precise functional criterion which can be used to verify the homogeneity of the specimen; this test is based on the biological activity of the tRNAs, namely their ability to give acyl derivatives with activated amino acids. The use of this criterion during fractionation led to the rapid development of methods of obtaining highly purified specimens of tRNA, and structural investigations are now carried out on specimens with an average purity of not less than 90%. Cells contain little tRNA, but sufficient to allow the necessary quantity of material for analysis to be obtained: about 10% of the total RNA content of the cell consists of tRNA, and the content of individual tRNAs in the total specimen may amount to several per cent (tRNA^{Val}, 5%; tRNA^{Ser}, 1–2%; etc.). A factor of considerable importance in this choice was the existence in tRNA of minor components, which act as natural reference marks, and enable particular sequences to be identified faultlessly. During the study of primary structure, yet another advantage of tRNA was discovered. This was that the tRNA molecule has a rigid secondary structure; because of this structure, it is possible to obtain products of incomplete enzymic hydrolysis, with a characteristic composition, and under milder conditions it is possible to split a single phosphodiester bond, approximately in the center of the molecule, thus dividing it into halves.

It would be difficult to overestimate the importance of the tRNAs from the standpoint of elucidation of the links between chemical structure and biological activity. As adaptors in protein synthesis they perform many functions and react with many cell components, i.e., they are highly polyfunctional. The transfer RNAs interact with aminoacyl-tRNA-synthetases (6.1.1) and with messenger RNA, and they form complexes with ribosomes; the tRNAs are substrates for C-C-A-pyrophosphorylase (2.7.7.20 and 2.7.7.21), methylases (2.1.1), and other modifying enzymes. Some of these functions are highly specific, while others are general and independent of the amino-acid specificity of the tRNA. Different functions are evidently performed by particular nucleotide sequences or by spatially localized active centers, which, to correspond to the function they perform, may be highly specific and individual or, on the other hand, may be common to all transfer RNAs. The problem of functional topography, i.e., the discovery of the structural groups responsible for a particular function in the tRNA molecule, can be solved only when the primary structure is known. Although the study of primary structure is interesting for its own sake, its ultimate goal lies in the possibility of functional investigations against the background of the known chemical structure.

Investigation of every new tRNA is interesting at the present time, because their comparison and the discovery of their common features and differences are an essential stage in the determination of correlation between chemical structure and biological activity.

A list of transfer RNAs whose structure was known in July 1971 is given in Table 1. The work began on tRNA from *Saccharomyces cerevisiae,* and it culminated in the discovery of the nucleotide sequence in alanine, serine, tyrosine, phenylalanine, and valine tRNAs. A comparison revealed the common structural plan of these tRNAs, which is considered to reflect not only their common functions, but also their common origin from some

TABLE 1. Transfer RNAs Whose Primary Structure Is Known[a]

Source	Amino-acid specificity	Reference
S. cerevisiae	Ala_1	Holley et al., 1965a
"	$Ser_{1,2}$	Zachau et al., 1966a
"	Tyr	Madison et al., 1966a
"	Val_1	Baev et al., 1967a
"	Phe	RajBhandary et al., 1967
"	Asp	Keith et al., 1970
"	Leu_3	Kowalski et al., 1971
E. coli	Tyr Su_{III}^- and Su_{III}^+	Goodman et al., 1968
"	fMet	Dube et al., 1968
"	Met	Cory et al., 1968
"	Val_1	Yaniv and Barrell, 1969
"	Phe	Barrell and Sanger, 1959
"	Leu	Dube et al., 1970
"	Tyr Su^- and Su^+	Hirsh, 1970
"	$Val_{2A,\ 2B}$	Yaniv and Barrell, 1971
"	$Gln_{I,\ II}$	Folk and Yaniv, 1972
"	Gly	Carbon and David, 1971
T. utilis	Val_1	Takemura et al., 1968a
"	Tyr_1	Hashimoto et al., 1969
"	Ileu	Takemura et al., 1969a
Rat liver	Ser	Staehelin et al., 1968
Wheat germ	Phe	Dudock et al., 1969

[a] Work on yeast tRNA[Ala] (Merril, 1968) and on tRNA[Tyr] (Doctor et al., 1969) and tRNA[Phe] (Gassen and Uziel, 1969) of *E. coli*, in which different results were obtained, are not included in the table; these papers are discussed in the text.

single proto-tRNA. Later, when the investigations had been extended to tRNAs from other sources, the concept of the unity underlying the structure of this class of nucleic acids became more evident.

However, the primary structure as a whole proved to be less informative than was initially supposed. Despite the accumulation of a wealth of material, permitting comparison from different points of view, the anticodon is still the only functional center which has been positively identified without any shadow of doubt. The structure and localization of other centers, namely the center for interaction with pyrophosphorylase, with aminoacyl-tRNA-synthetases, with modifying enzymes, and so on, are problems which still await solution.

Elucidation of the nucleotide sequences of a number of tRNAs has created further interesting problems: how are isoacceptor tRNAs formed and why do they exist; how are taxonomic differences brought about; do tRNAs isolated from different organs and tissues, and also from tissues in different functional states, differ from each other? Primary structure as such gives no answer to these questions, but its elucidation has provided us with the key to their solution. Once the covalent structure of the molecule is known, controlled cleavage becomes possible, particular chemical groups can be removed, and precisely identified nucleotides can be modified. In other words, the most direct approaches to establishment of correlation between structure and function can be used. Research into the primary structure of tRNA laid the foundations of what has been called "functional topography," which is now developing extraordinarily rapidly.

Evidence is now being increasingly obtained, however, to suggest that to understand the principles governing interaction between tRNA and proteins, for this is the essential crux of the problem of functional centers, it is not sufficient to know the primary structure of tRNA; its three-dimensional structure must also be known. Fresh opportunities for progress in this direction have been provided by the x-ray structural investigation of tRNA, which, although it only started recently, has made considerable progress. Crystallization of many tRNAs has been accomplished, and several large crystals of various forms have been obtained, including crystals of isomorphic derivatives.

X-ray structural analysis, which has proved so fruitful in the case of proteins, has thus also been found applicable to the study of tRNA. Most probably, therefore, the three-dimensional structure of tRNA will very soon be solved, and this, in turn, will be of great assistance to the study of the functional centers of tRNA.

Some years ago, Holley et al. (1965a) stated that the identification of the primary structure of every nucleic acid and, in particular, of tRNA^{Ala}, is the first step toward the synthesis of biologically active nucleic acids. Work in this direction is already proceeding. Khorana and his collaborators (Khorana, 1971) have synthesized a DNA fragment whose sequence is complementary to that of yeast tRNA^{Ala}; in other words, the cistron of yeast tRNA^{Ala}, the replica from which must be identical with tRNA^{Ala}, but which must not contain any modified nucleotides, has now been synthesized. The opportunities offered by these researches are far-reaching.

Advances in the study of tRNA have stimulated structural investigations on other RNAs, and considerable progress has already been achieved in this direction. A most important factor, which has played an important role in the success of these investigations, is that methods developed for the study of tRNA could also be applied to the study of other RNAs. In particular, limited enzymic hydrolysis, which has yielded decisive results in the elucidation of the primary structure of tRNA, is not the sole prerogative of this class of nucleic acids: it has been used to study the structure of 5S and 6S RNA, and also of the high-molecular-weight rRNAs and polycistron RNAs of bacterial viruses, giving evidence of the highly regular structure of these polymers.

Success in the study of the primary structure of tRNA has thus stimulated the development of adjacent fields in molecular biology.

METHODS OF STUDYING THE PRIMARY STRUCTURE OF TRANSFER RNA

Theoretical aspects of the study of the primary structure of nucleic acids were widely discussed early in the 1960s (Habermann, 1963; Tumanyan and Kiselev, 1963; Rice and Bock, 1963; Mirzabekov and Baev, 1965) and initially some workers were doubtful that the nucleotide sequence could ever be determined in a given nucleic acid. However, along with these theoretical investigations, methods of analysis of primary structure were intensively developed and improved, and considerable progress has now been achieved in this field. By the use of various methods of chromatography and of countercurrent distribution highly purified specimens of tRNA, suitable for structural research, have been isolated. Great advances have been made in the fields of purification of nucleases and the study of their specificity, with the result that these enzymes have been converted into a fundamental tool for the study of nucleic-acid structure. Parallel developments have taken place in methods of fractionation, identification, and study of the structure of the nucleic-acid fragments. The results obtained by these methods have put an end to the theoretical disputes, because the determination of the primary structure of tRNA has been shown to be a practical possibility.

A survey of these methods is given in the present chapter. Attention is concentrated on the block method and on problems directly related to it, because this method has so far played the decisive role in structural research.

METHODS OF ISOLATING SPECIMENS OF INDIVIDUAL tRNAs

Let us now briefly examine the isolation of specimens of individual tRNAs, the initial stage of every structural investigation.

Of the many different methods which have been suggested for the isolation of individual tRNAs, only those giving adequate yields of specimens of a high degree of purity have acquired real importance. Because they did not satisfy demands of yield and purity, many of the published methods were not further developed.

Countercurrent distribution played the main role in the isolation of the first specimens of tRNA. The system devised by Holley and collaborators (Doctor et al., 1961; Apgar et al., 1962), developed from the findings of Warner and Waimberg (1958), has been most widely adopted. This system, consisting of isopropanol, formamide, and phosphate buffer, has been used to isolate alanine-, valine-, serine-, tyrosine-, and phenylalanine-specific transfer RNAs from yeast, and also to fractionate tRNA from *E. coli* (Weisblum et al., 1962; von Ehrenstein and Dais, 1963; Goldstein et al., 1964). Another system has been developed and used for the isolation of valine-, alanine-, and serine-specific tRNAs by Zachau and collaborators (Zachau et al., 1961; Karau and Zachau, 1964). By means of the countercurrent method, much larger quantities of purified specimens of individual tRNAs can be isolated, and a second advantage is that, in contrast to the situation with chemical methods, after countercurrent distribution fractions of all tRNAs can be collected and subjected to further purification. Yet another undoubted advantage is that during countercurrent distribution isoacceptor tRNAs are separated from each other. However, this method is more laborious than others, and the apparatus required is fairly complex.

Another method, using differences in the solubility of specific tRNAs in biphasic systems, is partition chromatography (Kirby, 1960; Everett et al., 1960; Tanaka et al., 1962; Zachau, 1965). Bergquist and Robertson (1965) described the isolation of five fractions with activity relative to serine by partition chromatography on Sephadex G–25. The fractionation of tRNA on ion-exchange celluloses and dextrans (Ofengand et al., 1961; Bergquist et al., 1965; Muench and Berg, 1966a), and on hydroxylapatite (Muench and Berg, 1966b) also has been described.

Cherayil and Bock (1965) made a detailed study of the conditions of elution of tRNA from DEAE-cellulose and DEAE-Sephadex in relation to pH, temperature, urea concentration, and salt gradient.

Japanese workers fractionated tRNA from *T. utilis* on DEAE-Sephadex by means of a series of eluting systems (Miyazaki et al., 1966; Miyazaki and Takemura, 1966; Miyazaki et al., 1967). They obtained highly purified specimens of valine-, tyrosine-, and isoleucine-specific tRNA, which were then utilized for the study of their primary structure (Takemura et al., 1968a; Takemura et al., 1969a; Hashimoto et al., 1969).

On the basis of differences in their secondary structure, Baguley et al., (1965a,b) fractionated tRNAs on DEAE-cellulose. Jacobson and Nishimura

(1963, 1964) suggested fractionating tRNA on ion-exchange paper, but this method has not been developed.

MAK columns have been widely used by Sueoka and Yamane (1962), who described the existence of several isoacceptor tRNAs for most amino acids. The method is limited by the small capacity of the columns, but this drawback can be overcome by the use of the modification suggested by Okamoto and Kawade (1963).

Chemical methods of isolating purified specimens of individual tRNAs also enjoy considerable popularity. They are mostly based on the enzymic aminoacylation of the tRNA which it is intended to isolate, and on the subsequent oxidation of the cis-diol groups of the remaining tRNAs contained in the total preparation with sodium periodate. The tRNAs with a dialdehyde group were removed in different ways: some workers (Zamecnik et al., 1960; Stephenson and Zamecnik, 1961; Zubay, 1962) converted the dialdehydes into Schiff bases, while other workers used the reaction between the dialdehyde group and polyacrylic acid hydrazide (von Portatius et al., 1961; Zachau et al., 1961; Frolova et al., 1964; Mirzabekov et al., 1965b; Grachev et al., 1966). Modification of aminoacylated tRNAs, e.g., the building up of synthetic polymers on the free amino group of an attached amino acid, is another possibility (Mehler and Bank, 1963; Simon et al., 1964). A disadvantage of all these methods is the impossibility of separating the isoacceptor tRNAs, and for this and certain other reasons, they are often combined with other methods.

A successful combination which has been used to purify valine-specific tRNA from yeast was developed by Mirzabekov et al. (1965b); in this method, the tRNA is enriched by the method of countercurrent distribution before being purified on hydrazide–agar gel by the method of Knorre et al. (1964). Countercurrent distribution enables $tRNA^{Val}$ to be separated into three fractions and yields a specimen 4–6 times richer; subsequent periodate oxidation of the specimen aminoacylated by valine, and adsorption of all the tRNAs except $tRNA^{Val}$ on hydrazide–agar gel, yield a specimen of more than 90% purity.

Preparative quantities of $tRNA^{Val}$ of a high degree of purity are also obtained by a preliminary enrichment of the total $tRNA^{Val}$ preparation from yeast by ion-exchange chromatography as described by Kawade et al. (1963), followed by periodate oxidation and adsorption of the dialdehyde-tRNA on hydrazide–agar gel. The method yields a highly purified $tRNA^{Val}$, but without separation of the isoacceptor tRNAs (Grachev et al., 1966). Separation of a specimen obtained in this manner into 5 fractions specific for valine on TEAE-cellulose has been described by Vasilenko et al. (1970).

The method of reversed-phase chromatography, proposed by Kelmers et al. (1965), has been widely adopted. It has the advantage of high resolving power, especially with respect to isoacceptor tRNAs. In 1966, Kelmers used this method to isolate specimens of very high degrees of purity of tRNA^{Phe} (Kelmers, 1966a) and tRNA^{Leu} (Kelmers, 1966b). The method of reversed-phase chromatography is undergoing continuous improvement, and already six modifications are known. These modifications apply both to the organic phase (Weiss et al., 1968) and to the sorbent (Pearson et al., 1971). Until recently, quaternary ammonium salts of fatty acids, adsorbed on diatomaceous earth treated with hydrophobic dimethyldichlorosilane, with a particle size of 100–120 mesh, have been used (Weiss and Kelmers, 1967; Weiss et al., 1968). Attempts to improve the fractionation by reducing the particle size of the diatomaceous earth have been unsuccessful. A much clearer separation, especially of the isoacceptor tRNAs, together with the possibility of fractionating large quantities of material, have resulted from the use of a polychlorotrifluoroethylene resin instead of the diatomaceous earth (Pearson et al., 1971).

The DEAE-celluloses acylated with aromatic acids (e.g., BD-cellulose) are becoming increasingly popular. Substitution of the hydroxyl groups by benzoyl and other residues reduces the capacity somewhat and gives a higher yield of RNA (Gillam et al., 1967). The binding of the nucleic acids to these sorbents is largely dependent on the relative number of double-helical segments, and it accordingly varies with factors influencing secondary structure (bivalent cations, pH, temperature).

A later development of the method of fractionation on BD-cellulose is introduction of a phenoxyacetyl residue into the NH$_2$ group of the amino acid used for aminoacylation (Gillam et al., 1968). The hydrophobic substituent causes marked displacement of the elution point of the phenoxyacetylated tRNA from the BD-cellulose and its separation from other tRNAs which have not been aminoacylated.

Hense et al. (1969) recently described a simple modification of this method enabling highly purified specimens of tRNA$_f^{Met}$ and tRNA^{Met} to be obtained from E. coli. A single fractionation of the phenoxyacetylated specimen of tRNA from E. coli, containing methionyl-tRNA, at 4°C led to separation of tRNA$_f^{Met}$ and tRNA^{Met}, and to their preparation as specimens of 70 and 35% purity respectively; on repeating the procedures, the purity of the tRNA$_f^{Met}$ reached 100%, and that of tRNA^{Met} was increased to 70%. These workers point out that phenoxyacetylation changes the order of elution of these two tRNAs; N-acetylation of the methionine residue probably gives rise to different conformational changes in the methionyl-tRNA^{Met} and methionyl-tRNA$_f^{Met}$, which are exhibited as changes in their affinity for BD-cellulose. It is assumed that the suggested modification will be useful for separating isoacceptor tRNAs.

The first stage in the production of individual tRNAs from E. coli

(tyrosine-, formylmethionine-, and methionine-specific) was usually fractionation on DEAE-Sephadex (Nishimura et al., 1967b); subsequent enrichment was carried out either on BD-cellulose (Gillam et al., 1967) or by
reverse-phase chromatography (Kelmers et al., 1965). For the isolation of
valine-specific tRNA from E. coli, both simple fractionation on BD-cellulose
(Gillam et al., 1967) and enrichment of val-tRNA^{Val} on BD-cellulose after
phenoxyacetylation (Gillam et al., 1968), have been used. Phenylalanine-
specific tRNA of about 90% purity has also been isolated from wheat germ
by a combination of chromatography on BD-cellulose and reverse-phase
chromatography (Dudock et al., 1969).

By combining fractionation on BD-cellulose and on DEAE-Sephadex
with reverse-phase chromatography, not only has the clear fractionation of
preparative quantities of tRNA$_1^{Val}$ and tRNA$_2^{Val}$ from bakers' yeast become
possible, but it has also been shown that tRNA$_2^{Val}$ consists of 3 fractions;
a specimen of one of these fractions of tRNA$_2^{Val}$ has been isolated in a highly
purified state and in sufficient quantity for structural analysis; it is being
studied at the present time (Kryukov et al., 1971).

Electrophoresis on polyacrylamide gel, originally suggested by Ornstein
(1964), Davis (1964), and Raymond et al. (Raymond and Weintraub, 1959;
Raymond and Wang, 1960), has recently come to be extensively used for the
fractionation of nucleic acids and their fragments. The decisive factors in this
method are the molecular weight and shape of the molecule; the rate of
movement as a rule is inversely proportional to the sedimentation coefficient
of the RNA. The range of the method used is very great; since fractionation
is based on molecular filtration, the pore size can be matched to the molecular weight of the corresponding RNAs. Separation of small RNA molecules was described by Richards et al. (1965); partial hydrolysates of ribosomal RNA were fractionated by this method by Gould (Gould, 1966, 1967;
Gould et al., 1969), while Hindley (1967) used it to separate 5S from 4S
RNA. A modification of the method, making it possible to work with 10–15
μg tRNA, has been described by Simukova and Budovskii (1970). Adams
et al. (1969) fractionated partial guanyl-RNase hydrolysates of P^{32} RNA
from phage R17 and obtained up to 40 discrete bands, the highest of which
corresponded to Gp and the lowest to fragments containing about 300
nucleotides. Considerable improvement in the fractionation of high-
molecular-weight RNAs was achieved by Loening (1967), who obtained
transparent gels by using recrystallized acrylamide.

Electrophoresis in polyacrylamide gel has been used to isolate tRNA^{Leu}
from E. coli (Dube et al., 1970) and from bakers' yeast (Kowalski and
Fresco, 1971).

As Zachau (1969) points out in his survey, in none of these cases is it
known exactly which structural features of the tRNA molecule lie at the

basis of their separation. It may be that both in countercurrent and in chromatographic fractionation the basic factor is a difference in stability, with a consequent difference in the degree of preservation of the spatial structure of the different forms of tRNA.

TERMINAL ANALYSIS

Long before it was possible to study the structure of the whole tRNA molecule, attempts had been made to analyze part of the polynucleotide chain and, in particular, its 3'- and 5'-ends. The results of most of these investigations, conducted on total tRNA specimens, are nowadays only of limited interest, but many of the methods developed still retain their importance even today.

The simplest method of determining 3'- and 5'-terminal nucleotides of tRNA is by alkaline hydrolysis, for it results in the removal of a nucleoside diphosphate from the 5'-end and a nucleoside from the 3'-end. Separation of these fragments from each other and from the total mass of 3'-nucleotides of the inner segment of the molecule is not a difficult task. This method can be used to determine the length of the polynucleotide chain, to verify the integrity of the terminal nucleotides in different tRNAs, and also to establish the presence and number of ruptures in the molecule. The only difficulty is that the relative content of terminal fragments is small: they constitute approximately one-eightieth of the total quantity of uv-absorbing matter, and the accuracy of their quantitative determination is correspondingly low.

Exonucleases are also used for the terminal analysis of tRNAs and polynucleotides. Exhaustive hydrolysis of a dephosphorylated polynucleotide by snake venom PDEase (3.1.4.1) identifies the 5'-terminal nucleotide, and corresponding hydrolysis with spleen phosphodiesterase (3.1.4.1) identifies the 3'-terminal nucleotide liberated in both cases in the form of nucleosides (Fig. 1).

Endonucleases—pyrimidylic (2.7.7.16) and guanylic (2.7.7.26) RNases—split oligonucleotides of different length and composition from the ends of different tRNA molecules, and these can easily be distinguished from the

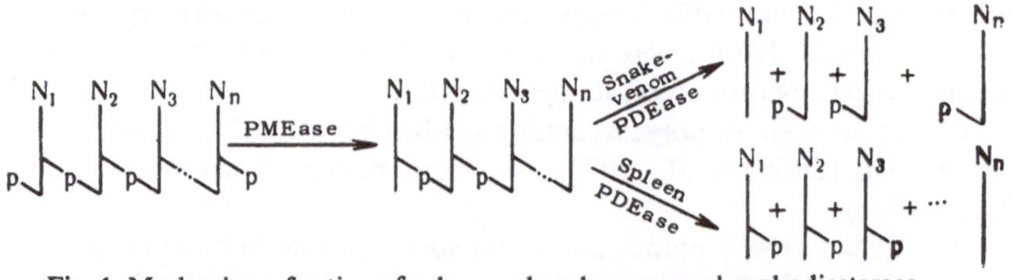

Fig. 1. Mechanism of action of spleen and snake-venom phosphodiesterase.

remaining oligonucleotides. The 5'-terminal oligonucleotide of the pancreatic hydrolysate has the general formula pN_1-N_2-N_3...Pyp, while that of the guanylic hydrolysate is pN_1-N_2-N_3...Gp; they differ from oligonucleotides of the same length in the inner part of the molecule only by their single surplus phosphate residue, which is used for their detection (Tomlinson and Tener, 1963b; McLaughlin and Ingram, 1965). The acceptor end of all tRNAs is formed by the trinucleotide C-C-A_{OH}, and for this reason pyrimidyl RNase splits adenosine from it if the end is intact, and cytidine if the end is broken off and the molecule now ends in the sequence C-C_{OH}, Guanyl-RNase fragments of the 3'-end of the tRNA molecule differ in length depending on the distance from the first guanylic-acid residue to the terminal adenosine; the longest oligonucleotide, consisting of 19 components, has been identified in a hydrolysate of tRNA^{Tyr} from *E. coli* (Goodman et al., 1968), and the shortest (C-C-A_{OH}) in tRNA^{Ser} from yeast (Zachau et al., 1966b) and rat liver (Staehelin et al., 1968) and in tRNA^{Asp} from brewers' yeast(Keith et al., 1970). A characteristic feature of the 3'-terminal oligonucleotides is that, unlike all other guanyl-RNase oligonucleotides, they do not contain the 3'-monophosphate of guanosine, inosine, or any methylated derivative of guanosine at the 3'-end.

To increase the sensitivity of terminal analysis, uniformly labeled RNAs have been used, and labeled reagents have been introduced into the terminal groups. In the first case, the need for detecting the small number of terminal residues against the background of the main mass of the nucleotides in the inner part of the molecule still remains, although at another level of sensitivity. It is therefore much better to introduce the label into the terminal sequences of the polynucleotide chain.

Some methods of identifying the 5'-terminal nucleotide are based on the fact that it has the only 5'-phosphate group in the molecule. ^{14}C-alanine (Ralph et al., 1963), $^{14}CH_3$-methylphosphomorpholidate (RajBhandary et al., 1964), and so on have been suggested for use in introducing the label into the 5'-terminal nucleotide. Ralph et al. (1963) converted 5'-phosphomonoester groups into the corresponding phosphoanilidates:

$$C_6H_5-NH-\overset{\displaystyle O}{\overset{\displaystyle \|}{\underset{\displaystyle OH}{\underset{\displaystyle |}{P}}}}-O\cdots$$

The next stage of the analysis was cleavage of the modified tRNA by spleen phosphodiesterase or alkali and fractionation of the hydrolysis products by ion-exchange chromatography. By this means, numerous fractions containing phosphoanilidate were isolated from the total tRNA. The chief conclusions obtained by this method were that most yeast tRNAs

(60–70%) have a terminal pG, some have a terminal pA, and a few a terminal pU. Many chains have a pG-C dinucleotide at the 5'-end. These conclusions were confirmed and expressed more precisely by Bell et al. (1964), who found different 5'-terminal nucleoside diphosphates in yeast tRNA in the following proportions: pGp, 78%; pUp, 10%; pAp, 7%; and pCp, 5%.

The advantages of the method using ^{14}C-methylphosphomorpholidate (RajBhandary et al., 1964) are the greater specificity of the reaction, the mildness of its conditions, and the stability of the radioactive group incorporated in the 5'-position (Fig. 2). Results obtained by this method for the 5'-terminal nucleotides of tRNA were as follows: pG, 75.5%; pU, 10.1%; pA, 4.1%; and pC, 4.1%; in a few cases, methylated pG was also found at the end of the molecules. Hence, even before individual tRNAs had been analyzed, it was concluded that the 5'-end of tRNA is formed chiefly by guanylic acid, but not by it alone. This was confirmed by analysis of individual tRNAs (Table 2).

It was suggested by Szekely and Sanger (1969) that 5'-OH-polynucleotide kinase be used to incorporate the label into the 5'-ends of the oligonucleotides of a guanyl-RNase hydrolysate of tRNA. By means of this method, the fingerprinting technique of Sanger et al. (1965) can be applied to nucleic acids which cannot be labeled *in vivo*. In this case, partial hydrolysis with snake-venom PDEase is used for subsequent analysis of the oligonucleotides.

The acceptor end of tRNA has been labeled principally by oxidation of

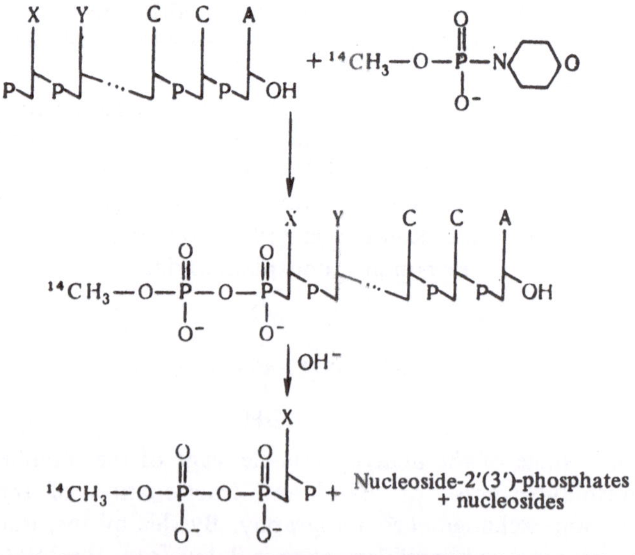

Fig. 2. Determination of 5'-terminal nucleotide with the aid of ^{14}C-methylphosphomorpholidate.

TABLE 2. 3'- and 5'-Terminal Sequences in Individual tRNAs

tRNA	Source	5'-Terminal oligonucleotides from		3'-Terminal oligonucleotides from guanyl-RNase hydrolysate
		guanyl-RNase hydrolysate	pyrimidyl-RNase hydrolysate	
Ala	S. cerevisiae	pGp	pG-G-G-Cp	U-C-C-A-C-C-A$_{OH}$
Ser	"	pGp	pG-G-Cp	C-C-A$_{OH}$
Tyr	"	pC-U-C-U-C-Gp	pCp	A-C-C-A$_{OH}$
Phe	"	pGp	pG-Cp	C-A-C-C-A$_{OH}$
Val	"	pGp	pG-G-Up	A-A-A-U-C-A-C-C-A$_{OH}$
Asp	"	pU-C-C-Gp	pUp	C-C-A$_{OH}$
Leu$_3$	"	pGp	pG-G-Up	C-A-A-C-C-A-C-C-A$_{OH}$
Tyr	E. coli	pGp	pG-G-Up	A-A-U-C-C-U-U-C-C-C-C-C-A-C-C-A-C-C-A$_{OH}$
fMet	"	pC-Gp	pCp	C-A-A-C-C-A$_{OH}$
Met	"	pGp	pG-G-Cp	C-C-A-C-C-A$_{OH}$
Phe	"	pGp	pG-Cp	C-A-C-C-A$_{OH}$
Val$_1$	"	pGp	pG-G-G-Up	U-C-A-U-C-A-C-C-C-A-C-C-A$_{OH}$
Gln	"	pU-Gp	pUp	C-C-A$_{OH}$
Gly	"	pGp	pG-Cp	C-U-C-C-A$_{OH}$
Leu	"	pGp	pG-Cp	C-A-C-C-A$_{OH}$
Trp	"	pA-Gp	pA-G-G-G-G-Cp	C-C-A$_{OH}$
Val$_2$	"	pGp	pG-Cp	C-A-C-C-A$_{OH}$
Val$_1$	T. utilis	pGp	pG-G-Up	A-A-A-U-C-A-C-C-A$_{OH}$
Tyr	"	pC-U-C-U-C-Gp	pCp	A-C-C-A$_{OH}$
Ilu	"	pGp	pG-G-Up	A-C-C-A-C-C-A$_{OH}$
Ser	Rat liver	pGp	pG-Up	C-C-A$_{OH}$
Phe	Wheat germ	pGp	pG-Cp	C-A-C-C-A$_{OH}$

the cis-diol group with periodate, followed by condensation of the resulting dialdehyde with various reagents, including those containing the label. The following compounds have been used for this purpose: ^{35}S-thiosemicarbazide (Dulbecco and Smith, 1960), ^{14}C-semicarbazide (Steinschneider and Fraenkel-Conrat, 1966; Mandels, 1967a,b), and isonicotinic acid ^{3}H-hydrazide (Hunt, 1965, 1970). Reduction of the 2',3'-dialdehyde groups with ^{3}H-NaBH$_4$ was found to be a more specific and quantitative method; the tritiated diol derivatives thus formed are stable in alkaline and weakly acid media (Sugiyama, 1965; Lee and Gilham, 1965).

Oxidation of RNA with periodate and subsequent reduction of the terminal dialdehyde with ^{3}H-NaBH$_4$ has been further improved by RajBhandary (RajBhandary et al., 1966; RajBhandary and Stuart, 1966a; RajBhandary, 1968):

The method was used to analyze the 3'-terminal sequence of tRNA^{Phe} (RajBhandary et al., 1968c) and also to analyze the structure of oligonucleotides isolated from it (RajBhandary et al., 1968a). Periodate oxidation followed by labeling the resulting dialdehyde has been successfully used to analyze the 3'-terminal sequences of high-molecular-weight RNAs, notably phage MS2 (De Wachter and Fiers, 1967; Glitz et al., 1968), phage f2, and TMV (Glitz et al., 1968).

Sequence studies in the part of the molecule next to the acceptor end are also based on reversibility of the condensation reaction of the terminal trinucleotide C-C-A. For this purpose, the C-C-A end is first removed by pyrophosphorylase; the preparation is then treated with the same enzyme in the presence of CTP (cytidine triphosphate) containing ^{32}P in the α-position; tRNAs in which the two terminal phosphodiester bonds are labeled (Lagerqvist and Berg, 1962) are obtained as a result:

$$\cdots pWpXpYpZpCpCpA$$
$$\downarrow \text{PPi}$$
$$\cdots pWpXpYpZ + 2CTP + ATP$$
$$\downarrow \text{CTP-}\alpha\text{-P*}$$
$$\cdots pWpXpYpZp^*Cp^*C + 2PPi$$

(where W, X, Y, and Z denote nucleosides next to the C-C-A end).

The tRNA is then degraded with alkali or with pancreatic or guanyl RNases. The 3'-terminal oligonucleotides (or nucleotides) contain ^{32}P, so that after fractionation they are easily distinguished from the oligonucleotides of the inner part of the molecule. Fragmentation of tRNA labeled in this manner gives essential information regarding nucleotides lying next to the C-C-A end:

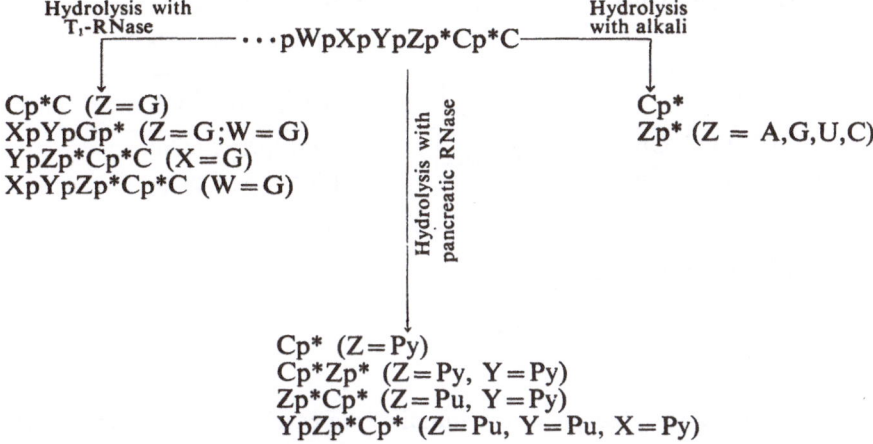

Hydrolysis with
T₁-RNase ——————— ···pWpXpYpZp*Cp*C——— Hydrolysis with alkali

Cp*C (Z=G)
XpYpGp* (Z=G;W=G)
YpZp*Cp*C (X=G)
XpYpZp*Cp*C (W=G)

Hydrolysis with pancreatic RNase

Cp*
Zp* (Z = A,G,U,C)

Cp* (Z=Py)
Cp*Zp* (Z=Py, Y=Py)
Zp*Cp* (Z=Pu, Y=Py)
YpZp*Cp* (Z=Pu, Y=Pu, X=Py)

Investigation of total tRNA showed that in approximately 60% of cases position 4 counting from the 3'-end is occupied by A, in 20–25% by G, and in 10–20% by U. Position 5 in 80% of the polynucleotide chains is a pyrimidine (Herbert and Wilson, 1962a,b).

A modification of this method is used to determine the 3'-terminal oligonucleotides of individual tRNAs without their preliminary isolation. The tRNA to be studied is aminoacylated; the specimen is then treated with periodate and an amine, with the result that the terminal adenine and ribose residues are removed from all the tRNAs except the one which is aminoacylated, leaving a 3'-phosphate group at the end. The amino acid is removed, and this is followed by pyrophosphorolysis, which affects only chains with an intact C-C-A group. The product is incubated with ³²P-labeled CTP, which is incorporated into the previously aminoacylated tRNA which has now lost its C-C-A group, thus leading to the formation of a mixture of labeled and unlabeled chains. After degradation and fractionation, the fragments formed from the tRNA specific for the amino acid used for aminoacylation can easily be identified by the label (Berg et al., 1962) (see scheme on page 22).

It was thus established that tRNAs binding isoleucine have an acceptor end of only one type, namely, G-C(U,C)A-C-C-A, whereas leucine-specific tRNAs have two different terminal sequences: G-C-A-C-C-A and G-U-A-C-C-A.

The functional properties of the C-C-A end are utilized in many methods. Total, partially enriched, or individual tRNA specimens are aminoacylated with a labeled amino acid and then subjected to terminal analysis. The

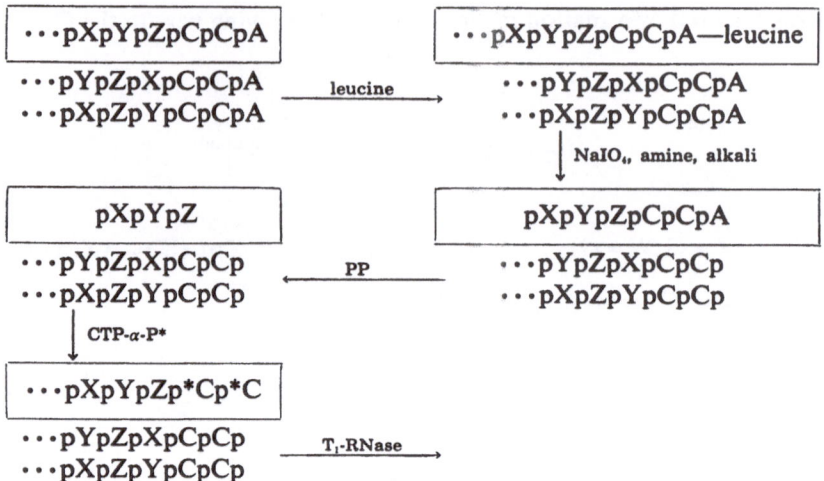

fragmentation and subsequent fractionation are carried out in a weakly acid medium, for the aminoacyl bond is unstable in an alkaline medium. Cleavage is usually carried out with guanyl RNase. The method is illustrated by the following scheme (Ishida and Miura, 1965):

$$...NpGp(Np)_nCpCpA\text{-}^{14}C \text{ (or } ^3H)\text{-amino acid} \xrightarrow{T_1\text{-RNase}}$$
$$(Np)_nGp + (Np)_nCpCpA\text{-}^{14}C \text{ (or } ^3H)\text{-amino acid}$$

where N is any nucleoside except G; $n = 0,1,2,3$, etc.

Herbert and co-workers (Herbert et al., 1964; Smith and Herbert, 1965) loaded total yeast tRNA with ^{14}C-amino acids and fractionated the guanyl-RNase hydrolysates by chromatography on DEAE-cellulose or by high-voltage electrophoresis. One terminal sequence was found for seven amino acids, and two for valine tRNA. The fragments were subsequently isolated and identified. The results for alanine, serine, and phenylalanine tRNAs agreed with those obtained on determination of the primary structure of these tRNAs. Bergquist (1966a) reported that all five yeast serine-specific tRNAs have identical acceptor ends, namely, G(A,C)C-C-A; these findings can no longer be accepted, because tRNA^{Ser} of bakers' and brewers' yeasts have a G-C-C-A$_{OH}$ sequence at their 3'-end (Zachau et al., 1966a; Neelon et al., 1967). At least two different 3'-terminal sequences were found in yeast tRNA^{Val} by aminoacylation with labeled amino acid (Smith et al., 1964; Ishida and Miura, 1963; Weil et al., 1964; Grachev et al., 1965). Ishida and Miura (1965) found two acceptor ends in valine-specific tRNA from rat liver. Several 3'-terminal sequences have also been found in tRNA^{Leu} from yeast, rat liver, and E. coli.

Aminoacylation with ^{14}C-valine and subsequent fractionation of the guanyl-RNase hydrolysate is used to determine the relative percentages of tRNA$_1^{Val}$ and tRNA$_2^{Val}$ in valine-enriched specimens (Grachev et al., 1945).

THE STEPWISE-DEGRADATION METHOD

The determination of primary structure by successive degradation of the polynucleotide chain, so apparently promising on theoretical grounds, in practice has proved to be difficult to carry out both chemically and enzymically. The attraction of this method was that by stepwise removal of monomers from either end of the molecule, their sequence could be determined both from the removed products of the reaction and from the undegraded residue of the polynucleotide chain.

The chemical method of stepwise degradation is based on the periodate oxidation of the *cis*-diol groups of nucleoside-5'-phosphate derivates with subsequent β-elimination of the dialdehyde in an alkaline medium or by the action of bases (Whitfield and Markham, 1953; Brown et al., 1955). It was assumed that by this method the tRNA bases could be determined by their stepwise removal from the 3'-end of the molecule:

$$R\text{-}C\text{-}C\text{-}A_{OH} \xrightarrow[\text{2) }\beta\text{-elimination}]{\text{1) NaIO}_4} R\text{-}C\text{-}Cp + \text{adenine}$$

$$R\text{-}C\text{-}Cp \xrightarrow{\text{PMEase}} R\text{-}C\text{-}C \xrightarrow[\text{2) }\beta\text{-elimination}]{\text{1) NaIO}_4} R\text{-}Cp + \text{cytosine}$$

$$R\text{-}Cp \xrightarrow{\text{PMEase}} R\text{-}C, \text{etc.}$$

This scheme has been studied by many investigators (Ogur and Small, 1960; Yu and Zamecnik, 1960; Khym and Cohn, 1961) in order to make the reaction more quantitative and to prevent the formation of byproducts. Their work showed that the glycine or ammonia, which were used originally, can be replaced by amines, under the influence of which β-elimination takes place through the formation of Schiff bases.

Yu and Zamecnik (1960) suggested that when cyclohexylamine is used for the treatment of oxidized RNA, the scheme of the reaction is as shown on page 24.

A useful modification of the method in which repeated cycles of periodate oxidation can be carried out was developed by Neu and Heppel (1964). These workers used their suggested modification of the method to determine four nucleotides from the 3'-terminal end of tRNA, while Whitfield (1965) used it to determine the terminal sequence in TMV RNA.

The use of cetyltrimethylammonium bromide (Cetavlon) for quanti-

RNA—O—P—O—CH₂ Base   RNA—O—P—O—CH₂ Base
        OH   OH   NaIO₄
                                                                  H₂N—C₆H₁₁
        HO OH   O O

RNA—O—P—O—CH₂ Base   RNA—O—P—O—CH₂ H
        OH   OH   +Base
     C₆H₁₁N NC₆H₁₁   C₆H₁₁N NC₆H₁₁
     Schiff base

tative precipitation of polynucleotides and their separation from the base and phosphatase has recently been suggested (Khym and Uziel, 1968). Aubert et al. (1967) used molecular filtration on dextran gel to analyze the reaction products after periodate oxidation of tRNA. Thus, many investigators have studied, and are continuing to study, ways of improving this method, by varying the pH, temperature, concentration of the reagent, and other conditions.

However, only one example of the use of stepwise degradation to study the primary structure of tRNA is so far known. By using the method of periodate oxidation, Uziel and Khym (1969) determined the 3'-terminal sequence of the tRNA^{Phe} of E. coli for the length of 19 nucleotides. The nucleotide sequence in the fragment which they studied coincided with the results obtained by degradation with pancreatic and guanyl RNases. However, these workers state that further oxidation cycles (20–26) gave unreliable results because of mechanical losses and asynchronous cleavage. It is interesting to note that the sequence in the 3'-terminal fragment established by periodate oxidation is identical to the structure of the same part of the tRNA molecule from E. coli suggested by Barrell and Sanger (1969), whereas the results for the remaining part of the molecule show considerable differences (Gassen and Uziel, 1969).

On the whole, the impression is gained that this method will not be widely used in research into nucleic-acid structure, for difficulties arise even during determination of the base sequence in oligonucleotides. One complication arising during work with tRNA is the slight breakdown of ψ (Tomasz and Chambers, 1965), for if the procedure is repeated many times, this breakdown becomes considerable in magnitude and may rupture the polynucleotide chain, leading to incorrect results. The behavior of H_2U during periodate oxidation has never been investigated, although it is unlikely to remain unaffected by it. Without question the method is suitable for determination and removal of one or two nucleotides from the 3'-end of the tRNA molecule or fragment, and this, in particular, is being used in structural and functional investigations (Chuguev et al., 1969, 1970).

The enzymic form of the stepwise-degradation method has been widely discussed in the literature (Vasilenko, 1963; Holley et al., 1963; Cantor and Tinoco, 1967). Exonucleases (PDEases) from snake venom and spleen can be used for this purpose. The structure of total yeast tRNA was studied by Vasilenko (1964, 1965) using snake-venom PDEase. The same enzyme was used by Melchers et al. (1965) to study yeast serine-specific tRNA. However, these investigations did not yield the expected result.

A clear example of the danger of using stepwise degradation with PDEase is the investigation of Nihei and Cantoni (1962). From their study of the kinetics of removal of nucleotides from total tRNA these workers concluded that the minor components are located in the central part of the polynucleotide chain. Examination of the structures of the tRNAs which they studied shows that this conclusion and the model proposed by Cantoni for the structure of tRNA on the basis of the results they obtained (McCully and Cantoni, 1962b) do not correspond to the facts.

The reason for these failures is that hydrolysis of the different chains, even of a homogeneous RNA specimen, takes place asynchronously and the residues of the molecules form a complex, heterogeneous mixture; oligonucleotides liberated at a given moment of time are therefore derived from different portions of the polynucleotide chain.

Snake-venom PDEase has been successfully used by a number of workers to study terminal sequences of transfer RNAs and also of high-molecular-weight RNAs (Preiss et al., 1961; Singer and Fraenkel-Conrat, 1963; Zubay and Takanami, 1964; Nihei and Cantoni, 1963). Hadjiolov et al. (1967b) used the method of stepwise degradation by this enzyme to study the structure of rRNA from rat liver.

The possibility of using spleen phosphodiesterase for stepwise degradation of tRNA has been investigated by Bernardi and Cantoni (1969) using an enzyme completely purified of endonuclease activity (Bernardi and Bernardi, 1968). However, this enzyme cannot be used for systematic stepwise degradation of tRNA because, as Philippsen and Zachau (1971) have shown, it attacks different tRNA sequences at different rates.

THE UNIT METHOD

The unit method of determining nucleotide sequence in nucleic acids is based on operations developed in connection with proteins: fragmentation of the polynucleotide chain by means of agents of varying specificity; preparation of fragments in an individual state and the study of their structure; and, finally, reconstruction of the molecule on the basis of overlapping of the units formed by various methods of degradation. Cleavage into units can be achieved, in principle, both enzymically and chemically; in practice, however, only the first method has so far been used.

More recently, and mainly because of the development of combined structural and functional research, various methods of specific cleavage of the polynucleotide chain at a number of minor components have been developed and tested on individual tRNAs (Philippsen et al., 1968; Wintermeyer and Zachau, 1970; Beltchev and Grunberg-Manago, 1970a). There is no doubt that these methods will be used in the future in structural research to obtain large fragments of tRNA.

The enzyme preparations used in structural investigations must be free from contamination by any other enzyme activity and must possess a high degree of specificity. Of the many known nucleases, pancreatic pyrimidyl RNase and guanyl RNase from takadiastase and actinomycetes have therefore been widely used. Only isolated examples of the use of other enzymes are known. For example, Brownlee et al. (1968) obtained useful information by splitting the 5S rRNA of *E. coli* with acid ribonuclease from the spleen (Bernardi and Bernardi, 1966) and with ribonuclease from *Bacillus subtilis*. The specificity of the other RNases, as a rule, is insufficiently strict, for they hydrolyze the polynucleotide chain at many nucleotides.

The stages by which the primary structure of a nucleic acid can be determined by the unit method are illustrated in the scheme. The specimen is first subjected to complete or exhaustive hydrolysis by two RNases, as a result of which two sets of oligonucleotides are obtained. Pyrimidyl RNase gives oligonucleotides ending in pyrimidines (P-oligonucleotides). As a result of the action of guanyl RNase, longer oligonucleotides are obtained, and depending on the specificity of the enzyme they end in Gp, Ip, or one of the methylated derivatives of Gp (T-oligonucleotides). The oligonucleotides of the pyrimidyl- and guanyl-RNase hydrolysates are fractionated and each oligonucleotide is isolated in an individual, chemically pure state. The individual oligonucleotides then undergo further structural investigation, i.e., the nucleotide composition and nucleotide sequence in the molecule are determined. In the course of subsequent analysis, the sequence, not of individual nucleotides, but of oligonucleotides in the tRNA molecule is determined. Each oligonucleotide is thus an independent unit which, in the subsequent stages of structural analysis, must be correctly identified.

By comparing the oligonucleotides of the P and T series and matching them with each other, longer segments of the polynucleotide chain can be reconstructed. However, this is insufficient to allow reconstruction of the whole molecule. The arrangement of the oligonucleotides is established by partial cleavage of the molecule by the same two RNases.

The possibility of partial degradation of the tRNA molecule is not an

Scheme of Determination of Primary Structure of Transfer RNAs by the Unit Method

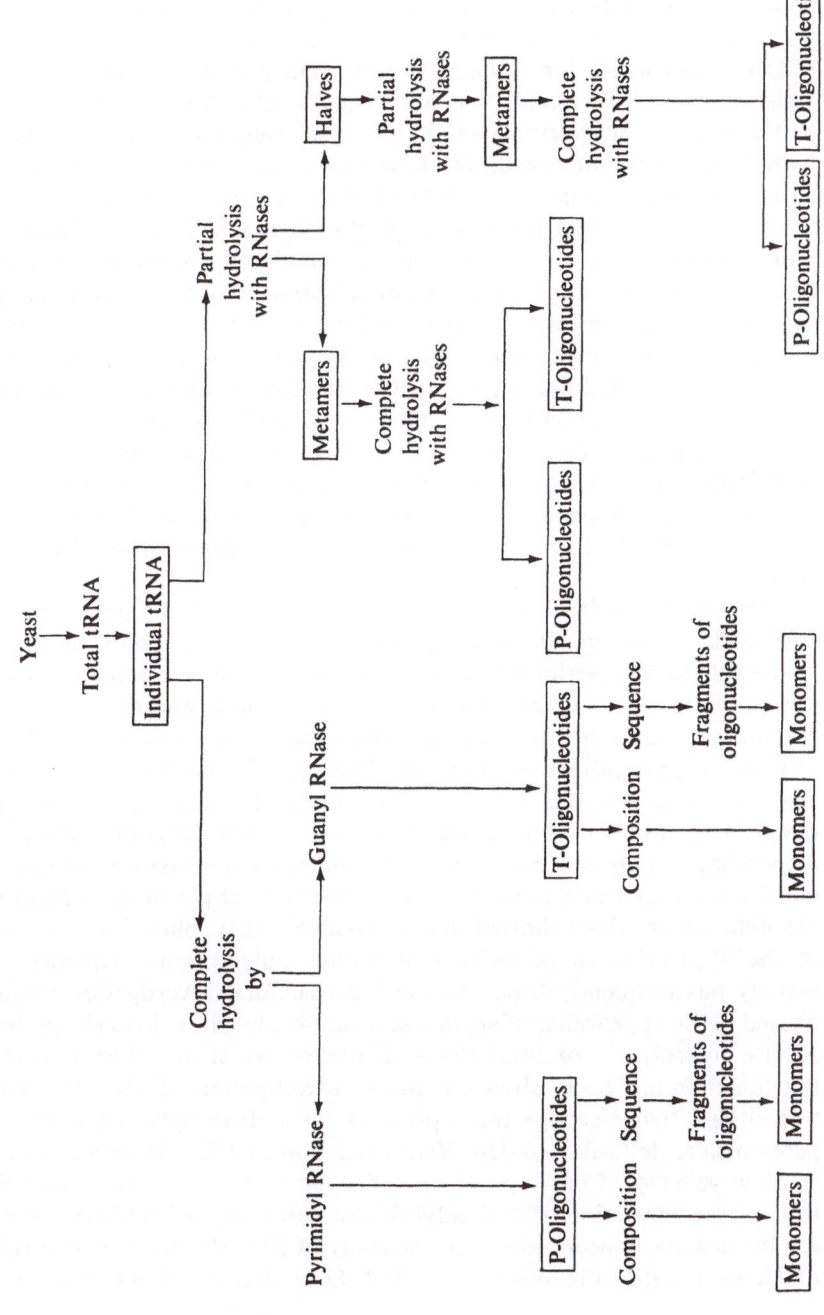

accidental phenomenon but it was discovered and has been used during the study of other RNAs, and it is evidently a characteristic feature of this class of nucleic acid, determined by their secondary structure. Under conditions unfavorable for RNases, only certain phosphodiester bonds accessible to the enzyme could be ruptured in the tRNA molecule; in this way metamers, i.e., fragments including several oligonucleotides of the total hydrolysate, were obtained. In some cases metamers were obtained from halves of the molecules into which tRNAs are broken up under the mildest conditions and in the presence of Mg^{2+}. Metamers obtained from whole molecules or from their halves are fractionated and isolated in a homogeneous state. Determination of the oligonucleotide composition of overlapping and abutting metamers by total hydrolysis with RNases gives sufficient information to allow reconstruction of the whole polynucleotide chain. A properly formulated hypothesis for the structure of each fragment plays an important role in structural research, for it enables the subsequent analysis of the fragment to be planned. The method of analogy and comparison is therefore extensively used in the determination of primary structure.

Let us now briefly consider the enzymes (pyrimidyl and guanyl RNases) which were the chief tools used to study the primary structure of the nucleic acids.

Pancreatic pyrimidyl RNase is a well studied enzyme belonging to the endonuclease group. It ruptures the phosphodiester bond between the 3'-phosphate of a pyrimidine (including most minor components which are pyrimidine derivatives) and the 5'-hydroxyl group of the next residue. The reaction products are pyrimidine-3'-phosphates and oligonucleotides with 3'-terminal pyrimidines of the $(Pup)_n$-Pyp type. From the point of view of structural research, this enzyme is not suitable because it gives a comparatively large number of mononucleotides and short oligonucleotides, corresponding to polypurine sequences. The longest pyrimidyl-RNase fragments consist of about 7 or 8 monomers. The specificity of pyrimidyl RNase is not absolute. Beers (1960) showed that it can also attack phosphodiester bonds in the 3'-position to adenylic acid within polyadenylic sequences. This activity has frequently been observed in structural investigations, where it has led to the appearance of small quantities of adenylic acid in the pyrimidyl-RNase hydrolysate of total tRNA (Mirzabekov et al., 1965a); a similar phenomenon has been observed during investigation of $tRNA_1^{Val}$, when it was shown that cleavage takes place at the polyadenylic sequence of the pentanucleotide G-A-A-A-Up (Venkstern et al., 1968a). In accordance with the low velocity of cleavage of polyadenylic sequences observed by Beers, this nonspecific action of pyrimidyl RNase can easily be abolished by reducing the enzyme concentration. In the study of $tRNA^{Phe}$ from E. coli (Barrell and Sanger, 1969), the oligonucleotide A-G-G-G-G-A-ψ was always found in small amounts, probably also because of the nonspecific action of pancreatic RNase. This enzyme evidently ruptures not only poly(A), but also poly(G) sequences. Pyrimidyl RNase has been used to produce both complete and partial degradation of tRNA. By the use of this enzyme, Dütting et al. (1966) obtained metamers from yeast tRNA[Ser], while Chang and RajBhandary (1968) obtained metamers and halves of the yeast tRNA[Phe] molecule; pyrimidyl RNase has also been used for the partial hydrolysis of tRNA[Asp] (Keith et al., 1970) and in other cases.

Guanyl RNase, which ruptures bonds of the GpN type and gives Gp and oligonucleotides with guanylic acid at the 3′-end as the reaction products, is more selective in its action. The enzyme T_1-RNase was discovered by Sato and Egami (1957) and is described by Egami et al. (1964). The analogous enzyme was isolated by Tatarskaya et al. (1964) from *Streptomyces aureoverticillatus* Kras. et Di Shen strain 1306. Guanyl RNases have also been found in other sources (Tanaka and Cantoni, 1963; Glitz and Dekker, 1964; Rushizky et al., 1964). Guanyl RNase is an enzyme which has played a decisive role in structural investigations; both total and partial hydrolysis with guanyl RNase, yielding metamers and halves of the molecules as products, have been extensively used to study the primary structure of all the transfer RNAs.

During the investigation of this enzyme much attention was naturally paid to its property of rupturing phosphodiester bonds formed by methylated derivatives of guanine (McCully and Cantoni, 1961, 1962b; Tanaka and Cantoni, 1963; Staehelin, 1964a; Dütting and Zachau, 1964b). Until recently this problem had not been completely settled, for different workers had obtained conflicting results. Determination of the primary structure of individual tRNAs shows that T_1-RNase ruptures all bonds formed by methylated derivatives of guanine except 7meGp and OmeGp. This is shown, for example, by the position of these nucleotides in the oligonucleotides 7meG-U-C-5meC-U-Gp and A-OmeC-U-OmeG-A-A-Y-5meC-U-Gp of the total guanyl-RNase hydrolysate of tRNA[Phe] from *S. cerevisiae* (RajBhandary et al., 1968b). All other methylated derivates of G occupy the 3′-terminal position in oligonucleotides of the total guanyl-RNase hydrolysate. For example, the dinucleotide U-1meGp has been identified among the hydrolysis products of tRNA[Ala] from *S. cerevisiae* and tRNA[Val] from *T. utilis* (Holley et al., 1965b; Takemura et al., 1968b); the dinucleotide C-2me$_2$Gp has been found in total hydrolysates of many yeast tRNAs; evidence for rupture of the bond formed by 2meG is given by the presence of the trinucleotide U-A-2meGp in the hydrolysate of RNA[Tyr] (Madison et al., 1967a) and of the pentanucleotide A-U-U-A-2meGp in yeast tRNA[Phe] (RajBhandary et al., 1968b). Admittedly all the guanine derivatives remain mostly as cyclic phosphates, suggesting that the corresponding bonds are ruptured more slowly than those formed by Gp, or that decyclization of cyclic phosphates of modified G derivatives is difficult.

The specificity of guanyl RNase from actinomycetes has been studied by Abrosimova-Amel'yanchik and co-workers (Abrosimova-Amel'yanchik et al., 1965); the study of oligonucleotides isolated form a pancreatic hydrolysate of total tRNA [(1meG, 2me$_2$G)Cp; (1meG, G)Cp; 2me$_2$G-ψp; 2me$_2$ G-Cp] showed that the actinomycete enzyme ruptures bonds formed by the methylated guanine derivatives, but that the rate of cleavage is much lower than with unmodified guanylic acid. This conclusion was confirmed by analysis of tRNA$_1^{Val}$ also, for the dinucleotide U-1meGp was identified in its hydrolysate although, unlike other oligonucleotides, it is partly preserved as the cyclic phosphate (Venkstern et al., 1968b).

Guanyl RNase (from takadiastase or from actinomycetes) also ruptures bonds formed by inosinic acid, and this has proved very useful in structural research. Yeast tRNAs are rich in Ip; this nucleotide is found in the anticodons of the alanine, serine, and valine tRNAs of *S. cerevisiae,* the valine and isoleucine tRNAs of *T. utilis,* and the serine tRNA of rat liver. Oligonucleotides with 3'-terminal Ip have been identified in exhaustive hydrolysates, namely: C-U-C-C-C-U-U-Ip in tRNA^{Ala}, C-ψ-U-Ip in tRNA$_1^{Val}$, A-ψ-U-Ip in tRNA^{Ser}, etc. T$_1$-RNase ruptures bonds formed by Ip more slowly than those formed by Gp (Whitfield and Witzel, 1963); actinomycete guanyl-RNase attacks bonds in the 3'-position relative to Gp and Ip at almost identical rates (Abrosimova-Amel'yanchik et al., 1965).

T$_1$-RNase does not hydrolyze the polynucleotide chain at methylated inosinic acid, as is shown by the presence of the tetranucleotide C-1meI-ψ-G in the total guanyl-RNase hydrolysate of tRNA^{Ala} (Holley et al., 1965b). The resistance of bonds formed by 1meI and 7meG to guanyl RNase is possibly not absolute, and with an increase in the quantity of enzyme or the reaction time its action might perhaps be exhibited. However, the use of a thousand times more of the actinomycete guanyl RNase than is necessary to produce cleavage of the GpN bond gave no indication of cleavage of the dinucleoside monophosphate 7meGpU, even after incubation for 2 h (Venkstern, unpublished data).

Guanyl RNase thus attacks the polynucleotide chain of the tRNAs not only at the guanosine residue, but also at its most widely occurring methylated derivatives, although at a slower rate; the exceptions are OmeGp, the structure of which does not permit the formation of a cyclic phosphate, an essential intermediate during the action of RNases, and 7meGp.

Experiments on poly(A), poly(U), and poly(C) have shown that the specificity of T$_1$-RNase relative to guanylic acid is not absolute (Irie, 1965). The nonspecific action of T$_1$-RNase has in fact frequently been observed, e.g., in the degradation of the oligonucleotide A-U-U-A-2meG>p to the trinucleotide A-U-U>p and the dinucleotide A-2meGp during the study of the structure of tRNA^{Phe} (RajBhandary et al., 1968b). With a shortening of the incubation times with the enzyme, the degree of this degradation was considerably reduced, but it could not be completely prevented.

Analysis of total tRNA hydrolysates obtained with the aid of the two RNases described above is one of the principal stages in the investigation of primary structure. However, the results have shown that notwithstanding the presence of minor components, enabling reconstruction of overlapping fragments to be carried out, it is impossible to reconstruct the whole molecule on the basis of the results of analysis of two total RNase hydrolysates. Cleavage at adenylic acid, by means of T$_2$-RNase as has been suggested (Sato and Egami, 1957), might provide the necessary additional information, but as has already been stated, the specificity of this enzyme is too wide (Rushizky and Sober, 1963). Because of the need to obtain additional overlaps, attempts were made to modify, in some way or other, the character of degradation with RNases in order to obtain longer fragments; modification of the substrate (Naylor et al., 1965; Seifert and Zillig, 1967; Budovskii, 1968), modification of the enzymes (Goldstein, 1967), and also a change in the conditions of their action (Litt and Ingram, 1964) have all been suggested for this purpose. For practical purposes, only the last method, i.e., the action of enzymes under unfavorable conditions, has been used to study the primary structure of the tRNAs, leading to the formation of incomplete or partial ribonuclease hydrolysates.

Incomplete, partial, or limited degradation of the polynucleotide chain is possible primarily because the rate of hydrolysis of the phosphodiester bond depends on the character of the neighboring bases. By varying the experimental conditions—the quantity of enzyme, the pH, or the temperature—it is possible to achieve rupture of only the most labile bonds while preserving the integrity of the more stable bonds. Kinetic constants for the formation of cyclic phosphates from dinucleotides by the action of pyrimidyl RNase (pH 6, 26°C), which were obtained several years ago (Witzel and Barnard, 1962; Witzel, 1963), are given below:

Substrate	K_2, sec^{-1}	Substrate	K_2, sec^{-1}
CpA	3000	UpA	1200
CpG	500	UpG	—
CpC	240	UpC	40
CpU	27	UpU	11
C > p	5.5	U > p	2.2

These results show that differences in the rates of hydrolysis are due not only to the pyrimidine, but to the right-hand component forming the phosphodiester bond. The dinucleoside monophosphate UpU is hydrolyzed the slowest of all, CpA the fastest of all; the difference between them is almost 300 times. The relative rates of hydrolysis of different substrates by T_1-RNase were determined by Whitfield and Witzel (1963):

Substrate	Relative rate of hydrolysis	Substrate	Relative rate of hydrolysis
GpCp	1100	GpU	250
GpC	800	IpC	150
GpA	550	G > p	2
GpG	450		

In this case the influence of the right-hand neighbor also is seen, but the differences in the velocities of hydrolysis are much smaller.

Analogous results have been obtained for guanyl RNase from actinomycetes by Abrosimova-Amel'yanchik et al. (1965) (Table 3).

Unlike T_1-RNase, the enzyme from actinomycetes hydrolyzes the G-Up bond fastest, followed in succession by I-Up and G-Cp bonds. Deamination and methylation of guanylic acid lead to considerable slowing of the reaction.

These results suggest that partial hydrolysis can be most easily carried out with pyrimidyl RNase, since the range of differences in this case is maximal. This property of pyrimidyl RNase has in fact been successfully used for the analysis of polypyrimidine sequences in guanyl RNase oligonucleotides of tRNA^{Val} from *S. cerevisiae* (Venkstern et al., 1968b).

In practice, however, it was not the unequal stabilities of the phosphodiester bonds, which depend on the nature of the nucleotides, that played the principal role in structural investigations, but conformational factors, i.e., the stereochemical accessibility of the particular bond to the action of the RNase. It has long been known that hydrolysis of nucleic acids by nucleases depends on their secondary structure (Singer et al., 1960; Grunberg-Manago, 1962; Cantoni et al., 1962; Nihei and Cantoni, 1963). The high stability of phosphodiester bonds in double-stranded parts of the molecule has been

TABLE 3. Rate of Hydrolysis of Dinucleosides by Guanyl Ribonuclease from Actinomycetes (in %)

Time, min	G-Cp	2me₂G-C	G-Up	G-ψp	2me₂G-ψ	I-Up
15	20	2	46	21	2	32
30	32	3	67	40	3	53
60	52	7	80	60	6	73
120	71	13	90	77	11	90
240	86	22	96	—	15	95
1440	95	71	100	—	19	95

clearly demonstrated in synthetic polynucleotides. Snake-venom PDEase, for instance, readily hydrolyzes single-stranded polynucleotides, such as poly(I), poly(C), and so on; hybrid complexes—e.g., [poly(A) • poly(I)], [poly(C) • poly(U)]—are much more difficult to hydrolyze (Cousin, 1963). Although T_2-RNase readily hydrolyzes poly-N^6-hydroxyethyladenylic acid, consisting of a single-stranded chain, it does not attack polyadenylic acid, which forms a double-stranded structure under the experimental conditions (Van Holde et al., 1965).

The inhibitory action of magnesium ions on degradation of nucleic acids by nucleases is mainly due to the increased stability of the secondary structure of the polynucleotides, as reflected, for example, in their increased melting temperature under these conditions. This effect of Mg^{2+} is seen particularly clearly in the case of tRNA (Nishimura and Novelli, 1963). That is why, if it is desired to preserve the secondary structure, partial hydrolysis of individual tRNAs is carried out by a small quantity of enzyme at $0°C$ and in the presence of Mg^{2+}.

The discovery that the action of RNases is highly selective in the presence of Mg^{2+} and at $0°C$ was a major event in structural research, and as a result of it the problem of determining the nucleotide sequence in tRNA was carried to a successful conclusion. Conditions could now be chosen under which RNases would rupture only one phosphodiester bond in the tRNA molecule, located in every case in the anticodon loop. The relative stability of this bond varies from one tRNA to another, but within a given molecule it is always minimal. Anticodon bonds are ruptured by guanyl RNase regardless of whether they are formed by Gp or Ip, and they are left intact only if they cannot be attacked by the enzyme at all (for example, in the tRNA^{Phe} containing OmeG in the anticodon). Hydrolysis in this manner is possible because the phosphodiester bonds located in the anticodon region in all tRNAs are most accessible to the action of RNases, probably because they are not protected by hydrogen bonds and by the screening effect of stereochemically close groups in the native molecule.

Cleavage of the tRNA molecule into two halves was first achieved in the case of yeast tRNA^{Ala} (Penswick and Holley, 1965) (Fig. 3). Total hydrolysis of the fragments I and II thus obtained with guanyl RNase demonstrated their complementary character in the sense that together they contained all the oligonucleotides of the guanyl-RNase hydrolysate of the whole molecule (Fig. 4). Cleavage took place at the phosphodiester bonds of the I-G-C anticodon. The halves of tRNA^{Ala} were not isolated in preparative quantities, and the subsequent structural investigation was carried out on metamers obtained from the whole molecule (Apgar et al., 1965).

The most labile bonds in yeast tRNA^{Ser} were found to be those of the I-G-A triplet, at which hydrolysis by T_1-RNase took place under the mildest

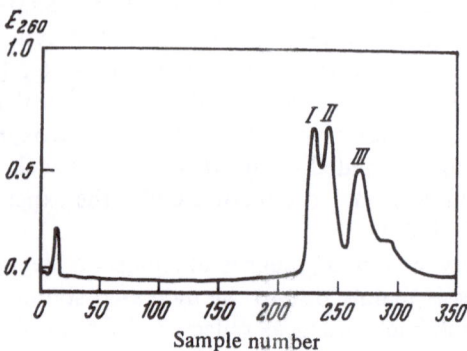

Fig. 3. Fractionation of yeast tRNA^{Ala} and its two halves on DEAE-cellulose (Penswick and Holley, 1965): (I) 3'-half; (II) 5'-half; (III) unsplit molecule.

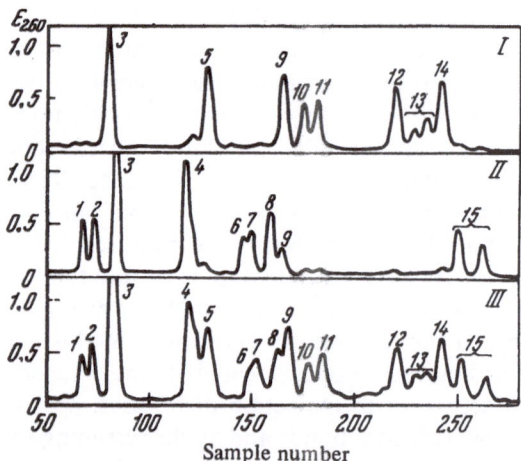

Fig. 4. Fractionation of total guanyl-RNase hydrolysates of yeast tRNA^{Ala} and of its two halves on DEAE-cellulose: (I) 3'-half; (II) 5'-half; (III) tRNA^{Ala}; (1–15) oligonucleotides.

conditions. Serine tRNA was found to be much more stable than tRNA^{Ala}. Under identical conditions of hydrolysis, when more than 50% of the tRNA^{Ala} was found as halves, 90% of the tRNA^{Ser} still remained as whole molecules. Appreciable formation of halves took place only when the incubation time was doubled or when the content of enzyme was increased by a factor of 3 or 4. It has been postulated that the reason for the greater stability of tRNA^{Ser} is its more compact three-dimensional structure. In this case also, halves were not used to determine the oligonucleotide sequences (Dütting et al., 1966).

Tyrosine tRNA from yeast, on the other hand, was hydrolyzed much more easily than tRNA^{Ala} by T$_1$-RNase (Madison and Kung, 1967). For instance, when the conditions were such that about 70% of the tRNA^{Ala} was still present as large fragments, more than half of the uv-absorbing substance of the tRNA^{Tyr} was found in the oligonucleotides. The 5'-half of the molecule was particularly labile; after hydrolysis of the whole molecule, virtually none of its oligonucleotides could still be left as large fragments. In order to determine its structure, these workers were therefore compelled first to isolate the 5'-fragment, and then to subject it to partial hydrolysis. Separation of the halves of the tRNA^{Tyr} molecule is illustrated in Fig. 5. Partial hydrolysis of the middle peak, representing the 5'-half of tRNA^{Tyr}, yielded the necessary overlaps for reconstruction. Madison and Kung consider the greater lability of tRNA^{Tyr} than of tRNA^{Ala} to be the result of the smaller number of hydrogen bonds in the helical part of the H$_2$U branch (three instead of four in tRNA^{Ala}); however, this can hardly be true because tRNA^{Ser}, which also has three hydrogen bonds in its helical part, is much more stable than tRNA^{Ala}.

As was mentioned above, yeast tRNA^{Phe} cannot be attacked by T$_1$-RNase at the anticodon because of the presence of a methyl group on the 2'-OH group of the ribose. In this case degradation into halves was achieved with pancreatic RNase. The possibility of splitting tRNA into halves with pancreatic RNase was first demonstrated by Litt and Ingram (1964). This problem was investigated more fully by Armstrong et al. (1966), who showed that the most labile bond in tRNA^{Ala} and tRNA^{Val}, so far as attack by

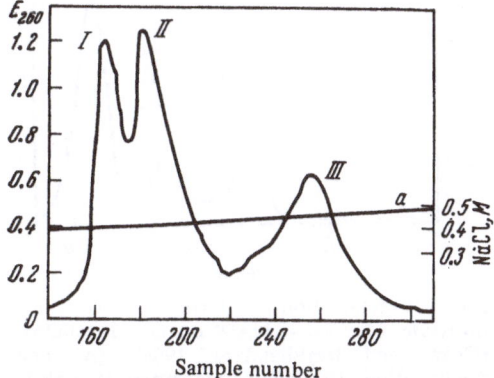

Sample number

Fig. 5. Fractionation of yeast tRNA^{Tyr} and its two halves on DEAE-cellulose (Madison and Kung, 1967): (I) 3'-half; (II) 5'-half; (III) unsplit molecule; (a) NaCl gradient.

pancreatic RNase is concerned, lies near the middle of the molecule. Zachau et al. (1966d) also stated that the phosphodiester bonds located in the anticodon region of tRNA^{Ser} are most labile relative to pancreatic RNase. Fractionation of a partial pyrimidyl-RNase hydrolysate of tRNA^{Phe} on DEAE-cellulose in a linear gradient of NaCl and 7 M urea at pH 7 gave the picture shown in Fig. 6A, while rechromatography under the same conditions, but at pH 3, gave the profile illustrated in Fig. 6B (Chang and RajBhandary, 1968). Analysis of the resulting peaks showed that Id represents the unsplit tRNA^{Phe} molecule which has lost its acceptor end. Peak Ic represents the 5'-fragment containing all nucleotides from pGp to the U residue bordering the anticodon. Peaks Ia and Ib correspond to the 3'-fragment of the molecule without the terminal pCpA, and terminating at the 5'-end in the hypothetical anticodon OmeG-A-A; these peaks differ only in that they contain different

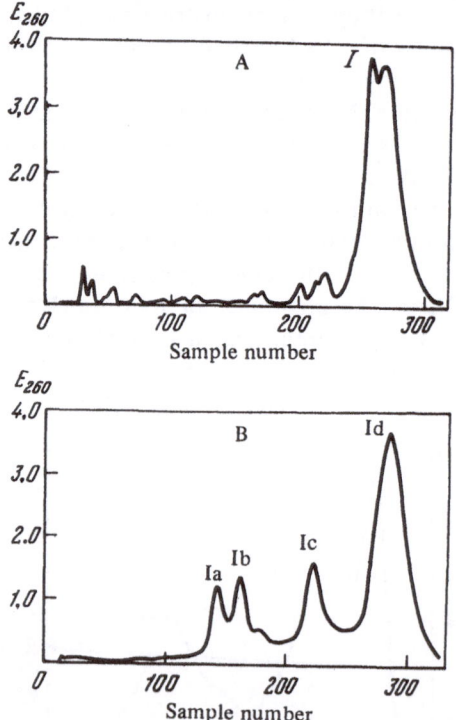

Fig. 6. Fractionation of partial pancreatic hydrolysate of yeast tRNA^{Phe} on DEAE-cellulose (Chang and RajBhandary, 1968): (A) first fractionation; (B) rechromatography of peak I; (Ia) 3'-half, nucleoside Y not modified; (Ib) 3'-half, nucleoside Y modified; (Ic) 5'-half; (Id) unsplit molecule of tRNA^{Phe}; (I, a, b, d) without acceptor end.

modifications of the nucleoside Y (see below). Peaks I (a and b) and Ic thus correspond to the 3′- and 5′-halves of the tRNA^{Phe} molecule respectively, and this was the first case in which an individual tRNA was split into two halves by pancreatic RNase. In this case also, the phosphodiester bonds of the acceptor end and the anticodon loop were most accessible to attack by pancreatic RNase.

Halves have been obtained from tRNA$_1^{Val}$ by means of guanyl RNase from actinomycetes (Baev et al., 1967b). Conditions for obtaining the optimal degree of hydrolysis were chosen empirically; the results showed that if about 70% of the tRNA$_1^{Val}$ is hydrolyzed, virtually only one phosphodiester bond in the molecule is ruptured. If a higher percentage of tRNA is hydrolyzed, products of more advanced hydrolysis are obtained, especially of the nonacceptor half, which is evidently less resistant to the action of RNases (Armstrong et al., 1966; Madison and Kung, 1967). The halves of the tRNA^{Val} molecule obtained under unfavorable conditions of hydrolysis could be completely separated from each other and from the unsplit molecule by two methods (Figs. 7 and 8) (Baev et al., 1967b). The halves of the tRNA$_1^{Val}$ molecule were isolated in preparative amounts, and the whole

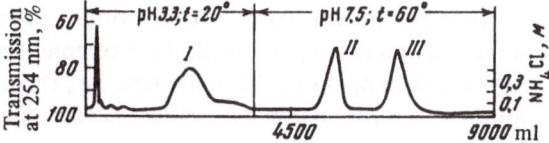

Fig. 7. Fractionation of tRNA$_1^{Val}$ and its two halves on DEAE-cellulose, after Aksel'rod et al. (1967): (I) 3′-half; (II) 5′-half; (III) tRNA$_1^{Val}$ (unsplit molecule).

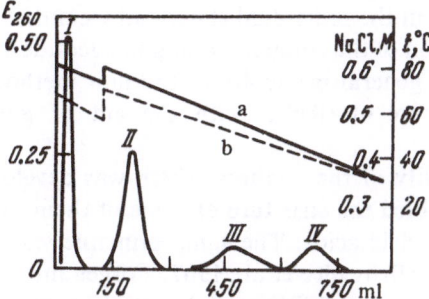

Fig. 8. Fractionation of tRNA$_1^{Val}$ and its two halves on DEAE-cellulose, after Mirzabekov et al. (1966): (I) 5′-half; (II) 3′-half; (III) unidentified hydrolysis products; (IV) tRNA$_1^{Val}$ (unsplit molecule); (a) temperature gradient; (b) NaCl gradient.

work of reconstruction was carried out with them (Mirzabekov et al., 1970).

One part of the tRNA molecule is thus most accessible to enzymic attack. Hydrolysis under mild conditions (Mg^{2+}, $0°C$, small quantity of enzyme) leads to rupture of a single phosphodiester bond in this part. Analysis of the resulting large fragments, using complete degradation with RNases, has shown that one of them is the 3'-half, and the other the 5'-half of the molecule, and together they contain all oligonucleotides identified in the native molecule; the only exception is the oligonucleotide where the hydrolysis takes place, whose fragments are shared among the different halves, and in the case of pancreatic RNase, the acceptor oligonucleotide, which does not contain the terminal CA_{OH}.

Under rather more vigorous conditions not one but a comparatively small number of bonds are ruptured in tRNA molecules. In this case also the ruptures are not at random, but occur at definite phosphodiester bonds located in accordance with the cloverleaf model in single-stranded parts of the molecule (see below). The fragments thus obtained, known as metamers, are smaller than halves but they each contain several oligonucleotides of the exhaustive hydrolysates. Metamers are isolated in the individual state and analyzed by exhaustive hydrolysis with RNases. The reaction products are oligonucleotides, already known from the previous analysis; they are identified, and in this way an idea is obtained of the oligonucleotide composition of the metamers. Overlapping of the metamers, i.e., the discovery of the same oligonucleotide in several metamers, enables the structure of the whole molecule to be rebuilt.

In some cases ($tRNA_1^{Val}$, the 5'-half of $tRNA^{Tyr}$), metamers were obtained, as was mentioned above, not from the whole molecule but from its halves, which is advantageous. First, it facilitates fractionation of the metamers and their isolation in the individual state, and second, it simplifies the procedure of identification of the component oligonucleotides.

These are the general principles of the unit method used to determine the primary structure of tRNA, with the aid of guanyl and pyrimidyl RNases.

The applicability of the method, which was developed for the tRNAs, and which has enabled the structure of some of them to be identified, is not limited to these nucleic acids. The same principle was successfully used to study ribosomal 5S (Brownlee et al., 1967; Forget and Weisman, 1967, 1969) and 6S RNAs (Brownlee, 1971), and led to the complete determination of their structure. The 5S RNA molecule is 1.5 times longer than the tRNA molecule and it contains no minor components. Neither of these factors represented an obstacle to determination of their structure. It can accordingly be concluded that the presence of minor components is not the deciding factor in the determination of primary structure, and that nucleotide sequence in a polymer can be elucidated by this method provided that the molecule contains four different structural elements.

During work with Sanger's methods (Sanger et al., 1965; Brownlee and Sanger, 1967, 1969) and using labeled RNAs, as is being done by an increasing number of workers, the absence of minor components and the presence of only four major nucleotides in the polynucleotide chain, on the contrary, considerably facilitates the task.

It is a more difficult task to determine the nucleotide sequence in high-molecular-weight (virus and ribosomal) RNAs. The difficulty is due to the much greater length of the polynucleotide chain—3000–6000 nucleotides for virus RNAs, 3200–5500 for 23–28S RNAs, and 1600–2500 nucleotides for 16–18S RNAs (Belozersky and Spirin, 1960; Spirin and Gavrilova, 1968)—and also to the absence of any adequate criterion for determining homogeneity of the specimens in this case.

However, it is now quite certain that it can be solved by existing methods. Recent experimental findings show that the RNAs of ribosomal subparticles are homogeneous and are therefore suitable for structural research. Evidence of this is given by the picture obtained on fractionation of the 18S and 28S rRNAs by electrophoresis in polyacrylamide gel (Hadjiolov et al., 1967a,b), by the results of analysis of terminal oligonucleotides, according to which each rRNA has one 3'- and one 5'-terminal oligonucleotide (Takanami, 1967a,b), and by other observations. The RNAs of the E. coli bacteriophages (MS2, f2, R17, etc.) have been found to be excellent models for studying the structure and function of natural messenger RNAs; they are single-stranded RNAs, 3300–3500 nucleotides long, and they can be isolated in a homogeneous state.

Information now available also indicates that the method of partial enzymic hydrolysis, which has played a decisive role in decoding the structure of tRNA, can also be used with the ribosomal and phage RNAs, since they too evidently possess a highly regular secondary structure.

The methods principally used until now to study the structure of high-molecular-weight RNAs have been those developed on tRNA, but some new and extremely interesting technical discoveries have recently been made and these must be briefly mentioned here. I shall not attempt to survey the literature in full on this problem, but I shall merely outline the present state of the situation.

Research into the structure of high-molecular-weight RNAs began as did research on tRNA: attempts were made to obtain a general idea of the structure of the molecule and to determine the sequences at its 5'- and 3'-ends, which in this case also were naturally more accessible for study than the inner segments of the polynucleotide chain. High-molecular-weight RNAs were hydrolyzed exhaustively with guanyl and pyrimidyl RNases; the 5'-

and 3'-terminal fragments were identified by their characteristic triphosphate and nucleoside terminal groups (De Wachter and Fiers, 1967, 1969; Weith and Gilham, 1967, 1969; De Wachter et al., 1968a,b; Dahlberg. 1968; Glitz et al., 1968; Fellner et al., 1970a; Bishop et al., 1968; Young and Fraenkel-Conrat, 1971) or by the use of various labels (Midgley and McIlreavy, 1966, 1967; Mandels, 1967a,b; Takanami, 1967a,b; Glitz et al., 1968; Glitz and Sigman, 1970; Hunt, 1965, 1970). From all the facts which have now been obtained it is possible to compare the 3'- and 5'-terminal sequences determined by different investigators for the same RNAs (Adams and Cory, 1970). An essential step was the production of catalogs of oligonucleotides of total pyrimidyl- and guanyl-RNase hydrolysates (Thirion and Kaesberg, 1968; Beck et al., 1970; Fellner et al., 1970b; Min Jou and Fiers, 1969).

Again, isolation of larger fragments was also an essential step in the case of the high-molecular-weight RNAs. To begin with partial enzymic hydrolysis was used for this purpose, but because of the rigid secondary structure of the ribosomal and virus RNAs, cleavage was possible only at certain parts of the molecule, and as in the case of the tRNAs, it yielded large units of characteristic composition (Fiers, 1968; Min Jou et al., 1969, 1970, 1971; Ehresmann et al., 1971). An interesting development of the method of partial hydrolysis has been to make a deliberate search among the resulting partial hydrolysis products for particular fragments. The *E. coli* phages have genomes with only a few cistrons and with a correspondingly small number of products of gene action. The RNA of these phages has been shown to code the synthesis of three proteins: the phage envelope protein, the maturation protein, and the enzyme replicase; the amino-acid sequence of one of them, the envelope protein, has been completely established. Determination of the nucleotide sequence in the RNA of these phages at the same time had the object of elucidating the process of coding of their proteins. For this purpose, starting from the known amino-acid sequence and using the genetic code, units corresponding to particular segments of this protein were isolated from the partial hydrolysis products of the phage RNA.

In this way Adams et al. (1969) identified and determined the structure of a fragment of the RNA of phage R17 which was 57 nucleotides long and which corresponded completely to the amino-acid sequence of a particular segment of the protein of phage R17. This work is of great theoretical significance, because for the first time the nucleotide sequence in a messenger RNA was determined chemically, and the validity of the genetic code was confirmed by the most direct method. These workers showed that the code used by the phage is in fact degenerate, because some of the amino acids which occur repeatedly are coded by different triplets. Later, fragments of the RNA of the same phage, corresponding to the first six amino acids of its envelope protein (Robinson et al., 1969) and the six terminal amino acids of this protein (Nichols, 1970), were identified. Adams and Cory (1970) have isolated and studied a fragment of the RNA of phage R17 which is 74 nucleotides long and which evidently is concerned not with the synthesis of specific protein, but with RNA replication.

An interesting approach to the production of large fragments of the high-molecular-weight RNAs is the treatment of RNA–ribosome complexes with RNases. The stability of certain segments of RNAs associated with ribosomes to the hydrolytic action of nucleases has been known for a long time (Shakulov et al., 1962; Takanami et al., 1965). By making use of this observation and obtaining R17 RNA–ribosome complexes under the conditions of initiation of synthesis of the polypeptide chain, the subsequent action of pancreatic RNase made it possible to remove the whole of the messenger RNA with the exception of that part which was in the initiator complex (Steitz, 1969); this part was isolated and studied. The same method was used to isolate, and later to study, sequences of 26 nucleotides including the initiator part of the cistron coding the envelope protein of phage Qβ (Hindley and Staples, 1969). The stability of RNAs associated with ribosomes has been used in Ebel's laboratory (Ehresmann and Ebel, 1970; Fellner et al., 1970c) both to obtain large fragments of 16S and 23S rRNA and also to examine their role in the association of subparticles. Yet another approach allowing systematic determination of nucleotide sequences in phage RNAs must be mentioned; it is based on the ability of Qβ replicase to use minus chains as templates for the formation of the infectious RNA of phage Qβ. During brief incubation under synchronization conditions in the presence of α-^{32}P-nucleoside triphosphates, homogeneous specimens of short radioactive fragments of the phage RNA of desired length from 5 to 300 nucleotides could be obtained. In this way the 5′-terminal sequence (175 nucleotides long) of the RNA of phage Qβ was determined (Billeter et al., 1969).

These last elegant investigations not only provided new methods of obtaining large RNA fragments for structural research and enabled the nucleotide sequence in considerable segments of phage RNA molecules to be determined, but they also gave a much deeper insight into the various stages of protein synthesis. So far as the study of the structure of these large fragments is concerned, it continues to proceed on the basis of the same unit principle whose applicability on such large scale has become possible only thanks to the series of methods proposed by Sanger and his collaborators (see below).

THE USE OF CHEMICAL MODIFICATIONS

An important approach to the study of nucleic acids is their chemical modification which, in principle, can be used to isolate the nucleic acids, to study their primary and macromolecular structure, and also for functional investigations. The use of chemical modifications in the case of nucleic acids is complicated by the similarity between the chemical properties of the different nuclear bases and the lability of the phosphodiester bonds. Nevertheless, the problem has proved solvable, and many selectively acting chemical agents which do not disturb the integrity of the polynucleotide chain are now known.

In the study of primary structure, chemical modifications can be used either to make the internucleotide bond next to the modified component more labile or to increase its stability to the action of nucleases.

One of the most thoroughly studied agents used to modify nucleic acids is hydroxylamine, which hydrolyzes uridine at pH 10 and modifies cytosine at pH 6 (Verwoerd et al., 1961; Kochetkov et al., 1962; Zillig et al., 1962; Cerna et al., 1964). At pH 10 and 0° C hydroxylamine reacts only with uridylic acid residues, which it converts into ribosylurea. Pyrimidyl RNase does not act on phosphodiester bonds formed by ribosylurea, and it ruptures the polynucleotide chain only at cytidine residues. RNA modified in this manner can be split into much larger fragments than native RNA (Kochetkov and Budovskii, 1964, 1968).

Pyrimidyl RNase can be converted into cytidyl RNase also by treatment of the RNA with cyclohexyl-β-[N'-(N-methylmorpholinium)]-ethylcarbodi-imide-p-tosylate (water-soluble carbodiimide) (Gilham, 1962; Knorre, 1966). On hydrolysis of the RNA by pyrimidyl RNase modified with water-soluble carbodiimide, oligonucleotides having only a cytidyl residue at the 3'-end are formed (Lee, J.C., et al., 1965). This modification also prevents degradation of the polynucleotide chain by PDEase (Naylor et al., 1965). A valuable property of water-soluble carbodiimide, and one which compels investigators to come back to this compound again and again, is its sensitivity to the spatial organization of nucleic acids. Girshovich and Shubina (1969) have recently isolated a product of interaction between uridine and water-soluble carbodiimide and have studied its properties.

RNA can be made resistant at its cytidyl residues by treatment with acethydrazide pyridinium chloride [Gerard's reagent (Ukita et al., 1964; Kikugawa et al., 1967a,b)], with semicarbazide (Hayatsu and Ukita, 1964), or with hydrazine. After treatment with hydroxylamine (modification of Up) and hydrazine (modification of Cp), pyrimidyl RNase hydrolyzes tRNAs only at Tp and ψp, for these nucleotides are resistant to treatment with both compounds (Verwoerd and Zillig, 1963; Seifert and Zillig, 1967).

Several RNA modifications at the guanyl residues have been described. Glyoxal "limits" the action of T_1-RNase as the result of which the enzyme splits the polynucleotide chain only at inosine and dimethylguanyl residues (Kochetkov et al., 1967). Bonds formed by guanylic acid become resistant to guanyl RNase as the result of the action of trinitrobenylate (Azegami and Iwai, 1964) and Kethoxal (Litt, 1969).

Most of the reactions listed above have been well studied from the point of view of their chemical characteristics, the integrity of the polynucleotide chain, the quantitative character of the yield, and their effect on the vulnerability of the corresponding phosphodiester bonds to enzymic attack. However, there are still very few cases in which these methods have been used to study the primary structure of nucleic acids. Brownlee et al. (1968), for instance, used two of them to obtain longer fragments in order to study the structure of 5S rRNA. Water-soluble carbodiimide has been used to modify U (Gillam, 1962), so that on subsequent hydrolysis with pyrimidyl RNase, cleavage occurred only at C (Lee, J.C., et al., 1965). The action of guanyl RNase has been limited by partial methylation of RNA with dimethyl sulfate, leading to conversion of some of the G residues into 7meG; as already mentioned, phosphodiester bonds formed by 7meG are resistant to the action of guanyl RNase (Lawley and Brookes, 1963; Brimacombe et al., 1965). Both reactions have been used to determine sequences in the single-stranded parts of the molecule which are most difficult to analyze, for these parts of the molecule were modified before the rest.

Modification by water-soluble carbodiimide, followed by the action of pancreatic RNase has also been used to study the structure of RNA of phage R17 (Adams et al., 1969) and of tRNA[Met] from E. coli (Cory and Marcker, 1970).

Attempts have been made in the past to use chemical methods for specific cleavage of polynucleotide chains. Thus the use of hydroxylamine appeared promising for this purpose, for by its action at 37° C an oxime group is formed from the ribose of uridylic acid, and this can be removed by treatment with cyclohexane at pH 4.0. As a result an aldehyde group is formed, the ribose ring is opened, and β-elimination takes place, with rupture of the phosphodiester bond at the 3'-position. Hence, in principle, rupture of the polynucleotide chain can be achieved at uridyl residues only.

A number of methods of specific chemical degradation of the polynucleotide chain at several minor components have now been developed for individual tRNAs. They are all based on the removal by some means or other of a particular base (from the Y nucleotide in tRNA[Phe], 7meG, and H_2U) and subsequent rupture of the labilized phosphodiester bond by a β-elimination mechanism (Philippsen et al., 1966; Wintermeyer and Zachau, 1970; Beltchev and Grunberg-Manago, 1970a).

The method of chemical modification suggested by Ziff and Fresco (1969) is of great importance in connection with the study of the primary structure of E. coli tRNAs rich in thiopyrimidines. The polynucleotide is treated with sodium periodate, which leads to the conversion of 4thioU (**1**) into 2-oxypyrimidine-4-sulfonate (**2**). This is followed by ammonolysis, for example, with methylamine; the fairly stable N^4-methylcytosine (**3**) is thus formed, and if labeled CH_3NH_2 is used, this contains the label. The principal nitrogenous bases remained completely unaffected by this procedure, and the polynucleotide chain remains intact.

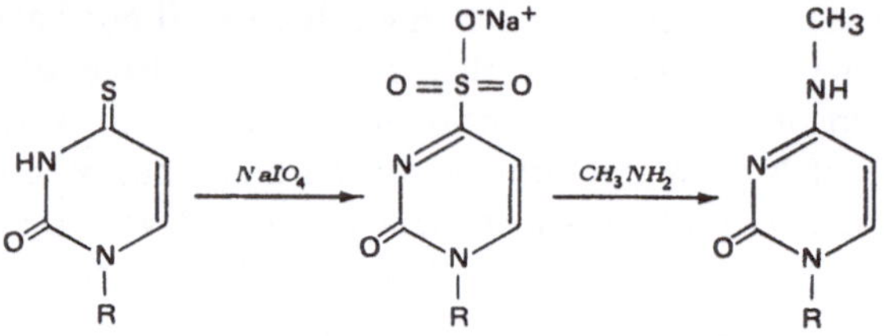

This modification thus leads to stabilization of the labile minor component and to specific introduction of the label. Subsequent degradation of the polynucleotide chain is carried out in the usual manner.

An interesting modification of tRNA, altering its vulnerability to attack by pyrimidyl RNase, has been described by Zachau (1964). According to his observations, if RNA is irradiated with uv light, neighboring uracil residues undergo dimerization:

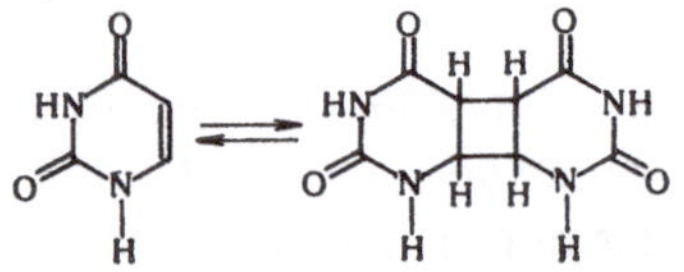

The dimers are not hydrolyzed by pyrimidyl RNase, and oligomers of considerable length are obtained through its action. A disadvantage of the method is the formation of a certain number of dimers through deamination of the cytidylic acid, as well as the formation, even by irradiation of an individual tRNA, of a variety of different dimerized oligonucleotides as the result of the random character of radiation injuries in the molecule.

Chambers and co-workers (Tomasz and Chambers, 1968; Schulman and Chambers, 1968) have suggested degradation of polynucleotide chains at ψ residues by the method of ultraviolet irradiation. For example, according to the results obtained by these workers, the photochemical degradation of T-ψ-C-Gp takes place in accordance with the following reaction:

$$\text{T-}\psi\text{-C-Gp} \xrightarrow{h\nu} \text{T-Y} + \text{5-Formyluracil} + \text{pC-Gp}$$

On irradiation of a purified specimen of tRNA^{Ala}, cleavage took place only at the 2 ψ residues in the molecule of this tRNA.

Chemical modifications are being increasingly used at the present time in structural and functional investigations and also to study the conformation of the tRNA molecule whose primary structure has already been identified. It is outside the scope of this monograph to discuss these matters in detail, and only a few general remarks and examples will therefore be given.

The role of individual nucleotides or nucleotide sequences in tRNA has been established by modifying them and investigating the functional activity of the modified tRNA. This approach was until recently confined mainly to the investigation of terminal sequences and minor components, which can be selectively modified.

For example, by acetylating the terminal hydroxyl groups of tRNA, Knorre et al. (1969) showed that the terminal adenosine residue takes part in the formation of the tRNA–aminoacyl-tRNA-synthetase complex. Modification of I in the anticodon leads to loss of the adaptor activity but has no effect on the acceptor activity of yeast tRNA^{Ala} (Yoshida et al., 1968b). Modification of the minor components will be examined in more detail later, in connection with their functions. A promising approach, which will probably allow modification to be carried out at any part of the polynucleotide chain, is to attach modifying reagents to oligonucleotides; by means of complementary interactions with certain sequences, these reagents can direct the modification to segments close to these sequences (Grineva et al., 1968). The first compounds of this type have already been synthesized (Belikova et al., 1967).

During the investigation of macromolecular structure the basic assumption is that some nucleotides, potentially capable of reacting with a given reagent, are concealed and do not react, while other are exposed and do react. It was found in this way that not only bases linked by hydrogen bonds in accordance with the clover-leaf model can be protected, but also some of those bases which, according to this model, lie in unpaired parts of the molecule. Despite some complications connected with the possible change in conformation of tRNA by the influence of the reagent and medium (Girshovich et al., 1968; Bekker et al., 1969), certain distinguishing features of the macrostructure of tRNAs have been established. For example, the results obtained by bromination and deamination of tRNA^{Ala} (Nelson et al., 1967), by treatment of tRNA^{Ala} with water-soluble carbodiimide (Brostoff and Ingram, 1967), etc., indicate that the segment containing the sequence T-ψ-C-G is concealed, while the anticodon loop and both terminal sequences are exposed.

Treatment with monoperphthalic acid has been used to study the tertiary structure of yeast tRNA^{Ser} and tRNA^{Phe} (Cramer et al., 1968; Erdmann et al., 1969). The localization of the oxidized adenosine residues was determined and it was concluded that the A residues in the CCA end, in the H$_2$U loop, and in the anticodon loop do not participate in the formation of the tertiary structure.

Similar results were obtained recently by Bollack et al. (1969) after methylation of yeast tRNA^{Phe} with dimethyl sulfate. These workers showed that in the presence of 0.02 M Mg^{2+} only five methyl groups are incorporated into tRNA^{Phe} and they are localized mainly in the 3′-terminal portion, in the anticodon and H$_2$U loops, but not in the loop of the "universal oligonucleotide."

Litt (1969) modified yeast tRNA^{Phe} with Kethoxal; under conditions allowing preservation of the secondary structure only two G residues, localized in the H$_2$U and anticodon loops, reacted.

On modification of yeast tRNA$_1^{Val}$ with O-methylhydroxylamine (Zhilyaeva and Kiselev, 1970; Zhilyaeva et al., 1970) three segments with exposed cytidine residues were found, namely, the anticodon and H$_2$U loops and the acceptor end of the molecule. The inaccessibility of the 1meA and 7meG residues in the intact yeast tRNA^{Phe} molecule to the reducing action of NaBH$_4$ suggested that these two positively charged minor components participate in the formation of intramolecular bonds, and that their role is possibly one of stabilizing the three-dimensional structure through interaction with the negatively charged phosphate groups (Igo-Kemenes and Zachau, 1971).

Chemical modifications of widely different types, resulting from the use of different chemical agents, thus yield uniform results indicating that both terminal sequences, the H$_2$U loop, and the anticodon loop in tRNA molecules are exposed, while the Tψ loop is concealed.

Judging from the results of experiments with 7meG, the extra branch or at least some of its nucleotides is also screened. This conclusion was also reached by workers who used not only the method of chemical modification to detect free and screened nucleotides, but also the opportunities afforded by the investigation of mutant RNAs (Smith et al., 1970, 1971; Cashmore, 1971). The reactivity of the C residues toward methoxamine was studied (Kochetkov et al., 1963, 1969) in tRNA^{Tyr} Su$_{III}^{+}$ and various mutants in which, because of substitution of certain nucleotides, ability to form several pairs was modified. Replacement of the A15 by the G15 of the H$_2$U loop in tRNA^{Tyr} Su$_{III}$ A15 in fact led to an increase in the reactivity of C57 of the extra branch, as would be expected from Levitt's model (Levitt, 1969): formation of the pair G15–C57, postulated by this model for all tRNAs, is impossible in the case of tRNA^{Tyr} Su$_{III}$ A15 because of the G15→A15 substitution. Kochetkov and co-workers conclude that the H$_2$U nucleotides in the loop and accessory branch interact with each other in the three-dimensional structure of tRNA^{Tyr} Su$_{III}^{+}$.

Electron microscopy is a promising method of study of the primary structure of nucleic acids, especially those of high molecular weight (Beer and Moudrianakis, 1962; Ulanov et al., 1967). In this case the selective chemical modification must make certain nuclear bases visible in the electron microscope. Beer and co-workers have suggested selective markers for G in DNA (Moudrianakis and Beer, 1965a,b), G and U in RNA, G and T in RNA, G and T in DNA (Fiskin and Beer, 1965), for C (Gal-Or et al., 1967), and also for T (Highton et al., 1968).

For electron-microscopic investigation of the secondary and tertiary structure of tRNA, Fröholm and Olsen (1969) used the method of staining with 1 % uranyl acetate and 4 % sodium silicotungstate. However, the method of electron microscopy has not yet been fully perfected and as yet it has hardly been used.

FRACTIONATION OF OLIGONUCLEOTIDES

One of the most important tasks in the determination of primary structure is separation of the fragments formed by specific degradation of tRNA molecules. Satisfactory methods of fractionation must possess high resolving power and must yield individual substances in sufficient amount for further analysis. In the study of primary structure it is essential to have methods capable of fractionating the following tRNA fragments: mononucleotides (including minor components), oligonucleotides of total ribonuclease hydrolysates, and products of incomplete enzymic hydrolysis, namely metamers and halves of the molecule.

When structural investigations began, only the first of these problems had been solved for practical purposes. To separate mononucleotides on paper several systems of solvents were used. As a rule they separated the principal components well and the few minor components known at that time more or less completely (Dunn et al., 1960). Cantoni et al. (1962) developed a method of fractionation of an alkaline hydrolysate of tRNA on Dowex-1 and, in various modifications, this has been used to determine the nucleotide composition of tRNA.

Satisfactory methods of fractionation of oligonucleotides, especially the higher members, were virtually absent, but they have been developed during the course of structural investigations and considerable progress has been achieved in this direction.

Ribonuclease hydrolysates of total tRNA are usually used as the model system. Fractionation of the more complex assortment of oligonucleotides obtained from total tRNA guarantees good results of fractionation of hydrolysates of individual tRNAs containing only a limited number of higher oligonucleotides.

Methods which can be used for fractionation of oligonucleotides include chromatography on columns, electrophoresis, chromatography on paper, and thin-layer separation.

Paper chromatography, provided that certain working methods are adopted, is a quantitative micromethod possessing great simplicity and clarity. This is particularly true of chromatography of the nucleic acids and their components which, of course, absorb uv light. Because of this, they can be detected by examination of the chromatogram in uv light without any additional treatment. Paper chromatography has now been developed for all components of nucleic acids from bases up to large oligonucleotides. An essential advantage of paper chromatography over the column method is that the fractionation can be checked in the course of the experiment and the resolution improved if necessary by further application of the solvent.

Complex mixtures of oligonucleotides obtained by enzymic hydrolysis of nucleic acids are usually fractionated on two-dimensional chromatograms, using electrophoresis in one direction and partition chromatography on paper in the other direction (Rushizky and Knight, 1960; Ingram and Pierce, 1962; Ingram and Sjöquist, 1963; Holley et al., 1963; Miura, 1964) or partition chromatography in both directions (Shapiro and Chargaff, 1963; Krutilina et al., 1964). If survey chromatograms, or fingerprints, of enzymic hydrolysates of individual tRNAs are obtained by using different principles of fractionation in two directions, adequate resolution can be obtained not only of the lower, but also of the higher oligonucleotides. The method of two-way separation gives very clear results and it is used for the direct comparison of oligonucleotide sets of isoacceptor tRNAs (Krutilina et al., 1970; Kryukov et al., 1971); to detect taxonomic differences between tRNAs with the same amino-acid specificity (Takemura and Miyazaki, 1969); to detect changes produced in individual tRNAs by chemical modifications (Brostoff and Ingram, 1967; Zhilyaeva and Kiselev, 1970), and also to determine the oligonucleotide composition of halves of tRNA molecules and also of any smaller fragments obtained by partial enzymic hydrolysis of the molecule.

Two-way chromatograms of the guanyl-RNase hydrolysate of the whole tRNA^{Val} molecule and of its two halves are shown in Fig. 9 (Mirzabekov et al., 1970). The distribution of oligonucleotides between the 3'- and 5'-halves of the molecule can easily be determined by comparing these chromatograms. The same systems of chromatographic solvents were used by Harada et al. (1969b) for thin-layer fractionation of the oligonucleotides of tRNA$_1^{Val}$ from E. coli. An advantage of two-way chromatography is that the position of each oligonucleotide is characterized by two parameters, and the reliability of identification by the position on the chromatogram is therefore much greater than that based on the position on the profile after column fractionation. Enzymic hydrolysates can be fractionated on two-way chromatograms without any form of preliminary treatment. The small

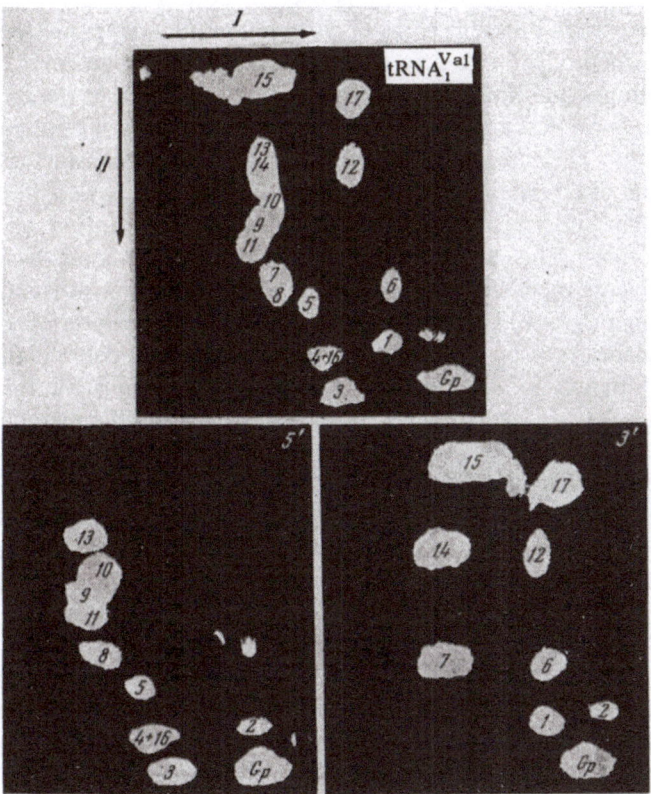

Fig. 9. Two-way chromatograms of guanyl-RNase hydrolysate of yeast $tRNA_1^{Val}$ and its two halves (contact print): (I) first direction, isobutyric acid–0.5 N NH OH (10 : 6), pH 3.7; (II) second direction, tertiary butanol–0.75 N HCOOH (1:1), pH 4.8; (1) C-Gp; (2) A-Gp; (3) U-1meGp; (4) U-Gp; (5) H_2U-C-Gp; (6) C-A-Gp; (7) T-ψ-C-Gp; (8) C-ψ-U-lp; (9) H_2U-H_2U-A-U-Gp; (10) U-C-ψ-A-Gp; (11) U-U-U-C-Gp; (12) A-C-A-C-Gp; (13) C-A-ψ-C-U-Gp; (14) 1meA-U-C-C-U-Gp; (15) A-A-C-7meG-H_2U-5meC-C-C-C-A-Gp; (16) pGp; (17) A-A-A-U-C-A-C-C-A_{OH}; (2, 4) breakdown products.

quantities of buffer and protein which are applied to the paper along with the mixture of oligonucleotides have no effect on the separation. A great advantage of fractionation on paper is that the oligonucleotides can be eluted from the paper after fractionation by any convenient solvent, including water, in any volume. By contrast with column chromatography, if the fingerprint method is used difficulties associated with large volumes and the need for desalting do not arise. This is very important with respect to subsequent spectrophotometric and, in particular, enzymic analysis of the oligonucleotides. Two-way chromatography is used in both analytical and preparative forms, for which different types of chromatography paper are used.

The method of two-way chromatography is not however, without its disadvantages and limitations. These include the fact that the error of determination is not less than $\pm 10\%$. However, the use of several parallel chromatograms can increase the reliability of the results considerably. Another important disadvantage is the inadequate resolving power of the method, resulting in superposition or insufficiently clear separation of certain oligonucleotides. This disadvantage can be partly overcome by passing the same solvents repeatedly or by additional fractionation of isolated areas of the chromatogram. The most rational method of analysis of a mixture of oligonucleotides is by the combined use of paper and column chromatography, e.g., by obtaining pure 3′- and 5′-halves in columns and then analyzing their total RNase hydrolysates on paper. However, this method requires a much larger amount of material.

The possibilities afforded by the method of two-way fractionation on paper are by no means exhausted, as its continuing development and improvement show. Attempts have been made to combine ion-exchange and paper chromatography and in this way to enjoy the advantages of both. Tyndall et al. (1964) described a method of two-way chromatography of pyrimidyl-RNase hydrolysates of ^{32}P-labeled RNA from HeLa cells on DEAE-paper.

A system of elegant methods, which have become widely used and which can yield much information on the transfer and high-molecular-weight RNAs, has been developed by Sanger and his collaborators.

The first method which was suggested was intended for fractionation of the products of complete degradation of RNA by ribonucleases (Sanger et al., 1965; Sanger and Brownlee, 1967). Oligonucleotides are fractionated in one direction by low-voltage electrophoresis on cellulose acetate at pH 3.5, and in the other direction on DEAE-paper, also at acid pH values. This method thus uses the forces both of ion exchange and of electrophoresis, so that good resolution of the oligonucleotides can be achieved. An essential feature of the method is that all the work is done on ^{32}P-labeled RNA; both detection and analysis of all the degradation products are carried out on the basis of radioactivity (autoradiography or counting). Correlation between the composition of an oligonucleotide and its position on the chromatogram is widely used to identify compounds. Accordingly, the quantity of a substance required for analysis was considerably reduced.

Fractionation of the oligonucleotides of a pyrimidyl-RNase hydrolysate of tRNA^{Phe} from E. coli by this method is illustrated in Fig. 37 (page 142).

Next, to separate partial hydrolysis products each containing 15–25 monomers, the use of chromatography on DEAE-paper in the second direction and a mixture of nonradioactive oligonucleotides as the eluent (homochromatography) has been used. Radioactive oligonucleotides are displaced from the origin by oligonucleotides of higher molecular weight and are distributed over the paper in accordance with their length, which determines their affinity for DEAE-paper (Brownlee et al., 1968).

To fractionate the longer oligonucleotides incorporating up to 50 monomers, the method of homochromatography was modified (Brownlee and Sanger, 1969). After ionophoresis on cellulose acetate at pH 3.5 in the first direction, fractionation in the second direction is carried out in a thin layer consisting of a mixture of cellulose and DEAE-cellulose. In this case also a mixture of nonradioactive oligonucleotides is used as the eluent. By transferring the homochromatography to a thin layer containing DEAE-cellulose, it is possible to fractionate larger fragments, and the greater compactness of the spots compared with those on paper leads to better resolution.

The methods described above have been adapted for use with RNAs which can be labeled in vivo. A further development will be to extend these methods to nonradioactive nucleic acids.

The first steps in this direction have already been taken (Szekely and Sanger, 1969). They suggested introducing the label into the 5'-hydroxyl group of the oligonucleotides of RNase hydrolysates of unlabeled RNAs by means of 5'-OH-polynucleotide kinase, and then carrying out the fractionation as before (Sanger et al., 1965) by two-dimensional ionophoresis.

By means of Sanger's method, the structure of 5S rRNA (Brownlee et al., 1967, 1968; Brownlee and Sanger, 1967) and 6S RNA from *E. coli* (Brownlee, 1971) and of the 5S rRNA from carcinoma KB (Forget and Weissman, 1967) has been completely elucidated; Hindley (1967) has used it to analyze fractions obtained by fractionating total tRNA from *E. coli* on polyacrylamide gel. The structures of all the tRNA from *E. coli* which have so far been studied were determined by the use of methods developed by Sanger and his collaborators. This method has also been used to determine the primary structure of asparagine tRNA from brewers' yeast (Keith et al., 1970).

The same method has been responsible for the enormous progress made recently in the study of the structure of high-molecular-weight nucleic acids, especially the polycistronic RNAs of viruses. It would be difficult to overestimate the importance of the method suggested by Sanger and his collaborators and of the researches undertaken on its basis. Suffice it to say that although the complete elucidation of the primary structure of any tRNA once took several years and required hundreds of milligrams of material, it can now be done in one month on micrograms of material. The only reservation which must be made is that by working blindly, i.e., purely by determining radioactivity, it is impossible to investigate the minor components, especially those hitherto unknown. Their identification requires the special isolation of adequately large amounts of the substance. As a result, many of the minor components of the tRNA from *E. coli* have not yet been identified.

A recent development in fractionation of oligonucleotides is the use of thin-layer chromatography and electrophoresis, which results in considerable saving of time and in great economy of material (Staehelin et al., 1968; Mirzabekov et al., 1969; Gassen, 1969). I have already mentioned homochromatography in a thin layer mixture of cellulose and DEAE-cellulose, by means of which nucleotides containing up to 50 monomers can be fractionated (Brownlee and Sanger, 1969).

The sensitivity of thin-layer chromatography is increased by combining it with a radiochemical technique. Randerath et al. (1968), for instance, have developed a method of ultramicroanalysis of ribonucleosides based on periodate oxidation with subsequent reduction of the nucleoside dialdehydes with ^{3}H-NaBH$_4$. By contrast with the most sensitive and precise method of nucleic acid analysis, which is based on biological labeling (with ^{32}P) (Midgley, 1962; Sanger et al., 1965), Randerath's method can be used to determine the base-composition of material which cannot be labeled with the isotope in vivo, and which is available only in minimal amount (for example, human nucleic acids). By means of this method, as little as 1 picomole (10^{-12} mole) of nucleoside can be determined with great accuracy, whereas the lower limit of spectrophotometry is 1 nanomole (10^{-9} mole).

Thin-layer fractionation of tritiated ribonucleosides was later used with nucleic acids and oligonucleotides containing minor components. If the radioactivity of the ^{3}H-NaBH$_4$ is high enough (2 Ci/mmole), a minor component present in the proportion of one per 3000 nucleotides can be detected on the chromatogram; these workers (Randerath et al., 1969) even suggest that this figure can be increased to one minor component per 100,000 nucleotides.

When oligonucleotides are fractionated on ion-exchange columns, advantage is taken of the polyanionic nature of these compounds. Depending on the conditions used, fractionation can be carried out with respect either to chain length or to nucleotide composition. The most difficult task is fractionation of structural isomers, i.e., of oligonucleotides of identical composition but with a different nucleotide sequence.

Ion-exchange chromatography of the oligonucleotides was introduced by Cohn (1949, 1951), who used the anion-exchange resin Dowex-1 for this purpose. An advantage of Dowex-1 is that it can be used not only to fractionate oligonucleotides with respect to their chain length, but also, by varying the fractionation conditions, to isolate individual oligonucleotides. A number of other investigators (Asano, 1961; McCully and Cantoni, 1962a) have used fractionation on Dowex-1 in the chloride and formate form. However, because of the high capacity of the strongly basic Dowex-1 small oligonucleotides can be eluted only at low pH values and in a high salt concentration, while the longer oligonucleotides are eluted poorly or not at all. Weaker anion-exchangers, containing DEAE-groups on cellulose or Sephadex, are therefore being used more often.

Aminoalkyl-substituted celluloses were first described by Peterson and Sober (1956). DEAE-cellulose was used by Rushizky and Sober (1963) and also by Staehelin (1964a) to fractionate ribonuclease hydrolysates of tRNA. The subject of chromatography on DEAE-cellulose is analyzed in detail by Staehelin (1963) in his survey.

When DEAE-cellulose is used to fractionate synthetic homologous oligonucleotides, they are separated in accordance with their chain length. However, separation of the degradation products of nucleic acids containing a heterogeneous mixture of oligonucleotides is complicated by secondary interactions between the cellulose matrix and the purine or pyrimidine bases, which vary in their affinity for it. Tomlinson and Tener (1962, 1963a) discovered that concentrated solutions of urea and other agents which prevent hydrogen bonding reduce this adsorption interaction of polynucleotides with DEAE-cellulose to a minimum. Essentially, only the electrostatic interaction between the phosphate groups and the quaternary ammonium groups of the cellulose is present in 7 M urea, and because of this the order of elution of the oligonucleotides is a function of the total negative charge, which corresponds to their chain length. Fractionation on DEAE-cellulose and DEAE-Sephadex in 7 M urea yield fractions containing isopliths, i.e., fragments with the same number of monomers.

It would be no exaggeration to say that the method of Tomlinson and Tener played the decisive role in the successful elucidation of the structure of the first transfer RNAs. Nowadays, when complex mixtures of oligonucleotides and polynucleotides are to be fractionated, it is customary to use a combination of different methods of fractionation, but most of the methods used are based on Tomlinson and Tener's principle.

On theoretical grounds Budovskii and Demushkin (1964) concluded that complex mixtures of oligonucleotides must be fractionated by the two-way method. They carried out the first cycle of chromatography at the most acid pH possible (pH $\sim$ 3) on Dowex-1 in formic acid, resulting in separation into groups of oligomers; the second cycle of fractionation, on the other hand, was carried out at alkaline pH values (pH $\geq$ 9) on DEAE-Sephadex in triethylammonium carbonate. Substances showing similar properties in the first system acquired substantial differences in the second because of the unequal degree of the changes in ionic and absorption interaction with the matrix of the ion-exchanger.

Rushizky and Sober (1964) used double chromatography on DEAE-cellulose in the Tomlinson–Tener system, and conducted the first fractionation in an almost neutral medium and the second in an acid medium.

To obtain individual oligonucleotides from RNase hydrolysates, Aksel'rod et al. (1965) combined consecutive chromatography on DEAE-Sephadex by the Tomlinson–Tener system at pH 7.5 in an ammonium acetate gradient with chromatography on Dowex-1 at pH 1.2–2.0 in a formic acid–ammonium formate gradient. In the first system the decisive factor is the number of phosphate residues, for the urea abolishes nonionic interactions, while dissociation of the heterocyclic purine and pyrimidine bases is ruled out in a neutral medium. On chromatography on Dowex-1 in an acid medium protonation of the bases takes place, leading to appreciable differences in the net charges and in the distribution coefficients depending on the base composition. In particular, the influence of the terminal pyrimidine, ψp, and other minor components is revealed. By means of Aksel'rod's method it has been possible to isolate practically all the oligonucleotides of total RNase hydrolysates of $tRNA_1^{Val}$ from *S. cerevisiae* in a pure state (Fig. 10).

Vasilenko et al. (1969) fractionated RNase hydrolysates of $tRNA^{Val}$ on TEAE-cellulose columns in 7 *M* urea solutions. To fractionate total RNase hydrolysates of yeast $tRNA^{Phe}$, fractionation on DEAE-cellulose in the chloride form at neutral pH and in 7*M* urea has been used (RajBhandary et al., 1968a,b).

The use of the Tomlinson–Tener system played an important role not only in the fractionation of total hydrolysates of individual tRNAs, but also in the fractionation of their incomplete hydrolysis products. In every case, 7 *M* urea was used for fractionation of metamers and half-molecules of individual tRNAs (Penswick and Holley, 1965; Dütting et al., 1966; Baev et al., 1967c; Madison and Kung, 1967; Chang and RajBhandary, 1968; RajBhandary and Chang, 1968), with differences only in minor details.

An important innovation, giving greatly improved results, was fractionation by the Tomlinson–Tener method at a raised temperature, a modification introduced by Goldthwait and Kerr (1962) and subsequently used by Baguley et al. (1965a,b) for the fractionation of tRNA, and also by Sedat and Sinsheimer (1964) to detect the polyadenylic sequences in DNA from phage φX174. The chief effect of temperature is to reduce the possible aggregation and secondary interactions with the matrix of the ion exchanger; the result is a much clearer separation of the higher oligonucleotides, including the half-molecules (Baev et al., 1967b).

A combination of different principles of separation has also been used successfully to fractionate large fragments of tRNA. Three modifications of fractionation in 7 *M* urea have been used by Zachau's group (Dütting et al., 1966), viz., (1) on DEAE-cellulose or DEAE-Sephadex at pH 7.5; (2) on

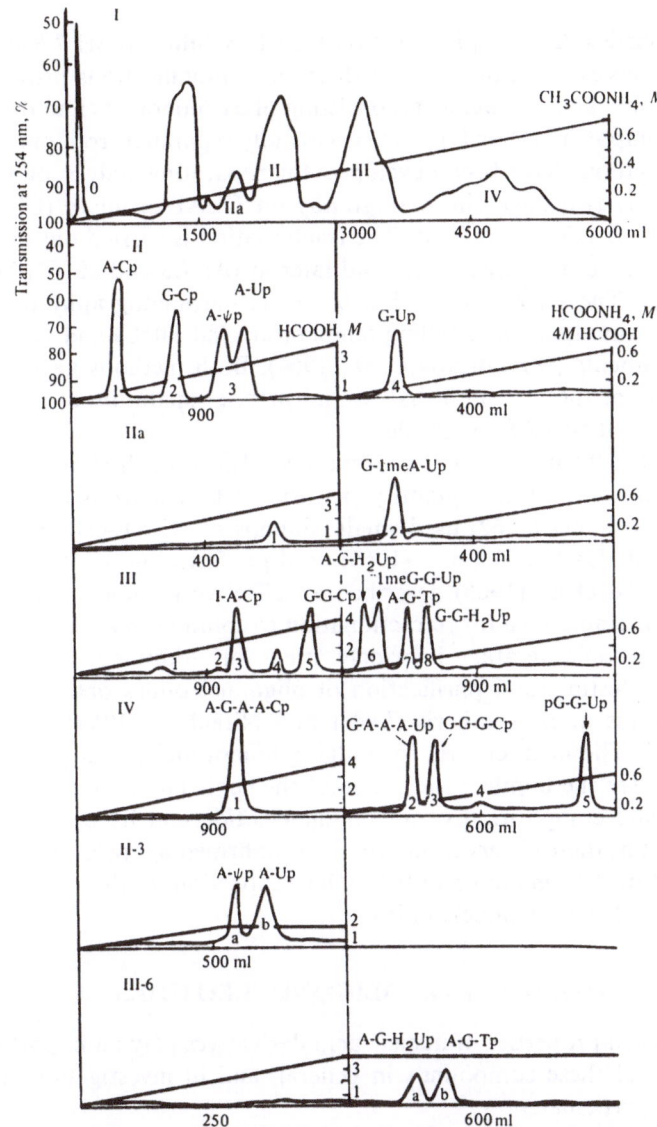

Fig. 10. Ion-exchange chromatography of oligonucleotides of pyrimidyl-ribonuclease hydrolysate of tRNA$_1^{Val}$. (I) Chromatography on DEAE-Sephadex (A-25, medium) in an ammonium acetate gradient in 7 M urea, pH 7.5; 1232 A$_{260}$ units absorbed; column 1.6 × 27 cm; mixing chamber 3000 ml 0.02 M NH$_4$OOCCH$_3$ in 7 M urea, pH 7.5; reservoir 3000 ml 0.7 M NH$_4$OOCCH$_3$ in 7 M urea, pH 7.5; rate of flow 130 ml/h. (II, IIa, III, IV) Rechromatography of corresponding fractions of A after adsorption on Dowex 1 × 4. Dimensions of columns 1 × 15, 1 × 5, 1.2 × 12.5, and 1 × 15 cm respectively. Flow rate 50 ml/h. (II-3, III-6) Second rechromatography of incompletely separated oligonucleotides; column 1 × 27 cm, elution started in linear gradient, thereafter constant concentration 4 M formic acid.

DEAE-Sephadex A-25 at pH 3; (3) on DEAE-cellulose at pH 7.5 and at 50–55 °C. The successive application of these three modifications proved highly effective. For example, fragments consisting of 31 monomers and differing in only one component (C or U) were successfully separated from one another.

Two methods have been developed for separating halves of $tRNA_1^{Val}$. The first is by two-stage chromatography on DEAE-cellulose (Cl^-) in $7M$ urea in a linear NH_4Cl gradient. The fractionation is carried out initially at pH 3.3 and at room temperature, and later at pH 7.5 and 60 °C (Aksel'rod et al., 1967). The second method is based on chromatography on DEAE-cellulose in simultaneously falling temperature (80–30 °C) and NaCl (0.55–0.35 M) gradients (Mirzabekov et al., 1966). Both methods gave complete fractionation of the 5'- and 3'-halves and precise separation of these halves from the unsplit $tRNA_1^{Val}$ molecule.

A great difficulty when the Tomlinson–Tener method is used is the need to remove the large quantity of salts and urea from the fractions. Rushizky and Sober (1962) used small columns of DEAE-cellulose for this purpose, while Uziel and Cohn (1965) used gel filtration on Biogel P-2 (or P-4). Aksel'rod et al. (1965) described an effective method of transferring fractions containing urea for refractionation to columns in a salt-free system.

Paper chromatography is widely used for additional fractionation, homogeneity testing, and purification of oligonucleotides obtained on ion-exchange resins or by synthesis (Leder and Nirenberg, 1964; Aoyagi and Inoue, 1968). The main criteria for judging homogeneity of a compound in this method are the number and shape of the spots and agreement between the length and composition of an oligonucleotide and its position on the chromatogram; these observations are then confirmed by spectrophotometric analysis and by determination of the molar proportions of the components in hydrolysates of the oligonucleotides.

SPECTROPHOTOMETRY OF OLIGONUCLEOTIDES

The optical properties of nucleic acid derivatives play an important role in the study of these compounds in general, and in investigations of their structure in particular.

Spectrophotometry of nucleic acid derivatives is the principal method used to identify them in the final stage after chemical and chromatographic analysis. Because of its speed and simplicity, spectrophotometry has become the dominant method and has supplanted all other methods used initially. Until recently, however, uv spectrophotometry of the nitrogenous bases, nucleosides, and nucleotides was the only one to have been adequately developed. Spectrophotometry of oligonucleotides did not receive the attention it deserves from biochemists and molecular biologists before

1963–1964. Papers dealing with the theoretical and practical aspects of the problem began to appear later, and now spectrophotometry of the oligonucleotides is being widely used in every case calling for their identification.

The investigation of oligonucleotides embraces determination of type and number of nucleotide components and of their sequence in the chain. The usual method of determining the base composition of an oligonucleotide is by its chemical or enzymic degradation, fractionation of the resulting components on a column or paper, and subsequent spectrophotometric identification of the individual fractions. If, however, many oligonucleotides require analysis, as, e.g., in the study of the primary structure of a tRNA, this procedure becomes extremely laborious. Attempts have accordingly been made to find simpler ways whereby the composition of oligonucleotides can be analyzed.

Several workers have tried direct spectrophotometry of mixtures of nucleotides (Reid and Pratt, 1960; Pratt et al., 1964; Lee, S., et al., 1965). The use of what, in principle, is simple spectrophotometric analysis of a mixture of nucleotides is complicated in practice by two peculiarities of their spectra. First, the absorption spectra of nucleotides are not highly differentiated, i.e., they lack those characteristic details which facilitate the identification of substances for analysis. This lack of details is particularly conspicuous when ultraviolet spectra are compared with infrared spectra. Second the spectra of Ap and Up are exceedingly close together, so that the identification of these nucleotides by their absorption spectra is unreliable without the use of special techniques. Nevertheless some workers (Lee, S., et al., 1965) have concluded that multicomponent spectrophotometric analysis is approximately equal in accuracy to the usual methods of analysis of nucleotide mixtures, and that it could be used as the basis for their automatic analysis. However, the method has never been used in practice.

The difficulties mentioned above, which arise during analysis of multicomponent mixtures of nucleotides, also occur in the analysis of undegraded oligonucleotides, i.e., during their direct spectrophotometry. In this case, however, yet another difficulty arises which cannot be disregarded. Comparison of the absorption of solutions containing a mixture of free nucleotides and of nucleotides organized into a polynucleotide chain reveals a decrease in absorption in the uv region. The cause of this phenomenon, known as the hypochromic effect, is that oligonucleotides in solution possess a definite conformation due to interaction between parallel bases (i.e., stacking). In practice, the hypochromic effect is determined by the degree of hyperchromicity, i.e., the increase in absorption of the oligonucleotide (polynucleotide) at 260 nm after hydrolysis into monomer components, expressed as per cent of original absorption. The degree of hyperchromicity depends on the length of the polynucleotide chain, the composition and sequence of the bases, the character of the internucleotide linkages, the pH of the medium, temperature, and also the wavelength at which it is measured (Michelson, 1959, 1966). However, practical experience has shown that the absence of strict additivity (hyperchromicity) and the nonlinear character of the hyperchromicity (its dispersion) in the region of 220–300 nm do not preclude determination of the composition of lower oligonucleotides from their spectra. Neither visual inspection of the spectrum nor calculation of the ratio between the absorptions at different wavelengths can reveal those very slight differences which depend on deviation from additivity and on the nonlinearity of the hyperchromic effect. Comparison of the spectra of lower oligonucleotides and of their corresponding hydrolysates shows that the dispersion of hyperchromicity in the case of the lower oligonucleotides is very small. This has been confirmed for the trinucleotides A-A-Cp (Staehelin, 1961) and (A, G)Cp (Venkstern et al., 1963a) and the polynucleotide poly(U_{13}, 4thioU) (Scheit, 1967b) (Fig. 11), and also by comparison of the absorption spectra of dinucleoside phosphates with the summated spectra of the corresponding nucleoside and nucleotide (Warshaw and Tinoco, 1965).

The additivity, for all practical purposes, of the spectra of the oligonucleotides is clearly apparent from the fact that all spectral differences between the mononucleotides are visible at the level of the lower oligonucleotides (Venkstern et al., 1963a,c; Baev et al., 1963). In structural investigations the molar extinction coefficients of oligonucleotides are for the most part regarded as equal to the sum of the molar extinction coefficients of their component units, although strictly speaking this is not true. Spectrophotometric parameters for oligonucleotides, incorporating appropriate corrections,

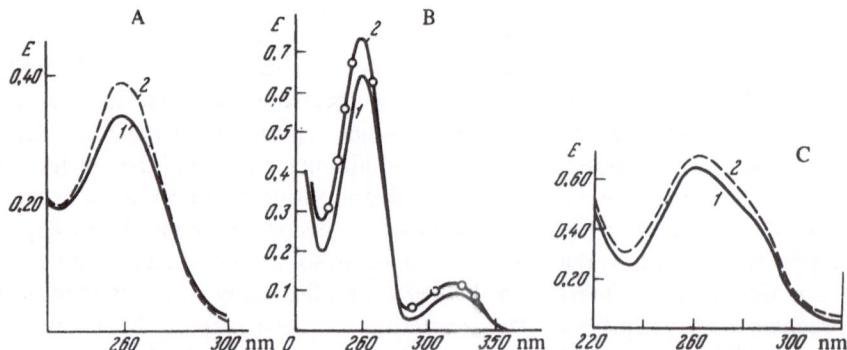

Fig. 11. Dispersion of hyperchromic effect of lower oligonucleotides. (A) Absorption spectrum of A-A-C: (1) before, (2) after hydrolysis with snake-venom phosphodiesterase; (B) absorption spectrum of poly (U_{13}, 4thoiU): (1) before, (2) after hydrolysis with pyrimidyl RNase; (C) absorption spectrum of (A, G) Cp: (1) before, (2) after alkaline hydrolysis.

have only recently begun to appear in the literature (Stanley and Bock, 1965; Aoyagi and Inoue, 1968).

The method of differential spectrophotometry of oligonucleotides, suggested by Vasilenko et al. (1965), is based on the virtual additivity of the spectra. As a result of partial hydrolysis of oligonucleotides with snake-venom exonuclease to determine their structure, a complete set of partial degradation products is obtained. After fractionation, the spectrum of each oligonucleotide is recorded against an equimolar concentration of another oligonucleotide differing from it by one nucleotide, so that the spectrum of the latter is obtained directly.

Reconstituted spectra obtained by graphic summation of the absorption curves of the component units are often used in work with oligonucleotides. For convenience of comparison, the reconstituted curve is equalized with respect to absorption at the maximum with the unknown spectrum, thus giving a reduced absorption curve. Clearly the use of these reconstituted spectra implies their additivity relative to the component mononucleotides.

Direct spectrophotometry of oligonucleotides is widely used in structural research, especially in the last stages of analysis in which the oligonucleotide composition of products of incomplete hydrolysis is determined, and in this case the oligonucleotides appear as a unique type of discrete structural units.

Absorption spectra provide a reliable characteristic of mononucleotides enabling them to be accurately identified. The spectrum of RNA is devoid of any characteristic features, for it is simply the averaged sum of the absorptions of many bases, and it cannot therefore be used to identify or determine the composition of the bases. Evidently a line can be drawn somewhere between these two limiting cases, so that the higher oligonucleotides with uncharacteristic spectra, unsuitable for analytical purposes, lie on one side and lower oligonucleotides, whose spectra have definite features of individuality, lie on the other side. This line cannot be accurately determined, but from experience gained by the study of the primary structure of $tRNA_1^{Val}$ from *S. cerevisiae,* the spectra of lower oligonucleotides containing two, three, four, and in some cases more monomers, are just as characteristic as the spectra of mononucleotides, and they can therefore be used for identification. The presence of a mononucleotide with a characteristic spectrum (e.g., Gp, ψp, etc.) or one which is repeated in the molecule, can also be established in the case of longer oligonucleotides. Individual features do not disappear completely in longer oligonucleotides, but they become latent in character and they can be used for identification only after the ratios between absorptions at characteristic wavelengths have been calculated. In that case the composition of the oligonucleotide is determined, its spectrum is obtained under various conditions, and its spectral characteristics are then used to identify that particular oligonucleotide in subsequent work. The characteristic parameters on which identification is based are the positions of the maximum, minimum, and inflections, the ratios between extinctions at different wavelengths (E_{250}/E_{260}, E_{270}/E_{260}, etc.) and, a very important piece of information, the values of all these parameters at at least two pH values and after bromination.

The additivity of the absorption spectra of the lower oligonucleotides is seen in the fact that any change in the spectrum of any of its components also affects the spectrum of the oligonucleotide as a whole. The sharper the change, the more suitable it is for analytical purposes. The change in absorption spectra of the nuclear bases and their derivatives with the pH of the solution is extensively used in the direct spectrophotometry of oligonucleotides. An example is the spectra of A-ψp and A-Up, which are very similar in an acid medium, and cannot therefore be used to identify the corresponding dinucleotides, but because of differences between the spectra of ψp and Up in an alkaline medium, these dinucleotides acquire sharply different spectral characteristics at pH values higher than 7 (Fig. 12).

The absorption bands of guanine, uracil, and cytosine in the 250–320 nm seen disappear as a result of addition of bromine, while the absorption spectra of adenine and hypoxanthine remain unchanged; this fact has long been used in the identification of nuclear bases and their derivatives (Wheeler and Johnson, 1907; Suzuki and Ito, 1958; Jones and Woodhouse, 1959). Bromination has also proved useful for the identification of oligonucleotides, because their component bases undergo the same changes by the action of bromine as in the free state. The effect of bromine on the spectra of oligoribonucleotides has been described (Baev et al., 1963b; Venkstern et al., 1963b). The presence of A in a given oligonucleotide can be accurately established from the spectrum of the bromination product, because the absorption band of any oligonucleotide not containing A (or its derivative) is lost as the result of bromination. The spectra of A-Gp and G-Up, which differ only a little in 0.1 N HCl, are given in Fig. 13 as an example; they can

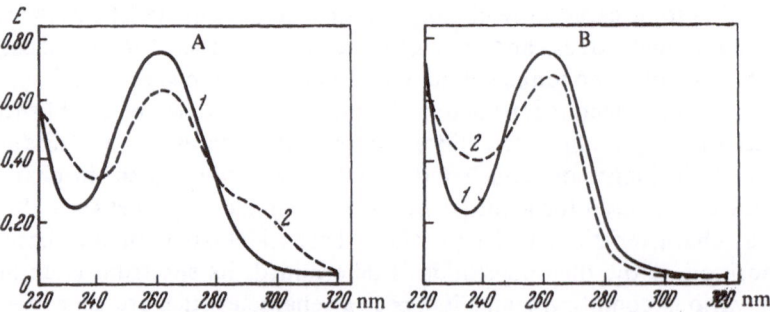

Fig. 12. Absorption spectra of A-ψp (A) and A-Up (B): (1) in 0.1 N HCl; (2) in 0.1 N KOH.

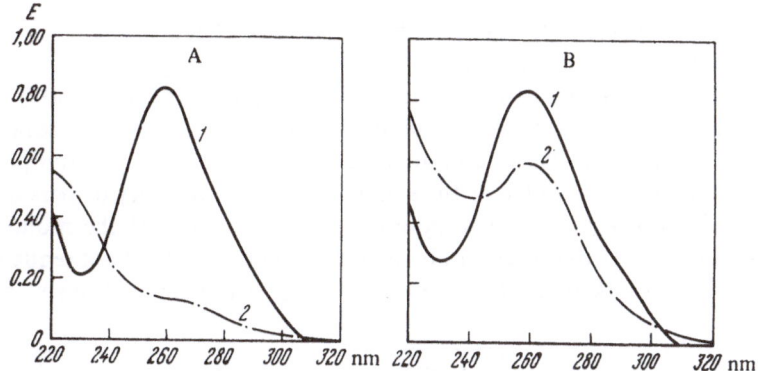

Fig. 13. Absorption spectra of G-Up (A) and A-Gp (B) in 0.1 N HCl: (1) before, and (2) after bromination.

be differentiated perfectly reliably by bromination, because the absorption bands of G-Up disappear while the spectrum of A-Gp is converted into the spectrum of Ap.

Oligonucleotides of pyrimidyl- and guanyl-ribonuclease hydrolysates obtained during the study of the primary structure of tRNA^{Val} have frequently been identified by direct spectrophotometry. The spectra of many oligonucleotides obtained in an individual stage during the study of the primary structure of yeast valine tRNA are given by Venkstern and Baev (1968) in their monograph. Stanley and Bock (1965) conducted a detailed spectrophotometric investigation of the dinucleotides and trinucleotides of a pyrimidyl-RNase hydrolysate of RNA. Spectrophotometric data, including hypochromic effects, for six dinucleoside phosphates and seven trinucleoside diphosphates have been published by Tinoco and co-workers (Warshaw and Tinoco, 1965; Cantor and Tinoco, 1965). Oligonucleotides of a guanyl-RNase hydrolysate of RNA have been fully investigated by Japanese workers (Aoyagi and Inoue, 1968).

The scope for spectrophotometric analysis of oligonucleotides is very wide. This method, so economical with material and time, has been successfully used to identify oligonucleotides of exhaustive pyrimidyl- and guanyl-RNase hydrolysates of many individual tRNAs. Lower oligonucleotides, which are identified without difficulty, play an important role in the identification of the structure of longer oligonucleotides, which is established by cleavage into progressively shorter fragments (partial hydrolysis with PDEase pancreatic RNase, etc.). The resulting fragments have to be identified, and this can be done most rationally by direct spectrophotometry.

Other promising methods for the study of nucleic-acid structure are circular dichroism and optical rotatory dispersion. A distinguishing feature of the single-stranded oligo- and polynucleotides which has been revealed by these methods is that only a slight cooperative effect is observed in these compounds, i.e., the stacking of any base pair of the polynucleotide chain is virtually independent of the stacking of its neighbors (Cantor and Tinoco, 1965; Cantor et al., 1966; Poland et al., 1966). It was accordingly concluded that optical rotatory dispersion and circular dichroism of oligo- and polynucleotides can be represented as the sum of the optical rotatory dispersion and circular dichroism of the corresponding dimers (but not of the monomers) and, consequently, that they can be used for identifying the base sequence in oligonucleotides (Cantor and Tinoco, 1965). Circular dichroism curves

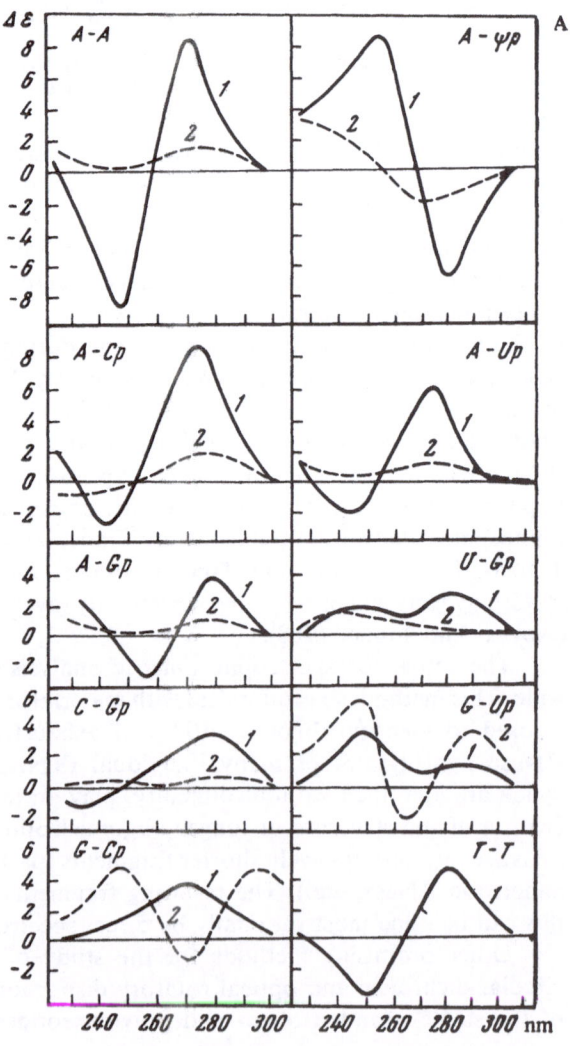

recorded for several oligonucleotides isolated from ribonuclease hydrolysates of yeast tRNA$_1^{Val}$ confirmed the promising nature of this method for structural research (Zavil'gel'skii et al., 1966; Zavil'gel'skii and Li, 1967). The shape of the circular dichroism curve is sufficiently characteristic for each pair (Fig. 14A) for it to be able to determine in the future not only the composition, but probably also the sequence of the nucleotides. For example, the curve for C-G differs appreciably from the curve for G-C. A change in the pH of the solution has a marked effect on the shape of the circular dichroism curve, and this also can be used for identification (Fig. 14A, B)

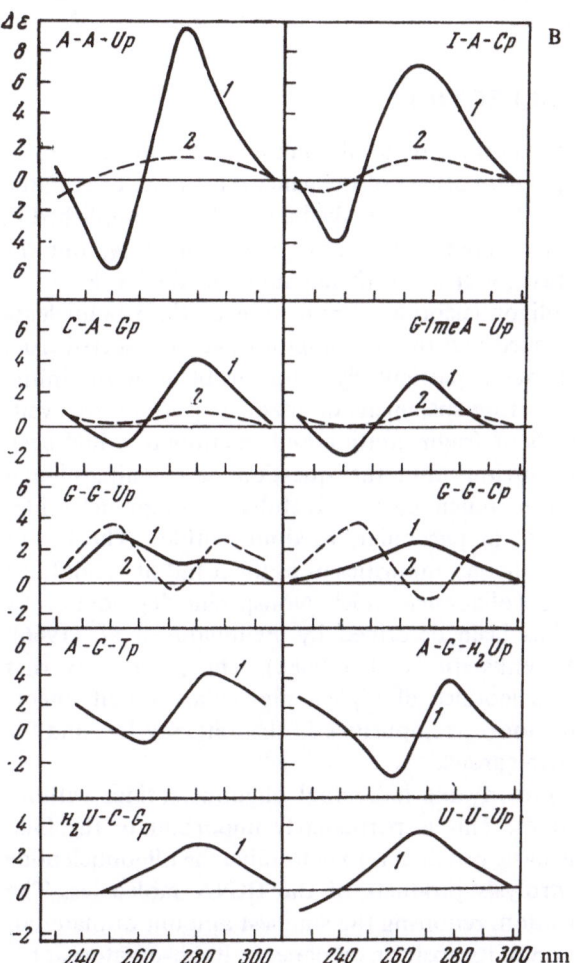

Fig. 14. Circular dichroism curves: (A) dinucleotides; (B) trinucleotides; (1) pH 5–6; (2) pH 1; $t = 5°C$.

Tinoco and co-workers (Warshaw and Tinoco, 1965; Cantor and Tinoco, 1965) investigated the optical rotatory dispersion of six dinucleoside monophosphates from a pyrimidyl-RNase hydrolysate of RNA and came to similar conclusions. Differences in the optical rotatory dispersion of structural isomers have also been reported by Japanese workers studying oligonucleotides of a guanyl-RNase hydrolysate of RNA (Inoue et al., 1967; Aoyagi and Inoue, 1968). The methods of circular dichroism and optical rotatory dispersion are undoubtedly of considerable promise; however, they are not yet sufficiently perfected, and they can only be used as adjuncts to existing methods. The field of their application is also limited, probably, to trinucleotides and tetranucleotides.

ANALYSIS OF OLIGONUCLEOTIDES

Analysis of the oligonucleotides of total ribonuclease hydrolysates of tRNA is a completely independent aspect of the general problem of research into the primary structure of tRNA. Three tasks have to be accomplished in oligonucleotide analysis; in practice this is done simultaneously, but the relative importance of the tasks varies at different stages of the work.

In the first stage the oligonucleotide composition of the ribonuclease hydrolysates must be established and the meaningful fragments sorted from the meaningless ones. This was a particularly acute problem at the initial stage of structural research, before specimens of adequate purity were available and before the methods of fragmentation and fractionation had been sufficiently perfected. It is obvious that the question of meaningful and meaningless fragments applies much more forcefully to oligonucleotides of low molecular weight and, in particular, to dinucleotides. The lowest oligonucleotides can appear either as impurities or as the result of breakdown of the most labile of the larger oligonucleotides. Nonspecific degradation by pancreatic and T_1-RNase has been described by Feldmann et al. (1966), Neelon et al. (1967), and Venkstern et al. (1968a). The probability that hydrolysates contain oligonucleotides of higher molecular weight, incorporating five monomers or more, as impurities is theoretically low and for practical purposes can be disregarded.

The second task is to identify the individual oligonucleotides without hydrolysis to mononucleotides. This is particularly important in the final stages of the work, when the basic problem is to determine the oligonucleotide composition of partial hydrolysis products of the tRNA molecules. The simplest method of identification, requiring the smallest amount of material, is that based on determination of its position on paper or its chromatographic profile after fractionation. Given good resolution and strict standardization of the fractionation procedure, these criteria give sufficiently reliable results.

If large quantities of material are available these results can be confirmed by spectrophotometric analysis, which has already been discussed. Often an additional method of identification can be used, one which does not require the complete analysis of the oligonucleotides, but consists of the determination of one characteristic component. For example, oligonucleotides containing H_2U can be detected by their characteristic yellow staining for the ureido group by Fink's method when the chromatogram is developed; oligonucleotides containing methylated guanine derivatives are easily identified by their specific violet fluorescence, etc.

The study of the structure of individual oligonucleotides, i.e., the composition and sequence of the bases in their molecule, is a most important and complex problem. If it is remembered that RNase oligonucleotides often contain 10–12 monomer units and, in some cases, as many as 19 units, the important role of determination of the structure of these large units in the general task of investigating the structure of the tRNA molecule, which consists of about 80 mononucleotides, will be at once realized. Determination of the nucleotide sequence in fragments such as these immediately identifies the structure of 1/7–1/8, or even 1/4 of the total polynucleotide chain.

Analysis of oligonucleotides is the most laborious part of the work entailed in determining the primary structure of tRNA. At the same time, it is an extremely important part, because the arrangement of the oligonucleotides, on the one hand, and their structure, on the other hand, are determined independently. Analyses of partial digests give little confirmation of the sequences of the oligonucleotides.

It is not, therefore, by chance that discrepancies found on repeated determination of the structure of some transfer RNAs have had to do principally with the structure of oligonucleotides, and not with their arrangement in the polynucleotide chain (Merril, 1968; Ebel and Baev, unpublished work).

Chromatographically pure oligonucleotides of RNase hydrolysates obtained either directly from fingerprints or tested for homogeneity and purified on paper from columns, are always investigated spectrophotometrically to begin with. This allows the reliability of the hypothesis regarding the nature of the oligonucleotide, based on the position of the stain on the chromatography paper or of the peak on the chromatographic profile, to be assessed. If it is proposed to use enzymes for the analysis, the spectra of the oligonucleotides are determined in water and the solution is used directly for subsequent analytical work.

The study of the structure of oligonucleotides is a simplified version of analysis of long polynucleotide chains, and for this reason many of the suggested methods are applicable to both cases.

As a rule the composition of the bases is determined by chemical hydrolysis, and this part of the problem could be considered to be solved

were it not for the complications introduced by minor components present
in many oligonucleotides of tRNA. Alkaline hydrolysis is frequently used,
although some investigators have found that this causes marked degradation
of cytidylic acid (Fritz and Röttger, 1963; Budovskii and Klebanova, 1967).
Acid hydrolysis is used in several versions. Hydrolysis with 1 N HCl at 100 °C
for 1 h is widely used; it yields purine bases and pyrimidine nucleotides, which
can be easily fractionated by various systems of chromatographic solvents.
A disadvantage of this method is the slight dephosphorylation of the pyrimidine nucleotides. Hydrolysis of oligonucleotides to bases, using 70 % HClO₄
in concentrated HCOOH has also been used. In 18 h at room temperature,
1 N HCl hydrolyzes oligonucleotides chiefly to mononucleotides, but some
degree of apurinization is unavoidable.

When the nucleotide composition of oligonucleotides of tRNA is
determined it must be remembered that many of the minor components are
highly labile: 1-methyladenosine is converted in an alkaline medium into
N^6-methyladenosine (Elion, 1962); 7-methylguanosine at alkaline pH values
forms 2-amino-6-hydroxy-5-N-methylformamide-6-ribosylaminopyrimidine
(Haines et al., 1962; Jones and Robins, 1963); acetylcytidine (Feldmann et
al., 1966) and, in particular, dihydrouridine, which is converted into β-
ureidopropionic acid (Batt et al., 1954), also are labile in an alkaline medium.
The high lability of the minor component Y, which is on the 3′-side of the
anticodon in yeast tRNA^Phe, in an acid medium and in 7 M urea has been
reported by RajBhandary et al. (1968a). Almost any form of acid hydrolysis
will cause some degree of decomposition of ψ. The thionucleosides presents
certain difficulties for experimental work, because they readily undergo
oxidation–reduction conversions. During chemical hydrolysis, minor components may thus be isomerized and converted into other minor components
(1meA→6meA), lose their characteristic grouping (acC→C), or become
decyclized and, consequently, lose their characteristic properties of nitrogenous bases (H₂U→ureidopropionic acid). In all these cases, determination
of the nucleotide composition by chemical hydrolysis yields incorrect results.

In order to preserve the minor components, milder agents such as
enzymes must be used for the hydrolysis. The commonest enzyme used for
this purpose is T₂-RNase, which ruptures phosphodiester bonds formed by
all nucleotides (Sato and Egami, 1957; Rushizky and Sober, 1963). In some
cases snake-venom PDEase also can be used, but if this is done it must be
remembered that it selectively attacks certain phosphodiester bonds (Baev et
al., 1963a; Venkstern, 1966; Feldmann et al., 1966). Exonuclease A5, which
is relatively insensitive to the base structure (Tatarskaya et al., 1970), can
probably be used for this purpose, and further details concerning this enzyme
will be given below.

The structure of oligonucleotides is studied chiefly by means of enzymes with subsequent fractionation of the reaction products by electrophoresis or by paper or column chromatography. Thin-layer chromatography has also been used successfully for the analysis of oligonucleotides and poly-nucleotides, especially where minor components are concerned (Björk and Svensson, 1967).

The assortment of nucleases available for use in the study of oligonucleo-tide structure is very wide. The same strict demands are made on enzymes used for this purpose as on those used for fragmentation of the tRNA molecule. Hydrolysis of the oligonucleotides must take place completely without the formation of byproducts. Only enzymes with strict substrate specificity and not possessing any enzymic side activity which could distort the results of the analysis can therefore be used. A matter of the utmost im-portance in this connection is the correct determination of the quantity of enzyme required for exhaustive hydrolysis. Substrate specificity of the nu-cleases and, in particular, of the RNases, as has already been mentioned, is not absolute, and if the enzyme preparation is used in high concentration, its nonspecific action usually begins to appear. On the other hand, if the enzyme concentration is low, products of incomplete hydrolysis such as cyclic phosphates may remain, thus multiplying the number of reaction products and confusing the picture. This is a particularly important matter in the case of partial hydrolysis of oligonucleotides by enzymes, when the result of analysis is entirely dependent on the correct choice of enzyme concentration and conditions for its action.

The 3'-terminal nucleotide is usually obvious because of the specificity of the RNase used to obtain the oligonucleotides. Otherwise it can be deter-mined by hydrolysis of the dephosphorylated oligonucleotide with alkali (OH⁻), acid (H⁺), pancreatic RNase (P), or spleen PDEase (Razzell and Khorana, 1961), in accordance with the following example:

$$(C, \psi, U) \, p \xrightarrow{\text{PMEase from } E. \, coli} (C, \psi, U) \xrightarrow[\text{Spleen PDEase}]{\text{OH}^-, \, \text{H}^+, \, \text{P}} Cp + \psi p + U$$

Total hydrolysis with snake-venom PDEase is used to determine the 5'-terminal nucleotide. In the case of trinucleotides with a known 3'-terminal nucleotide and in some special cases, e.g., of repeating monomers, exhaustive hydrolysis with PDEase can provide complete information concerning the sequence of the polynucleotides:

$$(A-, U-) \, Gp \xrightarrow{\text{PMEase}} (A-, U-) \, G \xrightarrow{\text{Snake-venom PDEase}} A + pU + pG$$

whence the sequence A-U-Gp is deduced; and

$$(C-, 3U-)Gp \xrightarrow{\text{PMEase}} (C-, 3U-)G \xrightarrow{\text{Snake-venom PDEase}} C + 3pU + pG$$

and consequently, the pentanucleotide has the structure C-U-U-U-Gp.

Quantitative results can be obtained when studying the structure of oligonucleotides labeled with ^{32}P by measuring radioactivity in the part of the paper containing the given nucleotide (or oligonucleotide). For this reason, the 3'- and 5'-terminal components in dephosphorylated oligonucleotides, split off as nucleosides, are determined in a somewhat unusual manner. By comparing the alkaline and phosphodiesterase (snake-venom) hydrolysates, the monomer absent in the alkaline hydrolysate is identified— this is the 3'-terminal component split off as a nucleoside. Correspondingly, the 5'-terminal component, also split off as a nucleoside, is absent in the phosphodiesterase hydrolysate.

The work of Zamir et al. (1965) showed that trinucleotides are hydrolyzed by snake-venom PDEase, purified as described by Keller (1964), without preliminary dephosphorylation, so that the position of all three nucleotides can be determined immediately:

$$(N_1\text{-},\ N_2\text{-},\ N_3\text{-})p \xrightarrow{\text{Snake-venom PDEase}} N_1 + pN_2 + pN_3p$$

A new enzyme, of evident promise for the elucidation of oligonucleotide structure, is actinomycete 5'-exonuclease, known as exonuclease A5 (Tatarskaya et al., 1970). This enzyme, like snake-venom phosphodiesterase, removes 5'-mononucleotides from the 3'-end of the chain, but it has the advantage over the other enzyme that the terminal 3'-phosphate is not a significant obstacle for it. Another difference is that this enzyme is virtually unable to hydrolyze dinucleoside monophosphates. The hydrolysis of, e.g., a tetranucleotide from a RNase hydrolysate thus leads to the formation of the following products:

$$(N_1\text{-},\ N_2\text{-},\ N_3\text{-},\ N_4)p \xrightarrow{\text{Exonuclease A5}} (N_1\text{-},\ N_2) + pN_3 + pN_4p$$

The trinucleotide is split correspondingly into mononucleoside diphosphate and dinucleoside monophosphate. The third feature distinguishing exonuclease A5 from snake-venom phosphodiesterase is its lower sensitivity to base structure. The enzyme does not discriminate against phosphodiester bonds formed by nucleotides with modified bases (Tatarskaya et al., 1970).

An effective method for use in oligonucleotide analysis is supplementary hydrolysis with RNase of a different specificity. Oligonucleotides obtained by the action of pyrimidyl RNase are subjected to total hydrolysis by guanyl RNase and vice versa. The advantage of this method is that oligonucleotides (di-, tri-, and tetranucleotides), and not monomers, are formed; these can easily be identified, so that the structure of fairly long sequences can often be unequivocally established. This method has proven very fruitful and has been used by all investigators studying the primary structure of oligonucleotides of individual transfer RNAs.

**TABLE 4. Determination of Base Sequence in Oligonucleotides of a
Pyrimidyl-Ribonuclease Hydrolysate by Exhaustive Hydrolysis with
Guanyl Ribonuclease**

Composition of oligonucleotides	Possible sequences	Hydrolysis products	Sequence established	Solution
(1meA, G) Up	1meA-G-Up G-1meA-Up	Gp, 1meA-Up	G-1meA-Up	Unequivocal
(A, A, A, G) Cp	A-A-A-G-Cp A-A-G-A-Cp A-G-A-A-Cp G-A-A-A-Cp	A-Gp, A-A-Cp	A-G-A-A-Cp	Unequivocal
(1meG, G) Up	1meG-Up G-1meG-Up	Gp, 1meGp, Up		No information

Examples of unequivocal determination of the structure of two oligonu-cleotides of a pyrimidyl-RNase hydrolysate are given in Table 4. The third example relates to a trinucleotide (1meG,G)Up, the structure of which cannot be determined by this method because guanyl RNase hydrolyzes it into mononucleotides, and the order of the 1meG and G remains unknown.

In the case of oligonucleotides of guanyl-ribonuclease hydrolysates, exhaustive hydrolysis with pyrimidyl RNase gives an unequivocal answer less frequently (the first two examples in Table 5). The difficulty of analysis of guanyl-RNase oligonucleotides is due not merely to the considerable length of their chain, but also to the fact that they consist mainly of pyrimidine residues. If the pyrimidine nucleotides do not alternate with Ap and form polypyrimidine units, on the action of pyrimidyl RNase these units are broken up into mononucleotides, and this degradation provides no informa-tion regarding the base sequence in the molecule (Table 5, example 5). In the case of oligonucleotides containing a certain number of adenylic acid residues as well as pyrimidine nucleotides, exhaustive hydrolysis by pyrimidyl RNase will always give some information about their structure, although it is not always complete. For example, hydrolysis of a pentanucleotide containing one Ap residue by pyrimidyl RNase reduces the number of possible structures from 24 to 6 (Table 5, example 3). The presence of polyadenylic sequences is also convenient from the standpoint of structural analysis, for they are degraded by pyrimidyl RNase into individual structural units with a 3'-ter-minal pyrimidine residue and usually possessing characteristic absorption spectra: A-A-Cp, A-A-Up, A-A-A-Up, and so on. If such units can be found in oligonucleotides, the number of alternative structures is considerably reduced (Table 5, example 4).

Comparison of the guanyl- and pyrimidyl-RNase digests plays an im-portant role in the determination of the structure of the oligonucleotides of the two total RNase hydrolysates, for the presence of minor components and

TABLE 5. Determination of Base Sequence in Oligonucleotides of Guanyl-Ribonuclease Hydrolysate by Exhaustive Hydrolysis with Pyrimidyl Ribonuclease

Composition of oligonucleotides	Possible sequences (number of possible structures)	Hydrolysis products	Established sequence (number of possible structures)	Solution
(C, A) Gp	C-A-Gp A-C-Gp (2)	Cp A-Gp	C-A-Gp (1)	Unequivocal
(2A, 2C) Gp	A-A-C-C-Gp A-C-A-C-Gp A-C-C-A-Gp C-A-C-A-Gp C-A-A-C-Gp C-C-A-A-Gp (6)	2A-Cp Gp	A-C-A-C-Gp (1)	Unequivocal
(U, C, ψ, A) Gp	(24)	Up Cp ψp A-Gp	U-C-ψ-A-Gp U-ψ-C-A-Gp C-ψ-U-A-Gp C-U-ψ-A-Gp ψ-U-C-A-Gp ψ-C-U-A-Gp (6)	Partial
(3A, 4C, H₂U, meC) Gp	(504)	A-A-Cp A-Gp 3Cp meCp H₂Up	A-A-C (H₂U, meC, 3C) A-Gp* (20)	Partial
(H₂U, C) Gp	H₂U-C-Gp C-H₂U-Gp (2)	H₂Up Cp Gp	H₂U-C-Gp C-H₂U-Gp (2)	No information

*The 5′-terminal position of A-A-C is determined by the presence of only one such sequence in the pyrimidyl-RNase hydrolysate, namely, in the composition of the pentanucleotide A-G-A-A-Cp

nonrepeating sequences in the molecule means that useful information can be obtained. For example, it will be clear that in a guanyl-ribonuclease oligonucleotide containing T the T must occupy the 5′-terminus if the trinucleotide A-G-Tp is found in the pyrimidyl-ribonuclease hydrolysate. In precisely the same way, identification of the trinucleotide A-A-Cp in an oligonucleotide of a guanyl-ribonuclease hydrolysate determines its 5′-terminal position if only one such sequence is found in the pyrimidyl-ribonuclease hydrolysate, e.g., in the pentanucleotide A-G-A-A-Cp. By comparing the oligonucleotides of two RNase hydrolysates, certain reconstructions are possible on the same basis, and these will be discussed below.

To determine the structure of longer oligonucleotides, methods of partial hydrolysis using PDEases, RNases, and polynucleotide phosphoryl-ase and micrococcal nuclease have been developed. As a rule partial hy-drolysis requires a much larger amount of material and greater skill on the part of the experimenter as regards choice of enzyme concentration and other reaction conditions and also as regards analysis of the products.

Partial hydrolysis with snake-venom PDEase is based on the fact that this enzyme, being an exoenzyme, splits off monomers one by one starting from the 3'-end of the molecule, and theoretically it must give the complete set of units of different lengths ranging from the original oligonucleotide down to mononucleotides. If the nucleotide composition of all the com-ponents is determined, or the 3'-terminal nucleotide of each of them is identi-fied by alkaline hydrolysis, the sequence in the original oligonucleotide can be restored:

$$(A,U,C,G) \xrightarrow[\text{with PDEase}]{\text{Partial}} (A,U,C,G) + (A,U,C) + (A,U) + (A) + \text{mononucleo-tides}$$

$$\downarrow \boxed{A}$$

$$Ap + \boxed{U}$$

$$Ap + Up + \boxed{C}$$

$$Ap + Up + Cp + \boxed{G}$$

Alkaline hydrolysis

The nucleosides are the 3'-terminal components of the stepwise-degrada-tion products of the oligonucleotide, and the original structure can be reproduced from them. By means of this method Holley et al. (1964b) determined the structure of several of the oligonucleotides of tRNA^{Ala}, while Vasilenko et al. (1965) determined the structure of the pentanucleotide A-A-C-G-Cp isolated from total yeast tRNA, and Bonnet et al. (1971) established the nucleotide sequence in an 11-member fragment of the guanyl-RNase hydrolysate of tRNA$^{Val}_2$ from brewers' yeast. However, snake-venom PDEase does not rupture all phosphodiester bonds at the same rate, so that a complete set of stepwise-degradation products is not always ob-tained. Feldmann et al. (1966), for instance, could not obtain all the theoreti-cally expected degradation products from the oligonucleotides U-C-C-U-Gp and C-U-C-U-Gp. Madison et al. (1967b) report that the sequence of three 5'-terminal nucleotides cannot be determined by partial hydrolysis with PDEase, because trinucleoside diphosphates and dinucleoside monophos-phates are hydrolyzed more rapidly than the larger oligonucleotides. As was pointed out above, PDEase attacks phosphodiester bonds formed by minor components much more slowly. The dinucleotide 2me$_2$G-Cp, is hydrolyzed by phosphodiesterase from the venom of *Vipera lebetina* much more slowly than the dinucleotide G-Cp (Baev et al., 1963a). Neelon et al. (1967) have observed that bonds of the N-ψ type are more resistant to the action of snake-venom PDEase. McLennan and Lane (1968) found a considerable number of 3'-terminal ψ residues among the products of partial hydrolysis of wheat-germ ribosomal RNA by PDEase. These workers suggest two possible alternative explanations: either PDEase possesses an endonuclease action, directed principally to ψ-rich areas, or the enzyme "stumbles" on the ψ residues. The second of these hypotheses seems more likely. It has also been shown that bonds of the OmeN$_1$-N$_2$ type are also relatively more resistant to the action of snake-venom PDEase (RajBhandary et al., 1968a).

The method of partial hydrolysis with PDEase suggested by RajBhandary and co-workers (RajBhandary et al., 1966; RajBhandary, 1968) has already been mentioned in connection with terminal analysis. In the case of oligonucleotides the method is also intended for small quantities of material and consists of introducing the radioactive label into the terminal 2',3'-diol groups of the products of partial PDEase hydrolysis by oxidation with periodate to dialdehydes and subsequent reduction with ^{3}H-NaBH$_4$ (Fig. 15). In this way the structure of the oligonucleotide G-G-G-A-G-A-G-Cp from yeast tRNA^{Phe} was established (RajBhandary et al., 1968a).

The properties of snake-venom PDEase are evidently dependent not only on the source, but also on the method of isolation of the enzyme, as is shown by the contradictory nature of results obtained by different workers for the rates of hydrolysis of oligonucleotides in relation to their length, composition, and presence of 3'-terminal P.

Sanger et al. (1965) developed a method of determining bases in ^{32}P-labeled oligonucleotides by means of limited degradation with spleen PDEase.

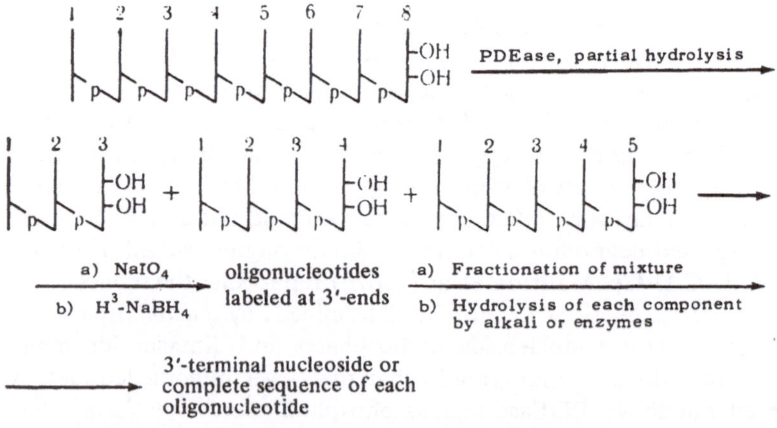

Fig. 15. The determination of the structure of oligonucleotides by the method of partial hydrolysis with snake-venom PDEase using ^{3}H-NaBH$_4$.

This enzyme produces stepwise degradation of oligonucleotides but, unlike snake-venom PDEase, it starts from the 5'-end of the molecule. Theoretically, the partial hydrolysate must contain all products of stepwise degradation of the oligonucleotide, e.g.,

$$A\text{-}U\text{-}C\text{-}G \longrightarrow A\text{-}U\text{-}C\text{-}G + U\text{-}C\text{-}G + C\text{-}G + G$$

The hydrolysate was separated electrophoretically on DEAE-cellulose at pH 1.9 and the reaction products were identified by their mobilities, assuming that the distance between two oligonucleotides differing by one component depends on the nature of that component. However, differences between the rates of removal of different nucleotides by the action of spleen PDEase were even greater than in the case of snake-venom PDEase. In practice, therefore, not all the potentially possible stepwise-degradation products are formed. On this account, the workers who developed and used this method to analyze oligonucleotides of 5S rRNA (Brownlee and Sanger, 1969) state that it can be used only in conjunction with some other method of analysis.

The enzyme micrococcal nuclease (3.1.4.7), which hydrolyzes both RNA and DNA, has been used to elucidate the structure of oligonucleotides. It was originally held that the chief products formed by the action of this enzyme are dinucleotides (Sulkowski and Laskowski, 1962; Ohsaka et al., 1964). However, subsequent investigations revealed a more complex pattern. In connection with the study of oligonucleotide structure, micrococcal nuclease has been used chiefly to determine the second nucleotide from the 5'-end, for on exhaustive hydrolysis the 5'-terminal dinucleoside remains unsplit, whereas the rest of the oligonucleotide is degraded to mononucleotides. As an example of the use of this enzyme, determination of the structure of the "universal" oligonucleotide carried out in Holley's laboratory (Zamir et al., 1965) can be cited. Both total and partial hydrolysis was used; the latter preserved the 5'-terminal trinucleotide the structure of which has been established by exhaustive hydrolysis with snake-venom phosphodiesterase:

$$T p\psi p C p C p \xrightarrow{\text{Micrococcal nuclease, exhaustive hydrolysis}} T p\psi p + C p + G p$$

TpψpCpCp $\Big\downarrow$ Micrococcal nuclease, partial hydrolysis

TpψpCp + Gp

By varying the degradation conditions, Feldmann et al. (1966) determined the structure of several of the oligonucleotides of yeast tRNA[Ser] by means of micrococcal nuclease. Later, Feldmann (1967b) made a more detailed study of the mechanism of action of this enzyme on seven oligonucleotides of known structure. He showed that under certain conditions crystalline micrococcal nuclease (Sulkowski and Laskowski, 1966) removes the 3'-

terminal dinucleoside monophosphates and 5′-terminal dinucleoside diphosphates from oligonucleotides. Tetranucleotides are split for the most part in
the center, while pentanucleotides give two nucleotides and the central component as a mononucleotide. In the case of long nucleotide sequences, a
more complex pattern is obtained. Feldmann concludes that micrococcal
nuclease can be used with advantage to study the structure of any oligonucleotide. However, the use of this enzyme to study the structure of the oligonucleotides of yeast tRNA^{Tyr} (Madison et al., 1967a) led to the conclusion
that the enzyme acts insufficiently reproducibly.

For this reason, Madison et al. (1967b) suggested that the second and
third nucleotides from the 5′-end of the oligonucleotide be identified by
means of polynucleotide phosphorylase, which, in the presence of inorganic
phosphate, removes nucleotides from the 3′-terminal end of the polynucleotide chain; the products of its action are nucleoside-5′-diphosphates and one
dinucleoside monophosphate. If small quantities of enzyme are used, the 5′-
terminal trinucleoside diphosphate remains unsplit:

$$\text{C-A-U-U-A-G-C} \xrightarrow[\text{phosphorylase}]{\text{Polynucleotide}} \text{C-A-U} + \text{ppU} + \text{ppA} + \text{ppG} + \text{ppC}$$

$$\downarrow \text{Alkaline hydrolysis}$$

$$\text{Cp} + \text{Ap} + \text{U}$$

It this way, on analysis of the heptanucleotide G-G-G-A-G-A-C and the
hexanucleotide C-U-C-U-C-G, respectively, the trinucleoside phosphates
G-G-G and C-U-C were obtained, and their structure could be established
without difficulty.

A method of partial hydrolysis with pyrimidyl RNase has been developed to determine the sequence in polypyrimidine units of guanyl-RNase
oligonucleotides (Venkstern et al., 1968b). The method is based on the fact
that the rate of cleavage of the phosphodiester bonds in oligonucleotides by
pyrimidyl RNase varies and depends both on the nature of the pyrimidine
and on the nature of its neighbor on the right with which the bond is formed.
As has already been mentioned, differences in the rate of cleavage of different
phosphodiester bonds by pyrimidyl RNase are sufficiently great for endonucleases to be suitable for partial hydrolysis; for the most labile CpA bond,
for example, $K_2 = 300 \text{ sec}^{-1}$, while for the most stable UpU bond $K_2 = 11$
sec^{-1}. It is therefore comparatively easy to select conditions under which
only some of the bonds potentially accessible to pancreatic RNase are ruptured; by analysis of the resulting assortment of hydrolysis products the
structure of the original oligonucleotide can be recreated. For example, by
partial hydrolysis of the tetranucleotide of the guanyl-RNase hydrolysate
of tRNA$_1^{Val}$ (C,ψ,U)Ip by pyrimidyl RNase, the trinucleotide (C,ψ)Up, was

obtained. The dinucleotides C-ψp and ψ-Up were found in the hydrolysate of the same oligonucleotide, thus unequivocally determining the structure of the trinucleotide as C-ψ-Up, and the structure of the original tetranucleotide as C-ψ-U-Ip. Partial hydrolysis by pyrimidyl RNase has been used to determine the structure of all oligonucleotides of the guanyl-RNase hydrolysate of tRNA$_1^{\text{Val}}$ of *S. cerevisiae* containing polypyrimidine sequences. The most important problem in partial hydrolysis by pyrimidyl RNase, just as in partial hydrolysis by any other enzyme, is establishment of the essential conditions for the optimal degree of degradation. If degradation goes too far, the units obtained are too short and contain too little information; analysis of long units is complex and does not always achieve its purpose. It is impossible to determine the theoretically optimal conditions because each oligonucleotide is an individual in the sense that it has its own assortment of phosphodiester bonds. Hence, in the analysis of every new oligonucleotide, a test degradation must be carried out. The preliminary experiment is also useful from the point of view of choice of chromatographic solvents for precise separation of the incomplete hydrolysis products on paper. In practice, partial hydrolysis by pyrimidyl RNase has been carried out with strict standardization of all the conditions (time, temperature, buffer), only the quantity of enzyme being varied.

The use of partial hydrolysis by guanyl RNase has been suggested for analysis of the oligonucleotides of pyrimidyl-RNase hydrolysates (Thirion and Kaesberg, 1968); in practice, however, this method has not yet been used. Vasilenko et al. (1969) analyzed the oligonucleotides of a guanyl-RNase hydrolysate by degradation with pyrimidyl RNase while adsorbed on TEAE-cellulose, a method with several advantages. More recently (Takemura et al., 1969; Adams et al., 1969; Cory and Marcker, 1970), the RNase U_2 (2.7.7), which ruptures phosphodiester bonds formed by purines (Arima et al., 1968), has also been used to study the structure of oligonucleotides.

The principal chemical method used to determine the structure of oligonucleotides is periodate oxidation. This method has been discussed earlier in the book in connection with the stepwise degradation of tRNA. At this point all that need be added is that this method, especially in the modification of Neu and Heppel (1964), can be used successfully to determine two or three residues from the 3'-end of the molecule, but the repeated application of the procedure is laborious and calls for a comparatively large amount of material. Further difficulties arise in the analysis of oligonucleotides containing ψ and H_2U in their composition.

The study of oligonucleotide structure is the subject of a detailed survey by Holley (1968).

In conclusion, two examples will be given of determination of the structure of oligonucleotides by the methods described above.

The course of determination of the structure of one of the hexanucleotides from the guanyl-RNase hydrolysate of RNA$_1^{Val}$ from *S. cerevisiae* is shown in the scheme. The results obtained by different methods of hydrolysis of the oligonucleotide show a gradual approach to the true structure, starting from 60 equally probably formulas, all of which would be possible with the nucleotide composition established by chemical hydrolysis and with guanylic acid in the 3′-position, determined by the specificity of the guanyl RNase.

Scheme of Determination of the Structure of a Hexanucleotide

		Possible structures	
Hydrolysis*	Hydrolysis products	Formula	Number
H⁺	2Cp + Up + ψp + Ade + Gua ⎱	(C, C, U, ψ, A) Gp	60
OH⁻	2Cp + Up + ψp + Ap + Gp ⎰		
PMEase + snake-venom PDEase	C + pC + pU + pψ + pA + pG	C(C, U, ψ, A) Gp	24
Pancreatic RNase (total hydrolysis)	2Cp + Up + A − ψp + Gp	C(C, U, A − ψ) Gp	6

```
                          ┌→ Pentanucleotide
                          │    │ PMEase +
                          │    ↓ Pancreatic RNase
Pancreatic RNase          │   2Cp + A − ψp + U ⟶ C (A − ψ, C) Up
(partial hydrolysis)      │                              ↓
                          │                       C-A-ψ-C-U-Gp
                          │                       C-C-A-ψ-U-Gp        2
                          └→ Dinucleotide
                               ↓ PMEase + OH⁻
                              Cp + U ─────────────→ C-Up
                                                    C-A-ψ-C-U-Gp
                                                    (final structure)
```

*In this and the following scheme, H⁺ denotes acid hydrolysis (1 *N* HCl, 100°C, 1 h); OH⁻ denotes alkaline hydrolysis (0.3 *N* KOH, 37°C, 18 h).

A scheme for the determination of the structure of the oligonucleotide OmeG-A-A-Y-A-ψ from tRNA^{Phe} of *S. cerevisiae* is given below (RajBhandary et al., 1968a).

Despite the fact that this oligonucleotide consists of only six monomers, the identification of its structure required the use of five enzymes under the conditions of total and partial hydrolysis. The study of this hexanucleotide is made more difficult by the fact that it contains three minor components which have a considerable effect on the rate of enzymic cleavage of the corresponding phosphodiester bonds.

The examples given on page 77 show that determination of the structure of each oligonucleotide is an independent problem, by no means easy, and requiring an individual approach.

Scheme for the Determination of a Hexanucleotide, OmeG-A-A-Y-A-ψ

(a) T$_1$-RNase ───────────→ No hydrolysis

(b) OH$^-$ ───────────────→ OmeG-Ap + 2Ap + Yp + ψ

(c) Spleen PDEase ─────────→ Slow hydrolysis

(d) $\dfrac{\text{Snake-venom PDEase}}{\text{Total hydrolysis}}$ ───→ OmeG (yield 20%)

(e) $\dfrac{\text{Snake-venom PDEase}}{\text{Partial hydrolysis}}$ ──→ $\begin{array}{l} \overset{\text{T}_2\text{-RNase}}{\longrightarrow}\text{OmeG-Ap + Ap + Y} \\ \text{OmeG-A-A-Y + pA + p}\psi \\ \quad\longrightarrow\text{micrococcal nuclease}\longrightarrow\text{OmeG-Ap+ A}-\text{Y} \end{array}$

(f) T$_2$-RNase ───────────→ OmeG-Ap + Ap + A-Y > p + ψ $\quad\begin{array}{l}\rceil \text{T}_2\text{-RNase} \\ \text{Spleen PDEase} \;\; \longrightarrow\text{Ap+Y} \\ \quad\longrightarrow\text{Ap +Y >p} \end{array}$

Chapter 3

THE PRIMARY STRUCTURE OF INDIVIDUAL TRANSFER RNAS

THE DEVELOPMENT OF RESEARCH INTO THE PRIMARY STRUCTURE OF tRNA

Alanine tRNA from Bakers' Yeast

Research into the primary structure of tRNA essentially began in 1958 when Holley began his systematic study of the fractionation of soluble RNA from tissue homogenates by means of the countercurrent method. The first experiments were sufficient to demonstrate the promising nature of this method, for the results showed wide distribution of the uv-absorbing material with the differentiation of separate maxima of acceptor activity relative to the various amino acids (Holley and Merrill, 1959; Holley et al., 1959).

In the subsequent years, the members of Holley's group have directed their efforts toward studying the method of countercurrent distribution, improving the sensitivity of the systems, and increasing the yield of material. They have achieved these aims by increasing the number of transfers and by choosing different systems of solvents (Holley and Doctor, 1960; Apgar et al., 1961; Doctor et al., 1961). Holley's group eventually settled on a system originally suggested by Warner and Waimberg (1958), consisting of concentrated phosphate buffer (pH 6), formamide, and isopropanol, in which the partition coefficients for the various tRNAs differ very considerably from each other (by ten times for valine and tyrosine tRNAs).

In the course of the work it was found that the main cause of inactivation of the specimen during distribution is the action of two nucleases, one of which is present as an impurity in the actual tRNA specimen itself, while the second is introduced from the experimenter's hands (Holley et al., 1961b). The procedure of countercurrent distribution was therefore modified so as to reduce contamination with these two enzymes to the minimum. Holley recommended repeated treatment of the tRNA sample with phenol in order to remove the RNase present as an impurity in the sample itself; the RNA was isolated from the countercurrent fractions by the method suggested by Kirby (1960), by concentrating the tRNA in a small volume to shorten the time needed for dialysis and evaporation of the fractions; dialysis against water saturated with chloroform to reduce the possibility of growth of microorganisms; work with rubber gloves to prevent contamination of the tRNA specimen with nuclease from the skin of the hands and, finally, precipitation of the tRNA as rapidly as possible with alcohol and subsequent maintenance in the dried state (Apgar et al., 1962). Meanwhile work was carried out to simplify the method of isolating the tRNA and the activating enzyme (Holley et al., 1961a), and soon after a method of large-scale isolation of the total tRNA preparation from yeast was suggested (Holley, 1963). This method became widely popular, for large quantities of the samples were required for structural investigations.

The final countercurrent system used by Holley consisted of 17,880 ml of a solution containing 2200 g K_2HPO_4 and 3400 g $NaH_2PO_4 \cdot H_2O$, 1200 ml formamide, and 5200 ml isopropanol; 4 g of the sample of yeast tRNA was used in the experiment. Fractions containing alanine, valine, and tyrosine tRNAs were subjected to countercurrent distribution a second time, resulting in specimens of 66%, 60%, and 45% purity, respectively.

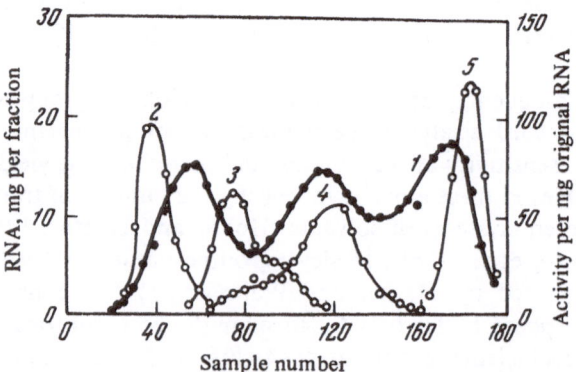

Fig. 16. Countercurrent distribution of yeast tRNA by Holley's method: (1) distribution of substance; (2–5) maxima of activity for alanine, valine, histidine, and tyrosine, respectively.

In 1962–1963, Holley had thus perfected his system of countercurrent distribution to such an extent that he could isolate large quantities of tRNA preparations of a high degree of purity (Fig. 16). The distribution coefficients for different tRNAs vary in this system from 0.12 to 4.25, i.e., by a factor of 36, while the distribution curves of the uv-absorbing material and of activity differ only slightly from theory. When discussing the purity of his specimens, Holley points out that the criterion used to assess the purity of the tRNAs, namely their acceptor activity, is in one sense too strict. The presence of small quantities of "denatured" or of very slightly altered molecules must have some effect on their acceptor activity, but in practice this is unimportant so far as the study of their primary structure is concerned. On this basis Holley considers that his specimens are suitable for structural research (Apgar et al., 1962).

As a result of the very first experiments on countercurrent distribution and of the comparative ease with which tRNAs specific towards different amino acids were separated from each other, the existence of substantial structural differences between them could be postulated. The first attempt to explain the nature of these differences was made by Holley as far back as in 1961 (Holley et al., 1961c). The results obtained with these relatively impure specimens are now of historical interest only; however, they did lead to the important conclusion that differences in nucleotide composition, although they exist, are not sufficiently great to account for the twentyfold differences in the distribution coefficients of the different tRNAs. It was accordingly postulated that the possibility of separating tRNAs depends, not so much on their nucleotide composition, as on their different nucleotide sequence; this was subsequently confirmed by Holley in the same investigation, for the fractionation profiles of oligonucleotides of the pyrimidyl-RNase hydrolysates of tyrosine, valine, and alanine tRNAs on DEAE-Sephadex proved to be completely different (Fig. 17). Hence, with comparatively unpurified specimens and with unimproved methods of fractionation of oligonucleotides, the conclusion was reached that the oligonucleotide composition of hydrolysates of different tRNAs is strictly individual and that, consequently, their primary structures are different.

Information on the nucleotide composition of tyrosine, valine, and alanine tRNAs was obtained in 1963 as the result of fractionation of their alkaline hydrolysates on two-way chromatograms (Holley et al., 1963); all the minor components except H_2Up were identified—this last was not discovered until much later (Madison and Holley, 1965).

At this time the method of fractionation of oligonucleotides of pyrimidyl- and guanyl-RNase hydrolysates in 7 M urea solutions was beginning to be introduced, and it led to a marked increase in the resolving power of the various systems. The isolated oligonucleotides were analyzed with

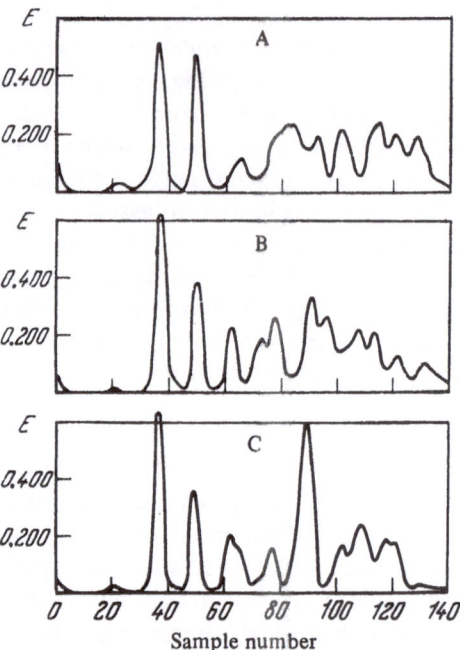

Fig. 17. Fractionation of pyrimidyl-RNase hydrolysates of tRNA on DEAE-Sephadex: (A) tRNA^{Tyr}; (B) tRNA^{Val} ; (C) tRNA^{Ala}.

respect to their composition (but not to their sequence), but the results on the whole were still far from the truth. Of the 14 oligonucleotides of the pyrimidyl-RNase hydrolysate of tRNA^{Ala}, only five agreed with the eventual results; in the case of tRNA^{Val} the situation was not much better, because correct results were obtained for only eight of the 16 oligonucleotides. The reason for this was partly that good resolution was possible at that time only for the lower oligonucleotides, and longer fragments could not be isolated in a homogeneous state.

Despite this objection, discussion of the next steps to be taken in analysis of primary structure began. There was no doubt that improvement in the methods of fractionation would ultimately enable all oligonucleotides to be obtained in a homogeneous state; the ways of studying their structure were also more or less clear. But how was the sequence of the oligonucleotides in the tRNA molecule to be established? This question confronted various groups of investigators almost at the same time, but it was first considered by Holley (Holley et al., 1963). He discussed three alternative ways of tackling the problem. Because of progress made in the analytical use of pyrimidyl and guanyl RNases, and also because of the discovery of T$_2$-RNase, wrongly regarded as adenyl RNase (Sato and Egami, 1957), hopes grew that new, specific endonucleases would be discovered and that they would provide a sufficient number of overlaps to enable reconstruction of the molecule. However, no such endonucleases have yet been found. The second way, which, like the first, has not been widely applied to analysis of the tRNAs, is the removal of segments of the molecule from either end by enzymic or chemical methods. Finally, the third possible way is by partial, incomplete, or restricted degradation of the molecule so as to obtain large overlapping units, by analysis of which the molecule can be reconstructed. As we know, this last method has been put to practical use, and it did in fact lead to the elucidation of the structure of tRNA^{Ala} in the beginning of 1965 (Holley et al., 1965a). Partial enzymic hydrolysis has not lost any of its importance even today. This method has been used to study all tRNAs whose primary structure have now been established, with the exception of tRNA^{Phe} of *E. coli* (Uziel and Gassen, 1969; Gassen and Uziel, 1969; but see Barrell and Sanger 1969). It has played an equally important role in the determination of the nucleotide sequences in 5S and 6S rRNAs and, judging from recent observations, it is also completely applicable to high-molecular-weight ribosomal and phage RNAs.

The group of workers under Holley's direction formulated the results of their work to determine the primary structure of purified specimens of alanine, valine, and serine tRNA as follows (Holley et al., 1964a): each of the three tRNAs has a characteristic nucleotide composition with a specific assortment of minor components. The sequence of the nucleotides is different in the tRNAs which have been investigated, as is shown by the fractionation profiles of their total pyrimidyl- and guanyl-RNase hydrolysates. Overlapping of the oligonucleotides of these two hydrolysates in some cases allows fairly long sequences to be obtained, but does not permit the structure of the complete molecule to be unequivocally reconstructed. In his paper Holley gives a structural formula of tRNA^{Ala}, but he stresses that it is only one of a thousand equally probable versions.

In the following months Holley succeeded in obtaining a sample of tRNA^{Ala} of 90% purity, with which his group completed the determination of the nucleotide composition and the study of oligonucleotides of exhaustive pyrimidyl- and guanyl-RNase hydrolysates (Holley et al., 1965b). Complete agreement was established between the oligonucleotides of the two hydrolysates, and in nearly all cases their structure was determined. The minor components and nonrepeating sequences found in the molecule allow certain reconstructions to be undertaken but, as Holley emphasized, only the positions of the 3'- and 5'-terminal fragments could still be taken for granted. The final task, that of determining the sequence of all the oligonucleotides in the polynucleotide chain of tRNA^{Ala}, was accomplished, as has already been mentioned, by partial enzymic hydrolysis. Degradation by T_1-RNase under the following strictly controlled conditions was used: a small quantity of enzyme, for a short time, at $0°C$, in the presence of Mg^{2+}. The distinguishing feature of T_1-RNase was the great selectivity of its action, and the distinguishing feature of tRNA was the rigidity of its secondary structure; for these reasons, this molecule could be degraded as a standard procedure into large fragments, the number of which varied depending on the experimental conditions (Penswick and Holley, 1965; Apgar et al., 1965, 1966). These fragments (metamers) were obtained in a homogeneous state by fractionating the oligonucleotides in 7 M urea on DEAE-cellulose (Tomlinson and Tener, 1963a,b). Exhaustive hydrolysis of the metamers with T_1-RNase gave oligonucleotides known from the previous analysis, and their overlapping demonstrated the sequence of the oligonucleotides in the polynucleotide chain.

When summing up his investigations, Holley points out that the final part of the work, the study of the structure of long nucleotide sequences in partial hydrolysis products, a hitherto completely unexplored field, was in fact comparatively simple and was completed surprisingly quickly. The most laborious part was the study of the structure of small fragments, namely oligonucleotides formed by exhaustive hydrolysis of tRNA by ribonucleases, as well as the identification of minor components.

In this part of the work Holley made several important discoveries which must now be mentioned. Holley identified two new minor components. One of them, dihydrouridine (Madison and Holley, 1965), is an unusual nucleoside; it does not absorb in the uv region and it is found (in comparatively large amounts, moreover) in almost all tRNAs so far studied. The problem of its role in tRNA is of great interest. The second minor component, MeI, also was found for the first time in a tRNA, and even today alanine tRNA remains the only one which possesses this compound in its composition.

An important event in the history of the study of tRNA structure was the identification of the tetranucleotide T-ψ-C-Gp (Zamir et al., 1965) (Fig. 18). This sequence was found in alanine, valine, and tyrosine tRNAs and also in total tRNA from yeast, E. coli, and rat liver. In view of the wide distribution of this oligonucleotide it was called "universal," and it is in fact found, as such or with slight modification, in all tRNAs so far studied. Holley deserves the credit for determining the structure of the "universal" oligonucleotide and also for the suggestion that it must participate in a function which is common to all transfer RNAs. Finally, Holley initiated the study of the structure of long oligonucleotides, using partial hydrolysis with PDEase (Holley et al., 1964b) and micrococcal nuclease (Zamir et al., 1965) for this purpose.

If Holley's work is considered chronologically, the logic with which

each successive step was taken and the importance of each published investigation are striking. The development and detailed study of the countercurrent distribution method, which with its various modifications lay at the
basis of isolation of many transfer RNAs; the development of a technique
for obtaining large quantities of total tRNA under laboratory conditions;
the identification of H_2Up and the "universal" oligonucleotide; and finally,
partial hydrolysis of the molecule with T_1-RNase with the production of
overlapping metamers and halves of the molecule—all these were technical
discoveries of decisive importance not only in elucidation of the structure of
alanine tRNA, but also in the development of research into the primary
structure of tRNA as a whole. Although every tRNA has its own special
features, it can be said that the determination of the primary structure of
a tRNA can be conducted as a standard procedure by following the scheme
first used by Holley to determine the primary structure of yeast $tRNA^{Ala}$.
Holley's final contribution was that he suggested a number of models for the
arrangement of Watson–Crick base pairs in tRNA; one of these, the clover-
leaf model, has been widely adopted and is at present the most firmly established model of a two-dimensional representation of the secondary structure of tRNA. Finally, Holley first showed the most probable location of the
anticodon (Holley et al., 1965; Holley, 1966).

It would be difficult to overestimate the importance of Holley's work;
as a pioneer in the field of study of tRNA structure he had no examples to

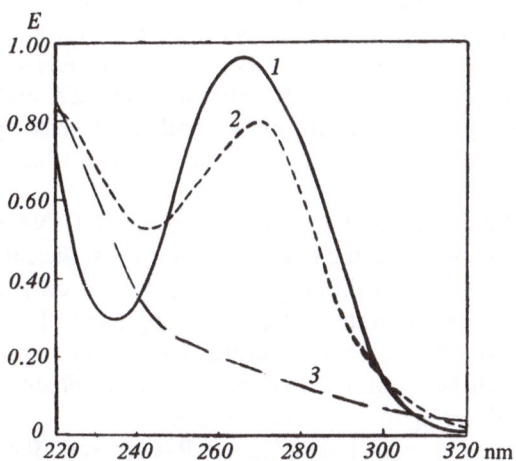

Fig. 18. Absorption spectrum of "universal"
oligonucleotide T-ψ-C-Gp; (1) in 0.1 N HCl;
(2) in 0.1 N KOH; (3) after bromination.

follow and was unable to proceed by comparison and analogy. He was the first to tread the difficult path to the determination of base sequence in nucleotide chains and he was the first to finish the course.

Serine tRNA from Brewers' Yeast

Systematic work to isolate purified specimens of this tRNA was commenced by Zachau's group in 1961. Zachau was convinced of the desirability of combining different methods of isolation of individual tRNA samples. This was perhaps because he and his collaborators tested a large number of different methods for this purpose, including the distribution of tRNA in aqueous solutions of natural and synthetic polymers forming biphasic systems (Tada et al., 1962); countercurrent distribution of tributylammonium salts of tRNA in a butanol–water–tributylamine–glacial acetic acid–dibutyl ether (100 : 130 : 10 : 2.5 : 24–31) system (Zachau et al., 1961); oxidation of total tRNA loaded with a given amino acid by means of sodium periodate, with subsequent fractionation on polyacrylic acid hydrazide (Zachau et al., 1961); chromatography on MAK columns (Melchers and Zachau, 1965); and partition chromatography on cellulose columns (Zachau et al., 1962). A particularly high degree of purification was obtained for serine tRNA both by Zachau and co-workers and by other investigators using this last method (Tanaka et al., 1962).

The first attempts at countercurrent distribution gave an indication that total yeast tRNA contains at least two forms of tRNA^{Ser}: the peak of serine activity during prolonged distribution had a shoulder (Thiebe and Zachau, 1965), and on repeated distribution it separated clearly into two peaks (Karau and Zachau, 1964). The presence of two forms of tRNA^{Ser} was also confirmed by chromatography of MAK columns (Melchers and Zachau, 1965). In addition, on repeated distribution a small third peak appeared (Thiebe and Zachau, 1965; Zachau et al., 1966b), the nature of which has subsequently been investigated by Feldmann (1967a).

Although the existence of at least two forms of tRNA^{Ser} has been confirmed by various methods, their preparation in a sufficiently pure form for determination of structural differences was achieved only at the beginning of 1966. It is interesting to note that, after testing most of the possible methods of isolation of tRNA, Zachau returned to the countercurrent method, by means of which he separated two polynucleotide chains consisting of 85 nucleotides and differing in only three of them.

The procedure by which samples of tRNA$_1^{Ser}$ and tRNA$_2^{Ser}$ of approximately 90% purity could be isolated was as follows: 40–50 g of the tRNA was distributed twice in a tri-n-butylamine system and then subjected to prolonged distribution in a system containing ammonium sulfate (Kirby, 1960).

As a result of this, serine-specific tRNA virtually free from other activities was obtained (Zachau et al., 1966b). The yield was about 70%, a result which these workers are inclined to ascribe to special precautions against contamination by nucleases, which they observed in accordance with the recommendations of Holley et al. (1961a), and to the use of the flotation method to isolate tRNA (Mirzabekov et al., 1964). Despite the by no means complete separation achieved, the side fractions contained $tRNA_1^{Ser}$ and $tRNA_2^{Ser}$ in a practically pure form.

The first attempt to study the difference between the primary structures of individual tRNAs was made by Zachau's group in 1962 on samples with an estimated purity of about 40% (Staehelin et al., 1962, 1963). The products of pyrimidyl-RNase hydrolysis of total, serine, and valine tRNA were fractionated on DEAE-cellulose. At that time results were obtained only for mononucleotides, dinucleotides, and some trinucleotides, including those containing minor components.

Much improved results for $tRNA^{Ser}$ were obtained in 1964, when three outstanding investigations were published simultaneously. One of these papers described the isolation of a purified sample of $tRNA^{Ser}$ by the countercurrent method and its separation by repeated countercurrent distribution into two fractions (Karau and Zachau, 1964), while the other two described the degradation of the isolated samples by pancreatic (Melchers and Zachau, 1964) and T_1-RNases (Dütting and Zachau, 1964a). The pyrimidyl-RNase hydrolysates of the two serine tRNAs were analyzed separately, but at this stage of the investigation it was not yet possible to establish the differences between them (Dütting et al., 1965). The oligonucleotides were characterized spectrometrically and by their electrophoretic mobility: the determination of their composition was almost completed and the study of their structure was commenced.

In 1964–1965 doubts were still held regarding the existence of isoacceptor tRNAs. Zachau's efforts were therefore directed toward proving that the two serine tRNAs actually exist and are not formed through secondary effects. This was a particularly crucial problem because the first analyses had revealed no differences between the two serine tRNAs. Zachau therefore continued his experiments on the fractionation of tRNA (Melchers and Zachau, 1965; Thiebe and Zachau, 1965) and concluded that the existence of several peaks of activity for certain amino acids (cysteine, valine, phenylalanine, etc.) is not an artifact but reflects the actual existence of several tRNAs specific for one amino acid but possessing a different primary structure and consequently a different secondary structure.

In later work by Zachau's group (Melchers et al., 1965; Dütting et al., 1965) attempts to detect differences between the two serine tRNAs were continued, but with a virtual lack of success mainly because they could not be separated from each other sufficiently. Analysis of the oligonucleotides at this period had made great head way, and the first reconstructions of the longer sequences could accordingly be made. To determine the base sequence in oligonucleotides, pyrimidyl and guanyl RNases, and micrococcal nuclease, as well as total and partial hydrolysis with snake-venom PDEase, were used (Feldmann et al., 1966).

Among the methods of obtaining longer sequences Zachau includes: (A) cleavage at minor components, which he used when determining differences in the rates of cleavage of phosphodiester bonds formed by Gp and its methylated derivatives by T_1-RNase (Dütting and Zachau, 1964b); (B) hydrolysis by pyrimidyl RNase after preliminary uv irradiation and the formation of Up dimers not attacked by this enzyme (Zachau, 1964); and, finally, (C) hydrolysis by enzymes under unfavorable conditions. Zachau used this last method in practice and he used both RNases for partial degradation (Dütting et al., 1966).

Experiments on the partition chromatography of tRNA were resumed in 1965 but this time the main object of this work was to show that oligonucleotides and polynucleotides formed by partial hydrolysis of tRNA could be separated by this method (Zachau, 1965; Schweiger and Zachau, 1965). The results showed that fragments of different composition can be separated particularly well from each other and that chain length plays a less important role. This could be significant in the investigation of metamers formed by partial hydrolysis of tRNA, but Zachau, like other investigators, used fractionation on DEAE-cellulose in 7 M urea for this purpose (Dütting et al., 1966).

The structure of $tRNA_1^{Ser}$ and $tRNA_2^{Ser}$ was finally elucidated at the beginning of 1966 (Zachau et al., 1966a,c), i.e., one year after publication of the structure of $tRNA^{Ala}$. Like Holley, Zachau later described experimental results confirming the structure as established, in the second half of 1966 (Zachau et al., 1966b; Feldmann et al., 1966; Dütting et al., 1966).

The elucidation of the structure of $tRNA_1^{Ser}$ and $tRNA_2^{Ser}$ was to some extent unexpected, for at the end of 1965 the work was still far from completion. The reason for this rapid success was that it was eventually possible to obtain pure samples of $tRNA_1^{Ser}$ and $tRNA_2^{Ser}$, uncontaminated by one another, and the ground had been prepared by all the previous work for the detection of the small differences existing between their structures.

An important success obtained by Zachau's group was the identification of two hitherto unknown minor components, acetylcytidine and isopentenyladenosine (Zachau et al., 1966c; Biemann et al., 1966), as well as guanosine and uridine methylated in the ribose residue, which were discovered for the first time in the composition of a tRNA.

The work of Zachau's group is thus the second example of elucidation of the structure of an individual tRNA and the first example of elucidation of the structure of two isoacceptor tRNAs.

Tyrosine tRNA from Bakers' Yeast

The study of the structure of tyrosine tRNA was completed almost simultaneously with that of the two serine tRNAs, i.e., at the beginning of 1966. Two of the authors describing the structure of $tRNA^{Tyr}$ (Madison and Everett) were also concerned with the elucidation of the structure of $tRNA^{Ala}$ and in a whole series of investigations carried out in the course of its determination. Naturally, therefore, the experience gained during the work on $tRNA^{Ala}$ was one of the chief reasons for the speed with which $tRNA^{Tyr}$ could be studied.

Following the tradition established after Holley's work, Madison began by publishing a structural formula and a minimal number of technical details (Madison et al., 1966a), and this was followed by the experimental material confirming the suggested structure (Madison et al., 1967a; Madison and Kung, 1967). The publication in which fractionation of $tRNA^{Tyr}$ was reported (Holley et al., 1963) was the only one which preceded the publication of the structural formula, if the work on the identification of H_2U (Madison and Holley, 1965), which applies equally to all tRNAs, and the work on partial hydrolysis by PDEase, used in conjunction with other methods to study the structure of the oligonucleotides of $tRNA^{Tyr}$ (Holley et al., 1964b), are disregarded. The methods used to analyze $tRNA^{Tyr}$ were the same as those used with $tRNA^{Ala}$: Holley's method of countercurrent distribution to obtain the individual sample of tRNA; total hydrolysis of the molecule by pancreatic and T_1-RNases with subsequent fractionation of the oligonucleotides on DEAE-cellulose in 7 M urea; finally, obtaining overlapping metamers by partial hydrolysis with T_1-RNase and subsequent determination of their oligonucleotide composition.

Judging from its oligonucleotide composition, the purity of the $tRNA^{Tyr}$ sample used for analysis was about 80%.

Besides the method used previously to analyze oligonucleotides, partial degradation by polynucleotide phosphorylase also has been used (Madison et al., 1967b). Instead of the 5'-terminal dinucleoside monophosphates formed by exhaustive hydrolysis by this enzyme (Singer, 1958), 5'-terminal trinucleoside diphosphates were obtained in this way, separated on DEAE-cellulose, and analyzed by alkaline hydrolysis. Chromatography and direct spectrophotometry have been extensively used in oligonucleotide analysis.

Half-molecules have been obtained by partial hydrolysis with T_1-

RNase, and much of the work on metamer formation and reconstruction of the molecule has been done on these products (Madison and Kung, 1967).

Phenylalanine tRNA from Bakers' Yeast

The study of the structure of phenylalanine tRNA, like that of tyrosine tRNA, has a comparatively short history.

The sample of tRNA^{Phe} used for structural investigations was obtained by countercurrent distribution in the scheme suggested by Holley et al. (1963). The first fractionation (200 transfers) was conducted at 25–26° C, and increased purity by a factor of 15; in the second cycle (780 transfers) the fraction containing phenylalanine activity was divided into two peaks, with an enrichment factor of 22 in the principal peak. During the third distribution (1200 transfers) the principal peak of phenylalanine activity was enriched further by a factor of 30 (Hoskinson and Khorana, 1965). It was subsequently found that this peak differs from the second peak only by the fact that it contains tRNA^{Phe} without a terminal adenosine residue (RajBhandary et al., 1968c).

From articles published in 1966 it was clear that work on the study of the primary structure of tRNA^{Phe} was nearing its end (RajBhandary and Stuart, 1966a,b). The fullest account was given in a paper at a Cold Spring Harbor Symposium (RajBhandary et al., 1966) in which a hypothetical structural formula for tRNA^{Phe}, reconstructed on the basis of incomplete experimental material and by analogy with the structures already known, was suggested. Reconstruction of the tRNA^{Phe} structure was based on the concept that the role of tRNA in protein synthesis requires a certain conformation of the molecule and that, consequently, like other tRNAs so far decoded, phenylalanine tRNA must conform to the clover-leaf type of model. The following principles consistent with his model must therefore be followed in creating the preliminary structure. (1) The polynucleotide chain of tRNA is folded, thus forming paired segments, starting from the fifth base from the acceptor end of the molecule which must be complementary to the first base at the 5'-end; next follow five or six pairs of nucleotides linked by hydrogen bonds. (2) The model has three principal loops (anticodon, Tψ-, and H$_2$U-loops), containing unpaired nucleotides. (3) The number of nucleotides in the anticodon and Tψ-loops is 7; the helical portions of these loops each contain five base pairs. (4) The 2me$_2$G residue, if present, lies at a distance of 9 nucleotides from the first letter of the anticodon. (5) The T residue occupies position 23 counting from the acceptor end of the molecule. (6) The anticodon is always followed by a modified base. With these principles as guide, and using the existing experimental data (complete analysis of the oligonucleotides of exhaustive pyrimidyl- and guanyl-RNase hydrolysates and their overlaps) it was possible to draw up a structural formula for tRNA^{Phe} in which only one correction was necessary on completion of the work. A further 6 months of work was necessary to verify the structure experimentally by partial hydrolysis with T$_1$-RNase, and the final version was published in January, 1967 (RajBhandary et al., 1967).

Khorana and his collaborators made an important contribution to the development of methods for determining the base sequence in oligonucleotides (Razzell and Khorana, 1961; Ralph et al., 1963; RajBhandary et al., 1964; RajBhandary, 1968). One of these methods was the introduction of a radioactive label into the terminal 2',3'-diol group. A combination of this method with PDEase degradation was used to determine the structure of several oligonucleotides of tRNA^{Phe}. Introduction of the label into the terminal nucleotides was also used to determine the 3'- and 5'-terminal sequences in tRNA^{Phe} (RajBhandary et al., 1968c).

Fourteen minor components have been identified in tRNA^{Phe}, including OmeC and 7meG, isolated from tRNA for the first time. The nature of one of the minor components of tRNA^{Phe} (Y) has only recently been discovered (Nakanishi et al., 1970), but its unusual properties have already been used to advantage in structural and functional investigations.

Experimental confirmation of the previously published structural formula of RNA^{Phe} appeared in 1968. In two papers (RajBhandary et al., 1968a,b) the results of determination of the composition and structure of oligonucleotides of total RNase hydrolysates of tRNA^{Phe} were presented. To analyze the oligonucleotides, extensive use was made of direct spectrophotometry. Besides the enzymes used previously to analyze oligonucleotides, T$_2$-RNase and spleen PDEase were used. The difficulties were due to the conversion 1meA→6meA, the lability of 7meG and nucleoside Y, and the much slower cleavage of the phosphodiester bonds formed by minor components by virtually all enzymes. Determination of the structure of the longest of the guanyl-RNase oligonucleotides, consisting of 12 monomers, was probably a very difficult problem, because this oligonucleotide contains five minor components, one of which (Y) is highly labile, while two of the others make the phosphodiester bond inaccessible to the action of RNase (OmeG and OmeC). By comparing oligonucleotides of RNase hydrolysates, it was possible to reconstruct longer sequences accounting for 55 of the 76 nucleotides constituting the tRNA^{Phe} molecule. The next two papers published by this group described the production of partial pyrimidyl-RNase (Chang and RajBhandary, 1968) and guanyl-RNase hydrolysates (RajBhandary and Chang, 1968) and analysis of the resulting metamers by total hydrolysis with the same enzymes; tRNA^{Phe} is the first individual tRNA from which halves have been obtained by means of pyrimidyl RNase.

Valine tRNA from Bakers' Yeast

Work to elucidate the primary structure of tRNA^{Val} started in the second half of 1962. It was not until the spring of 1964 that a sample of tRNA^{Val} sufficiently pure for structural research was isolated.

It was proposed to determine the structure of tRNA^{Val} by the unit method, and enzymes were used to hydrolyze it; the first effort were therefore directed toward the development of methods of fractionation and identification of the oligonucleotides. Initially it was intended to use paper chromatography only to fractionate the oligonucleotides, because of the small quantities of the sample available. For this purpose, a method of two-way chromatography was developed for use with the total tRNA sample (Krutilina et al., 1964), and this was later used to investigate an individual sample of the whole tRNA^{Val} molecule (Krutilina et al., 1965; Li et al., 1966) and its halves (Baev et al., 1966b,c), and also to compare the primary structures of tRNA$_1^{Val}$ and tRNA$_3^{Val}$ (Krutilina et al., 1970). However, it soon became clear that not all the problems connected with determination of primary structure could be solved by paper chromatography, and that fractionation on ion-exchange columns was necessary (Mirzabekov et al., 1965a; Aksel'rod et al., 1965). By working on columns it was possible to obtain comparatively large quantities of oligonucleotides sufficient for structural studies; paper chromatography was used, parallel with column chromatography, in order to obtain additional fractionation, to determine the homogeneity and purity of the specimens, and also to fractionate the hydrolysis products of the oligonucleotides during the study of their structure.

An important part of the first years of work on tRNA^{Val} was taken up with the development of a spectrophotometric method of identification and analysis of the oligonucleotides (Venkstern et al., 1963a,b; Baev et al., 1963b; Baev et al., 1963). Considerable attention was paid to the minor components (Venkstern et al., 1963b; Baev et al., 1963; Venkstern, 1964; Mirzabekov et al., 1965a). A number of investigations were carried out to study the behavior of the minor components in enzymic reactions which, although chiefly of practical importance, were nevertheless also of theoretical interest (Baev et al., 1963a; Abrosimova-Amel'yanchik et al., 1965; Venkstern, 1966).

The two most important events which led to success in the elucidation of the structure of yeast valine tRNA occurred in 1964. First was the discovery and isolation of guanyl RNase from actinomycetes, thus providing a vital tool for the study of the structure of valine tRNA from *S. cerevisiae* (Tatarskaya et al., 1964). A study of the specificity and other properties of this enzyme (Abrosimova-Amel'yanchik et al., 1965; Tatarskaya et al., 1966a,b) showed that it is a good substitute for T$_1$-RNase. Guanyl RNase from actinomycetes was used in both the initial and final stages of the investigation, in order to obtain halves and metamers of $tRNA_I^{Val}$; the enzyme has also been used to study the structure of pyrimidyl-RNase oligonucleotides.

The second achievement was the isolation of a specimen of an individual valine-specific transfer RNA from the yeast *S. cerevisiae* (Mirzabekov et al., 1965b). The $tRNA_I^{Val}$ was obtained from bakers' yeast as a specimen of more than 90% purity by the successive use of countercurrent distribution in a slightly modified Holley's system (Apgar et al., 1962) and oxidation of the purified specimen, aminoacylated with valine, by sodium periodate with subsequent fixation of the oxidized tRNAs on polyacrylic acid hydrazide, melted into agar as described by Knorre et al. (1964).

The general plan used to study the primary structure of $tRNA^{Val}$ differed only by the fact that the structural analysis was carried out almost entirely on half-molecules. It was known that, in principle, the tRNA molecule can be split at only one bond (this had been observed by Holley, Zachau, and Madison). A special feature of the work with $tRNA_I^{Val}$ was that the half-molecules were obtained as the only hydrolysis product, they were isolated in preparative amounts, and they were used for structural analysis. This was a great advantage and it enabled the oligonucleotides of the total RNase hydrolysates to be distributed between the acceptor (3'-) and nonacceptor (5'-) halves of the molecule, thus considerably reducing the number of alternative structures. However, this information was insufficient for unambiguous reconstruction of the molecule. A limiting assumption was then made, to serve as a working principle for reconstruction of the presumptive structure. On comparing oligonucleotides of exhaustive hydrolysates of $tRNA_I^{Val}$ and those of $tRNA^{Ala}$, the only transfer RNA whose structure had been elucidated at that time, units of identical or almost identical structure were found. It was postulated that the structure of all transfer RNAs is based on a single plan, and that identical and characteristic sequences occupy identical positions in their structure. Accordingly therefore, many of the oligonucleotides could be assigned to the 3'- and 5'-halves of the $tRNA_I^{Val}$ molecule, so that a preliminary and, to some degree hypothetical, structural formula was obtained. This was described at the Warsaw Conference of the Federation of European Biochemical Societies in 1966 (Baev et al., 1966b,c).

Experimental verification of the proposed formula and its correction were done on the basis of results obtained later. The final stage was analysis of the metamers obtained by partial hydrolysis, not of the whole molecule, but of its halves, so that the analysis was greatly simplified. Overlapping of the metamers enabled the two halves to be reconstructed, and by joining these together the structure of the whole molecule was obtained (Baev et al., 1967a).

Parallel investigations were carried out to study oligonucleotides of the total RNase hydrolysates. The structure of the pyrimidyl-RNase fragments was determined mainly by the use of guanyl RNase (Venkstern et al., 1968a). To study the structure of the longer guanyl-RNase oligonucleotides, both the standard methods of enzymic and chemical degradation and also the method of partial hydrolysis with pyrimidyl RNase developed for this purpose, were used (Venkstern et al., 1968b).

The known primary structure of $tRNA_1^{Val}$ and the isolation of halves of this molecule in preparative quantities provided a basis for functional investigations by the sliced-molecule method, at present undergoing intensive development (Baev et al., 1967b,c, 1968, 1969; Mirzabekov et al., 1969a, b,c,d, 1970).

Elucidation of the primary structure further enabled the anticodon of $tRNA_1^{Val}$ to be directly established (Mirzabekov et al., 1967).

The comparative study of the three isoacceptor valine tRNA, giving independent peaks on countercurrent distribution, is of great interest. The investigation of $tRNA_3^{Val}$ is now completed (Krutilina et al., 1970).

SUBSEQUENT DEVELOPMENT OF STRUCTURAL RESEARCH

The five yeast tRNAs discussed above were the vanguard for the study of the primary structure of nucleic acids. However, the material now available is by no means confined to them.

Since 1968 structural research has assumed a literally explosive character, and the number of completely elucidated structural formulas now exceeds twenty. No attempt will therefore be made to describe these investigations in detail, more especially because, like the earlier investigations, they are based on the unit method and the use of enzymes to break up the tRNA molecules. Only the general outlines of these investigations will be examined, and particular attention will be paid to refinements introduced, and to new prospects opened as a result of the accumulation of fresh experimental material.

The first fact to be mentioned is determination of the structure of serine tRNA from rat liver (Staehelin et al., 1968); this work is of interest because it is the first, and so far the only, example of elucidation of the structure of a tRNA from a mammalian tissue.

The structure of one tRNA of plant origin has also been determined: $tRNA^{Phe}$ from wheat germ (Dudock et al., 1969).

Two important centers for research into nucleic acid structure have arisen lately; each one has to its credit the determination of the complete nucleotide sequence of several transfer RNAs. The first of these centers is the Laboratory of Molecular Biology in Cambridge, England, where progress is being made in the study of the primary structure of RNA from *E. coli*. One of the main reasons for the rapid progress of research in this laboratory is that the structure of nucleic acids labeled with ^{32}P is studied by the method of Sanger and co-workers (Sanger et al., 1965; Brownlee and Sanger, 1967, 1969; Brownlee et al., 1968); as was mentioned above, this method is one of great precision and it requires much less material and time, because it is based on radiochemical techniques. Within a short time the structure of the tyrosine (Goodman et al., 1968), formylmethionine (Dube et al., 1968), methionine (Cory et al., 1968), valine (Yaniv and Barrell, 1969, 1971), phenylalanine (Barrell and Sanger, 1969), tryptophan (Hirsh, 1970), leucine (Dube et al., 1970), and two glutamine (Folk and Yaniv, 1972) tRNAs from *E. coli* have been determined at the Laboratory of Molecular Biology.

New advances, in the isolation of tRNA preparations have been the use of reverse-phase chromatography (Kelmers et al., 1965), of BD-cellulose (Gillam et al., 1967, 1968), and of electrophoresis in polyacrylamide gel (Adams et al., 1969), which enable the radioactive nucleic acids of high purity to be isolated in sufficient quantity for work with these new methods.

In research undertaken at the Laboratory of Molecular Biology, structural and genetic problems are closely interwoven. For example, the structure of the tRNA^{Tyr} of wild and amber-suppressor strains of *E. coli* has been elucidated (Goodman et al., 1968); the anticodon G-U-A of tRNA^{Tyr} of *E. coli* Su$_{\overline{III}}$ is converted as a result of the mutation into the triplet C-U-A, complementary to the terminating codon U-A-G so that tyrosine is incorporated into the polypeptide chain instead of chain termination occurring.

The nucleotide sequence has been determined in the tRNA^{Tyr} of many mutants and revertants of *E. coli* Su$_{\overline{III}}^{+}$ (Abelson et al., 1970; Smith et al., 1970, 1971). Comparison of accurately localized nucleotide substitutions and the accompanying changes in properties is a promising method of studying interaction between the various bases in tRNA and its functional topography.

A particularly important piece of research was the determination of the structure of *N*-formylmethionine tRNA (Dube et al., 1968), which initiates protein synthesis in bacterial systems (Clark and Marcker, 1965). This work is an essential step toward elucidation of the mechanism of initiation of protein synthesis in bacteria. Determination of the structure of valine and glutamine isoacceptor tRNAs is of geat importance; the results will probably shed light not only on the structural bases of recognition by the aminoacyl-tRNA synthetases, but also on the reason for the multiplicity of the tRNAs.

The use of Sanger's principle has led to the elucidation of the primary structure not only of many tRNAs, but also of two 5S ribosomal RNAs, isolated from *E. coli* (Brownlee et al., 1967) and from human carcinoma KB cells (Forget and Weissman, 1967). These investigations have demonstrated if not the universality, then at least the much wider applicability of the unit method than was hitherto apparent. In particular, it was found that not only transfer, but also 5S ribosomal, 6S ribosomal (Brownlee, 1971), and also the high-molecular-weight RNAs of ribosomes and bacterial viruses possess a rigid secondary structure. This meant that partial enzymic hydrolysis can be used to study them and the large fragments necessary for primary-structure determination can be obtained. For example, fragments from 20 to 200 nucleotides in length were obtained by restricted hydrolysis with T_1-RNase from the RNA of phage R17 which contains three cistrons (Adams et al., 1969). Some of these fragments were purified by electrophoresis in polyacrylamide gel, and then analyzed by Sanger's radiochemical method. Among the partial guanyl-RNase hydrolysis products of phage RNA, a fragment 57 monomers long was found and studied. Its nucleotide sequence corresponds to part of the amino-acid sequence in the phage coat protein (Weber, 1967) and it is consequently part of the cistron of this protein (Adams et al. 1969). This is the first case of determination of the structure of a messenger RNA by a chemical method, and it is the first direct confirmation of the validity of the genetic code. Examination of the sequence as thus established shows that some amino acids found more than once in the phage coat protein are coded by triplets differing in their third nucleotide; this is evidence of the degeneracy of the code used by the phage.

The study of the primary structure of high-molecular-weight RNAs is making rapid progress.

Among the structures which have recently been determined are large fragments of phage RNAs corresponding to different parts of the phage coat protein (Robinson et al., 1969; Nichols, 1970); terminal sequences of phage RNAs (Billeter et al., 1969); sequences composing the initiator segment of the cistron (Hindley and Staples, 1969), etc.

Large fragments of phage RNAs are now obtained not only by the method of partial hydrolysis, but also by the other methods mentioned earlier (Steitz, 1969; Billeter et al., 1969). The problem of determination of their structure has, in principle, been solved by the use of the unit method together with radiochemical techniques. It will probably not be long, therefore, before the first structural formula of one of the high-molecular-weight RNAs is known.

The second productive center for study of the primary structure of tRNA is Takemura's laboratory in Japan. Here the tRNAs of different amino-acid specificity have been isolated from *T. utilis,* mostly by the method of chromatography on DEAE-Sephadex using various systems of solvents (Miyazaki et al., 1966; Miyazaki and Takemura, 1966; Miyazaki et al., 1967). Elucidation of the primary structure of three of these tRNAs, namely, valine (Takemura et al., 1968a), tyrosine (Hashimoto et al., 1969), and isoleucine (Takemura et al., 1969b) is now complete. To study primary structure the workers of this group have used the ordinary methods of total and partial hydrolysis with pancreatic and T_1-RNases, followed by fractionation of the hydrolysis products on DEAE-cellulose in 7 M urea. Besides the enzymes used previously to study oligonucleotide structure (total and partial hydrolysis with PDEase, T_1-, T_2-, and pyrimidyl RNases), the enzyme U_2-RNase has also been used (Arima et al., 1968). During the study of tRNA^{Ile}, a new complex minor component, N-(purinyl-6-carbamoyl)-threonine nucleotide, has been identified in a tRNA.

Results, as yet only preliminary, have been obtained for the primary structure of tRNA^{Ala} from E. coli (Alvino and Clarke, 1968; Alvino et al., 1969).

Considerable progress has been made with the study of methionine tRNA of bakers' yeast, and its complete structural formula will evidently soon be established (RajBhandary and Ghosh, 1969). On countercurrent distribution two peaks of methionine activity were found, and the first was subsequently purified by chromatography on DEAE-Sephadex and by reverse-phase chromatography (Kelmers et al., 1965). Analysis of the products of guanyl-RNase hydrolysis showed that the 5'-terminal sequence consists of the dinucleotide pA-G, and the 3'-terminal sequence of the hexanucleotide C-U-A-C-C-A. This is the first case where A has been found in the 5'-position of a tRNA.

Of the other investigations which have been completed, mention must be made of the elucidation of the structure of asparagine tRNA from brewers' yeast (Keith et al., 1970) and leucine tRNA from bakers' yeast (Kowalski et al., 1971); the second of these investigations was carried out entirely by Sanger's method, for which a new technique of growing yeast on a medium rich in ^{32}P was developed and a tRNA^{Leu} with a high content of label was isolated. The structure of tRNA$_3^{Gly}$ from E. coli (Squires and Carbon, 1971) and of tRNA$_2^{Val}$ from brewers' yeast (Bonnet et al., 1971) has also been determined.

It was shown that met-tRNA$_1^{Met}$ of yeast, like the tRNA$_f^{Met}$ of E. coli, can be converted in the presence of formyltetrahydrofolic acid and methionyl-tRNA-transformylase from E. coli into formylmet-tRNA$_f^{Met}$. In the presence of the initiator codons G-U-G and A-U-G, tRNA$_f^{Met}$ of yeast, like the tRNA$_f^{Met}$ of E. coli, initiates protein synthesis. A study of the structure of the anticodon branch of yeast tRNA$_f^{Met}$ has shown that a modified adenosine, which is probably threonine carbamoyladenosine, is located on the 3'-side of the anticodon (RajBhandary and Kumar, 1970).

The structure of some transfer RNAs has been determined more than once. Merril (1968), for example, verified the structural formula established by Holley for tRNA$_2^{Ala}$ from bakers' yeast using a specimen purified by more sophisticated methods, and introduced one correction into it, although in principle nothing was changed. Differences in seven nucleotides were found between the results obtained for tRNA^{Tyr} from *E. coli* by Goodman et al. (1968) and Doctor et al. (1969). Admittedly, these investigations were carried out on different strains of *E. coli*. Parallel studies of the structure of tRNA^{Phe} from *E. coli* were made by Barrell and Sanger (1969) and by Gassen and Uziel (1969); the suggested structural formulas differ considerably from each other; their examination suggests that the version put forward by Sanger's group is more likely to be correct. Further tests by Harada et al. (1969b) confirmed the structure suggested by Yaniv and Barrell (1969) for tRNA$_1^{Val}$ of *E. coli*. The structural formula suggested by Baev et al. (1967a) has been verified by Vasilenko et al. (1969), and also jointly by the laboratories of Ebel and Baev (unpublished data). As a result of this confirmatory work, certain modifications were introduced into the structural formula of tRNA$_1^{Val}$ from bakers' yeast published in 1967 (see below).

The study of the structure of tRNA^{Ser} of bakers' yeast has been carried as far as the oligonucleotide-analysis level (Neelon et al., 1967); neither in the composition of the oligonucleotides nor in the minor components have any differences been found compared with the tRNA^{Ser} of brewers' yeast (Zachau et al., 1966b), thus suggesting that the primary structures of the serine tRNAs from these two sources are identical.

The structural formulae of all transfer RNAs so far studied, irrespective of their amino-acid specificity and source, can be represented by a clover leaf. This is an important argument in support of the view that this class of nucleic acids is built to a single common plan, and that the model which Holley suggested correctly reflects its basic features.

The list of transfer RNAs studied shows that the extensive material which has been collected provides a basis for their comparison from several different aspects: some results allow tRNAs of different and of the same amino-acid specificity from taxonomically close or distant sources to be compared: isoacceptor tRNAs; the tRNAs of different mutants of *E. coli*; tRNAs initiating protein synthesis and adding amino acids to the inner segments of the polypeptide chain; etc. It will be noted that primary structures themselves have proven less informative and have not justified the hopes which were initially placed on such a comparison. The primary structure does not tell the functional contribution of a particular nucleotide sequence. It is now obvious, therefore, that a known sequence of nucleotides is an essential but not adequate condition for full recognition of the functional topography of a tRNA. Structural and functional investigations based on a known primary structure are now being developed with extraordinary rapidity and in various directions. However, the detailed examination of these questions is outside the scope of the present monograph.

To complete this brief historical survey a few remarks of a general character may be made. All the research groups mentioned above, especially the first five, have studied the primary structure of tRNA by a method of gradual approximation to the truth. Determination of nucleotide sequence is basically a technical problem; the obtaining of experimental data corresponding more and more with the true state of affairs is therefore closely bound up with improvements in techniques, and every significant advance has followed directly after some technical innovation.

One of the chief problems arising during the determination of primary structure is the preparation of individual tRNAs; the work began with samples not exceeding 50–60% in purity, while the samples used for obtaining the final results were 90–100% pure. Reduction of degradation of tRNA to a minimum by eliminating the action of contaminating nucleases was of great importance to structural investigations; in this aspect of the problem such methods as treatment of material with phenol and bentonite, and isolation of Cetavlon salts of tRNA by flotation, played an important role in this achievement. An increase in the purity and a more detailed study of the properties of the enzymes used are also worthy of mention; nonspecific enzymic hydrolysis has been largely eliminated by the correct choice of enzyme concentration necessary for the required degree of hydrolysis.

An important aspect of the study of nucleotide sequence is fractionation of the tRNA fragments obtained as a result of different types of degradation of the molecule (different enzymes, different degrees of degradation by the same enzyme). Initially the fractionation of even total hydrolysates was a complex problem; only dinucleotides and mononucleotides could be separated successfully, and reliable analysis was restricted to these components. Methods which would allow fractionation of the higher oligonucleotides were nonexistent. An event of the first importance in this connection was the development of the method of fractionating oligonucleotides on DEAE-cellulose in 7 M urea. This system itself, and its different modifications and combinations with other methods have made it possible to obtain all the fragments into which individual tRNAs are split in the course of the study of their primary structure by the unit method, in an individual state.

Paper chromatography, a convenient method of fractionation and identification of the fragments of nucleic acids, has played an important role in structural investigations and it is without question the method of choice for the identification of minor components. The further development of chromatography and electrophoresis, the use of DEAE and acetate paper in these methods, the thin-layer modification, and conduct of the analysis at the radiochemical level have all proved extraordinarily fruitful. I have already mentioned above that a method of fractionating fragments including about 50 monomers in a thin layer has now been developed.

Another important factor has been the development of methods to investigate oligonucleotides. An important role in the solution of this problem has been played by the rational use of a wide range of enzymes, including the phosphodiesterases, ribonucleases, and polynucleotide phosphorylase, and more recently by the use of a combination of enzymic and radiochemical methods.

Such in brief are the lines of research whose development and perfection have made the determination of the primary structure of tRNA possible and which have laid the foundations of the study of nucleic acid structure in general. It only remains to be said that the road was longest for those investigators who first had to seek it; it was much shorter for those who set out upon it later and who were able to profit to the full from the experience gained by the pioneers.

It must also be noted that the rapid development of methods of investigating nucleic-acid structure has also enabled more precise data to be obtained, for all the stages involved in determination of nucleotide sequence are now being conducted at a higher experimental level. The verification of data obtained at the beginning of the period of structural research, by the more refined methods developed subsequently, is thus extremely useful, more especially because work of this type can now be carried out in an immeasurably shorter time.

We can now turn to the characteristics of the structure of individual transfer RNAs and to their comparison.

THE PRIMARY STRUCTURE OF INDIVIDUAL tRNAs

Alanine tRNA from Bakers' Yeast (Holley et al., 1965a)

Work on the determination of the primary structure of tRNA^{Ala} occupied altogether seven years, four of these years being taken up with purification of the tRNA. To determine the primary structure, about 1 g of tRNA^{Ala} was used; this was isolated from 200 g of a total preparation, to obtain which approximately 120 kg of yeast had to be processed (Holley, 1966).

The nucleotide composition of alanine tRNA from yeast is given below:

Nucleotides, principal	Moles/mole of tRNA	Nucleotides, minor	Moles/mole of tRNA	Nucleotides, minor	Moles/mole of tRNA
A	8	ψ	2	1meI	1
G	25	H_2U	2 (or 3)	1meG	1
U	12	T	1	2me$_2$G	1
C	23	I	1		

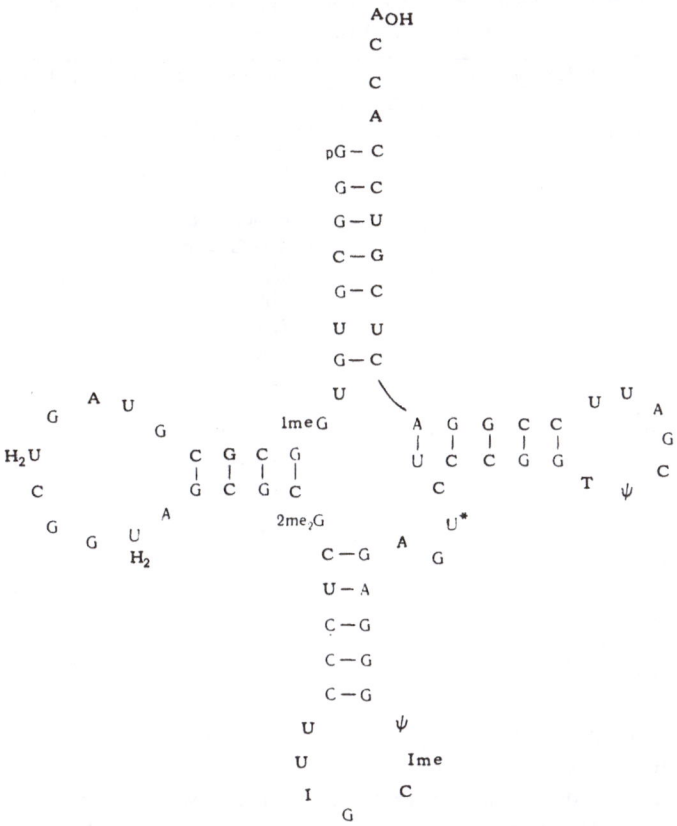

Fig. 19. Yeast tRNA^Ala in the clover-leaf configuration.

The molecule of tRNA^Ala (Fig. 19) consists altogether of 77 nucleotides, including nine minor components. Alanine tRNA differs from all other individual tRNAs in its lower content of A and the total absence of its derivatives. Other absentees from the composition of tRNA^Ala are 5meC and nucleotides methylated in the ribose residue.

On exhaustive hydrolysis with pyrimidyl RNase the molecule of tRNA^Ala breaks up into 40 fragments, including 21 mononucleotides, nine dinucleotides, seven trinucleotides, two tetranucleotides, and one octanucleotide (Table 6). The tetranucleotide pG-G-G-Cp forms the 5'-terminal end of the molecule. Adenosine is absent in tRNA^Ala, just as in most other tRNAs isolated from American commercial bakers' yeast, and in the usual preparation the molecule starts with the second nucleotide from the acceptor end, namely, C. The trinucleotide I-G-Cp is the anticodon of tRNA^Ala.

TABLE 6. Oligonucleotides of Total Ribonuclease Hydrolysates
of Bakers' Yeast tRNA[Ala]

Pyrimidyl-ribonuclease hydrolysate		Guanyl-ribonuclease hydrolysate	
Oligo(mono)nucleotides	moles/mole of TRNA	Oligo(mono)nucleotides	moles/mole of tRNA
C_{OH}	1	Gp	9
Cp	13	pGp	1
Up	6	C-me$_2$G > p	1
ψp	1	U-1meG > p	1
A-Cp	1	C-Gp	4
meI-ψp	1	A-Gp	2
me$_2$G-Cp	1	U-Gp	1
G-Cp	2	U-A-Gp	1
G-Up	4	H$_2$U-C-Gp	1
A-G-H$_2$Up	1	H$_2$U-A-Gp	1
1meG-G-Cp	1	C-meI-ψ-Gp	1
G-A-Up	1	T-ψ-C-Gp	1
I-G-Cp	1	A-C-U-C-Gp	1
A-G-Cp	1	U-C-C-A-C-C$_{OH}$	1
G-G-Tp	1	U*-C-U-C-C-Gp[a]	1
G-G-H$_2$Up	1	A-U-U-C-C-Gp	1
G-G-A-Cp	1	C-U-C-C-C-U-U-Ip	1
pG-G-G-Cp	1		
G-G-G-A-G-A-G-Up*	1		

[a]U* denotes mixture of U and H$_2$U.

On exhaustive hydrolysis of the molecule with T$_1$-RNase (Table 6), besides nine free Gp and pGp residues, 19 oligonucleotides are formed, including nine dinucleotides, three trinucleotides, two tetranucleotides, one pentanucleotide, three hexanucleotides, and one octanucleotide. The fact that tRNA[Ala] is degraded by guanyl RNase into almost the same small fragments as by pyrimidyl RNase is attributed to the relatively high content of guanylic acid in the molecule and to the fact that the total number of bonds at which the polynucleotide chain can be broken by pyrimidyl RNase is only a little larger than the number of bonds accessible to guanyl RNase.

The 3'-terminal fragment in the guanyl-ribonuclease hydrolysate is the hexanucleotide U-C-C-A-C-C$_{OH}$, and the 5'-terminal fragment is the nucleoside diphosphate pGp. The phosphodiester bonds formed by the methylated guanyl derivatives are more difficult to attack by guanyl RNase and some of them remain as cyclic phosphates. The bond formed by methylated inosinic acid is not ruptured by T$_1$-RNase.

The component U*, a mixture of U and H$_2$U, deserves special attention. As the workers who investigated it suggested, the sample was not homogeneous but consisted of two tRNAs, both alanine-specific and differing in their content of H$_2$U. Merril (1968) subjected a specimen of tRNA[Ala] obtained,

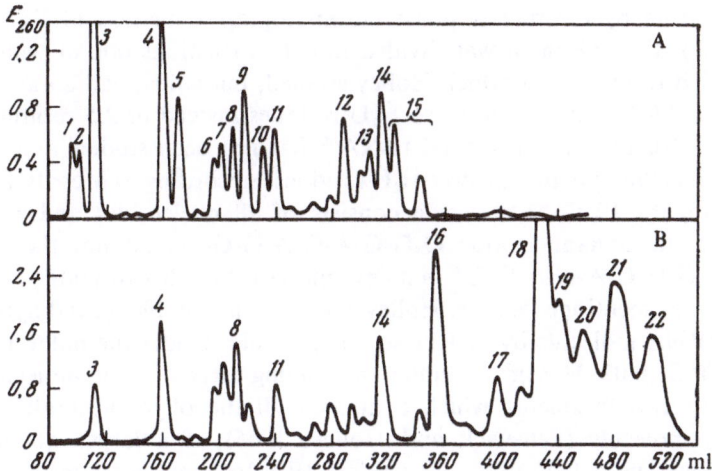

Fig. 20. Fractionation of guanyl-RNase hydrolysate of tRNA^{Ala} on DEAE-cellulose. (A) Total; (B) partial hydrolysis; (1–15) oligonucleotides; (16–22) metamers.

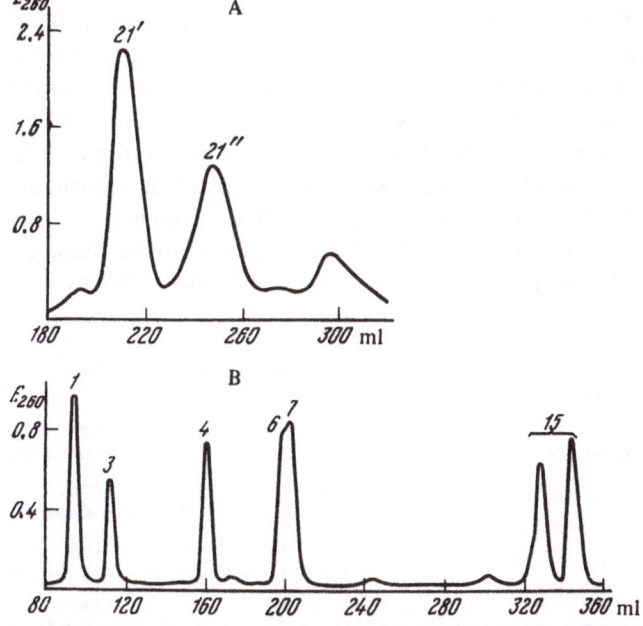

Fig. 21. A. rechromatography of peak 21 (Fig. 20B). B. fractionation of oligonucleotides forming metamer 21'; 21' and 21'' are metamers; 1, 3, 4, 6, 7, 15 are oligonucleotides.

like the sample studied by Holley by the countercurrent method, to further fractionation by the reverse-phase chromatography method of Kelmers et al. (1965); this specimen was divided into two fractions one of which corresponded to the tRNA which Holley studied, but which had an additional G in the 47–48 position and two H_2U residues instead of 2.5 residues. The primary structure of the second tRNA^{Ala} has not been studied and all that is known is that it contains three H_2U residues. According to Merrils' findings therefore, the tRNA^{Ala} molecule consists of 78 nucleotides and probably contains the nonanucleotide G-G-G-A-G-A-G-G-U and not the octanucleotide G-G-G-A-G-A-G-U* in its pyrimidyl-RNase hydrolysate.

As was mentioned above, Holley showed that specific partial hydrolysis of the tRNA molecule by T_1-RNase was possible. Under the mildest conditions (0° C, 0.02 M Mg^{2+}, 5 units enzyme/mg tRNA^{Ala}) the molecule was split into two fragments which contained all the oligonucleotides of the original molecule (Penswick and Holley, 1965); if a larger quantity of enzyme was used (225 units enzyme/mg tRNA^{Ala}) metamers were formed.

The profiles obtained on fractionation of total and partial guanyl-RNase hydrolysates of tRNA^{Ala} are shown in Fig. 20. Total hydrolysis (Fig. 20A) leads to the formation of oligonucleotides (peaks 1–15); on partial hydrolysis (Fig. 20B) most of the uv-absorbing material is concentrated on the right-hand side of the profile, i.e., in the region of compounds with higher molecular weight. Only a little uv-absorbing material is present in fractions 3–15, corresponding to oligonucleotides of total hydrolysates. Fractions 16–22 were rechromatographed to obtain the metamers in an individual state. As an example, the rechromatography of fraction 21 (see Fig. 20B) is shown in Fig. 21A, and fractionation of the oligonucleotides into which fraction 21′ was split by exhaustive hydrolysis with T_1-RNase is shown in Fig. 21B. This method was used to analyze the individual metamers.

The structural formula of tRNA^{Ala} and the metamers giving the essential information for its reconstruction (4) are given below (Holley et al., 1965a).

```
            Ime                        H₂      H₂       2me₂
pG-G-G-C-G-U-G-U-G̣-G-C-G-C-G-U-A-G-U̇-C-G-G-U̇-A-G-C-G-C-G̣-C-U-C-C-C-U-U-I-G-

            Ime                                         2me₂
pG-G-G-C-G-U-G-U-G̣-G-                     C-G̣-C-U-C-C-C-U-U-Ip!

                       H₂      H₂       2me₂
            U̇-C-G-G-U̇-A-G-C-G-C-G̣-C-U-C-C-C-U-U-Ip!

                       H₂      H₂       2me₂
      U-A-G-U̇-C-G-G-U̇-A-G-C-G-C-G̣-C-U-C-C-C-U-U-Ip!

            Ime                        H₂      H₂       2me₂
pG-G-G-C-G-U-G-U-G̣-G-C-G-C-G-U-A-G-U̇-C-G-G-U̇-A-G-C-G-C-G̣-C-U-C-C-C-U-U-I-Gp!
```

4

Examination of the structure of tRNA[Ala] shows that it has no long complementary sequences which could give G-C and A-U pairs. The longest such sequence contains only five nucleotides. This suggested that the double-helical segments in the molecule must be relatively short. With this in mind, and adhering to the principle of formation of the largest possible number of hydrogen bonds, three possible models for the arrangement of the Watson–Crick pairs in the tRNA[Ala] molecule were suggested (Fig. 22). The model known as the clover leaf (Fig. 22A) is most widely accepted.

Serine tRNA from Brewers' Yeast (Zachau et al., 1966a)

By triple countercurrent distribution, Zachau's group succeeded in separating two serine-specific tRNAs from yeast and analyzing them separately.

The nucleotide composition of the $tRNA_1^{Ser}$ and $tRNA_2^{Ser}$, calculated on the basis of the complete sequence of nucleotides, is given below:

Nucleotides, principal	Moles/mole of tRNA		Nucleotides, minor	Moles/mole of tRNA		Nucleotides, minor	Moles/mole of tRNA	
	$tRNA_1^{Ser}$	$tRNA_2^{Ser}$		$tRNA_1^{Ser}$	$tRNA_2^{Ser}$		$tRNA_1^{Ser}$	$tRNA_2^{Ser}$
A	17	15	ψ	3	3	5meC	1	1
G	23	25	H_2U	3	3	OmeU	1	1
U	13	14	T	1	1	OmeG	1	1
C	18	17	I	1	1	$2me_2G$	1	1
			acC	1	1	ipeA	1	1

The total number of nucleotides is the same in $tRNA_1^{Ser}$ and $tRNA_2^{Ser}$, namely, 85. The number and assortment of minor components in these two tRNAs are also identical. The difference lies in the replacement of two G and one U in the $tRNA_2^{Ser}$ molecule by two A and a C in $tRNA_1^{Ser}$ (Fig. 23).

Although the differences between the nucleotide compositions of the

1me
C-İ-ψ-G-G-G-A-G-A-G-Ů-C-U-C-C-G-G-T-ψ-C-G-A-U-U-C-C-G-G- A-C-U-C-G-U-C-C-A-C-C-A_{OH}

1me
C-İ-ψ-G-G-G-A-G-A-G-   A-C-U-C-G-U-C-C-A-C-C_{OH}

1me
C-İ-ψ-G-G-G-A-G-A-G-Ů-C-U-C-C-G-G-T-ψ-C-G-A-U-U-C-C-G-G-A-C-U-C-G-U-C-C-A-C-C_{OH}

1me
C-İ-ψ-G-G-G-A-G-A-G-Ů-C-U-C-C-G-G-T-ψ-C-G-A-U-U-C-C-G-

1me
C-İ-ψ-G-G-G-A-G-A-G-Ů-C-U-C-C-G-G-T-ψ-C-G-A-U-U-C-C-G-G-A-C-U-C-G-U-C-C-A-C-C_{OH}

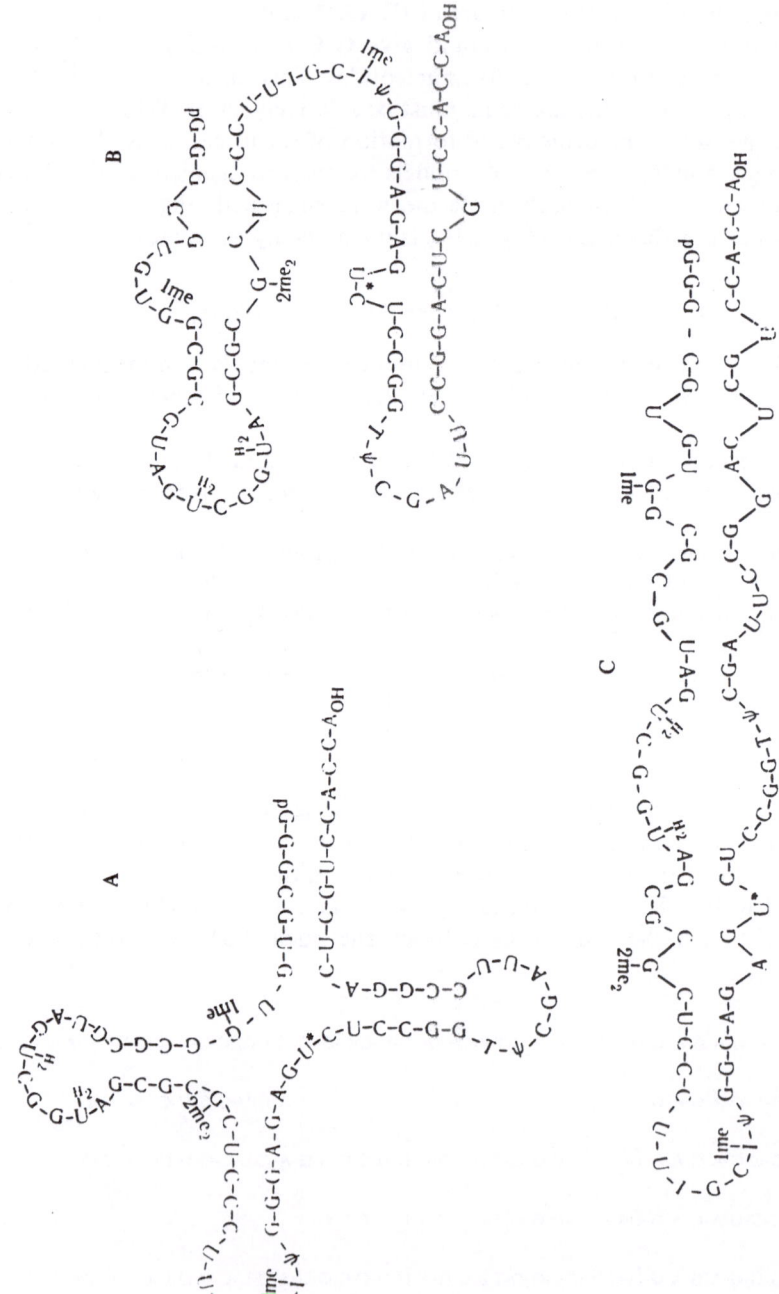

Fig. 22. Models of the secondary structure of tRNA[Ala] suggested by Holley.

two serine-specific tRNAs are slight, they have a marked effect on the oligonucleotides, particularly those formed on hydrolysis of the molecule by guanyl RNase. In the pyrimdyl-RNase hydrolysate of $tRNA_1^{Ser}$ and $tRNA_2^{Ser}$, 23 mononucleotides, six dinucleotides, five trinucleotides, three tetranucleotides, two pentanucleotides, one hexanucleotide, and one heptanucleotide have been identified. The number of Cp and Up groups split off in a free state, and also the composition of one tetranucleotide are different: the hydrolysate of $tRNA_1^{Ser}$ contains nine Cp, ten Up, and the tetranucleotide A-A-A-Up; from $tRNA_2^{Ser}$ pyrimidyl RNase splits off eight Cp and 11 Up, and the altered tetranucleotide has the composition G-A-G-Up (Table 7).

As a result of replacement of two G by two A, guanyl-RNase does not form the tetranucleotide T-ψ-C-Gp, the dinucleotide A-Gp, and the pentanucleotide U-C-C-U-Gp from $tRNA_1^{Ser}$; the corresponding altered oligonucleotides remain joined in one sequence consisting of 11 monomers: T-ψ-C-A-A-A-U-C-C-U-G. Replacement of U by C is manifested by the formation of

TABLE 7. Oligonucleotides of Total Ribonuclease Hydrolysates of $tRNA_1^{Ser}$ and $tRNA_2^{Ser}$ from Brewers' Yeast

Pyrimidyl-ribonuclease hydrolysate			Guanyl-ribonuclease hydrolysate[a]		
Oligo(mono)nucleotides	moles/mole of tRNA		Oligo(mono)nucleotides	mole/mole of tRNA	
	1	2		1	2
A_{OH}	1	1	G	8	8
Cp	9	8	5meC-G	1	1
acC > p	1	1	A-G	1	2
H_2Up	1	1	C-2me₂G > p	1	1
Up	10	11	C-C-A_{OH}	1	1
ψp	1	1	C-acC-G⎱	1	1
G-5meCp	1	1	C-C-G ⎰		
G-Cp	4	4	C-A-G	2	2
G-Up	1	1	U-C-G	1	1
A-A-Cp	1	1	H_2U-0meG-G	1	1
OmeG-G-H_2Up	1	1	U-U-G	1	1
A-G-Up	1	1	C-C-C-G	1	1
G-G-Cp	1	1	T-ψ-C-G	—	1
pG-G-Cp	1	1	A-A-A-G	1	1
A-A-A-Up	1	—	H_2U-H_2U-A-A-G	1	1
G-A-G-H_2Up	1	1	U-C-C-U-G	—	1
A-G-G-Tp	1	1	C-U-C-U-G	1	—
G-A-G-Up	—	1	C-U-U-U-G	—	1
2me₂G-A-A-A-G-A-ψp	1	1	C-A-A-C-U-U-G	1	1
A-A-G-G-Cp	1	1	A-ψ-U-I > p(+ A-ψ-U-I)	1	1
I-G-A-ipeA-A-ψp	1	1	A-ipeA-A-ψ-C-U-U-U-		
OmeU-G-G-G-Cp	1	1	OmeU-G	1	1
			T-ψ-C-A-A-A-U-C-C-U-G	1	—

[a]The oligonucleotides of the guanyl-RNase hydrolysate were dephosphorylated.

two different pentanucleotides, viz., C-U-C-U-Gp from tRNA$_1^{Ser}$ and
C-U-U-U-Gp from tRNA$_2^{Ser}$. The 5'-terminal oligonucleotide of the
pyrimidyl-RNase hydrolysate of the serine tRNAs is the trinucleotide
pG-G-Cp, and of the guanyl-RNase hydrolysate it is pGp. The trinucleotide C-C-A in the guanyl-RNase hydrolysate is the only oligonucleotide
not containing a 3'-terminal G or I and, consequently, it must form the 3'-
end of the molecule.

The nucleotide sequences of tRNA$_1^{Ser}$ and tRNA$_2^{Ser}$ are shown in
Fig. 24 (Feldmann et al., 1966). It is clear from this figure that the 5'-halves
of the two serine tRNAs are identical as far as the fiftieth nucleotide, and
the differences between them are concentrated in the acceptor half of the
molecules.

Zachau's group experienced technical difficulties in their analysis of the
oligonucleotides which could be of importance in other cases and which must

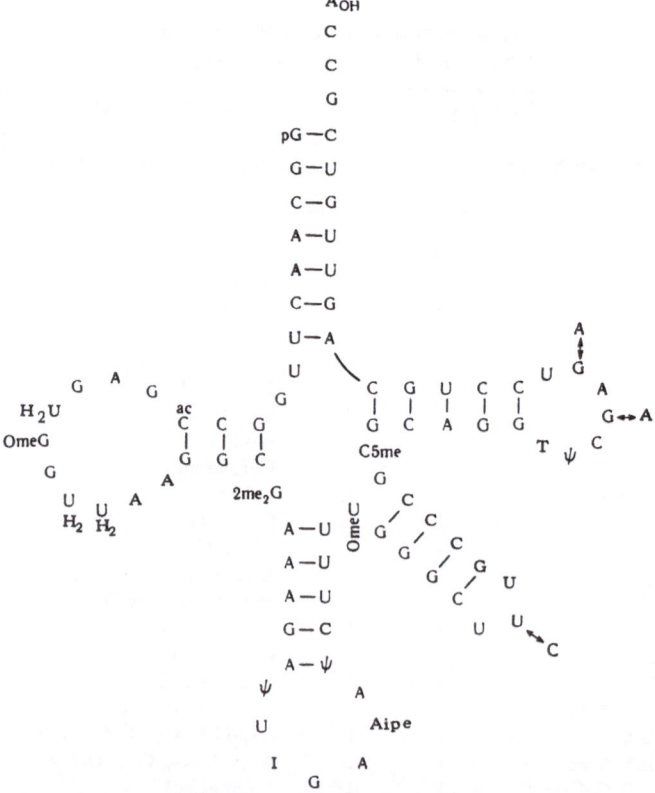

Fig. 23. Yeast tRNA$_2^{Ser}$ in the clover-leaf configuration. Arrows denote three nucleotides
distinguishing tRNA$_1^{Ser}$ from tRNA$_2^{Ser}$

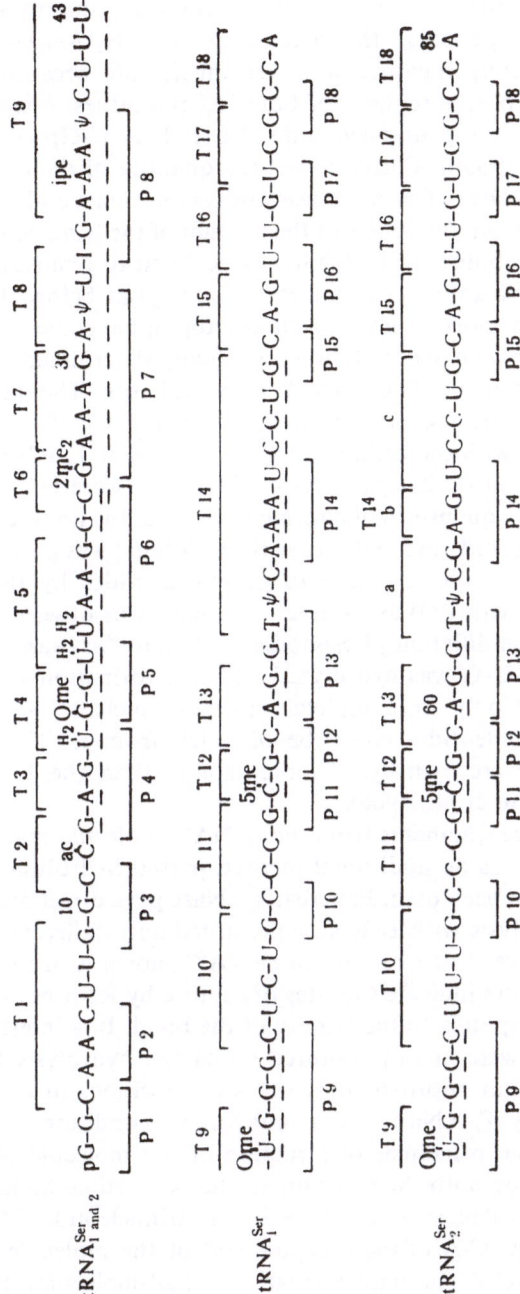

Fig. 24. Structural formulas of yeast tRNA$_1^{Ser}$ and tRNA$_2^{Ser}$: P = pyrimidyl-, T = guanyl-RNase oligonucleotides; broken line under-lines sequences reconstructed on the basis of oligonucleotide analysis.

therefore be mentioned. These difficulties were due primarily to the lability of the minor components: the pentanucleotide H_2U-H_2U-A-A-Gp, for instance, was initially regarded as a tetranucleotide because of the total removal of one or partial removal of both H_2Up residues. A similar observation was made on the pentanucleotide H_2U-H_2U-A-U-Gp of valine tRNA (Venkstern et al., 1968b). Furthermore, the quantity of the trinucleotide C-acC-Gp in hydrolysates of tRNA^{Ser} was always too small, and reached 1 mole per mole of tRNA only if added to the amount of the trinucleotide C-C-Gp; this result was attributed to the lability of acC and to removal of the acetyl group in a weakly alkaline medium. It was a long tme before the ipeA could be detected, for in most systems of chromatographic solvents it moves together with the solvent front. It was eventually determined in a system of ethyl acetate–n-propanol–H_2O (4:1:2; upper phase). The phosphodiester bonds formed by inosine and methylated derivatives of guanine were a nuisance because, with the ordinary concentrations of T_1-RNase used, they were not hydrolyzed (C-2me$_2$G-A-A-A-G) or they were left as cylic phosphates. When large quantities of enzyme were used, the nonspecific action of T_1-RNase was observed, and in this case the C-A bond was particularly labile.

In the study of the structure of oligonucleotides by the method of partial hydrolysis with PDEase a great difference was observed in the rates of hydrolysis of the different phosphodiester bonds; for example, the pentanucleotide U-C-C-U-G was hydrolyzed with the formation of the tetranucleotide U-C-C-U only or completely to mononucleotides; neither dinucleotides nor trinucleotides could be obtained from it. The dinucleoside monophosphates were hydrolyzed more rapidly than the longer oligonucleotides (Feldmann et al., 1966).

To obtain large fragments from the tRNA^{Ser} molecule mainly T_1-RNase has been used, but as an additional method partial hydrolysis with pyrimidyl RNase has also been used. Pancreatic RNase gave direct overlap in some cases where hydrolysis with T_1-RNase permitted only indirect conclusions to be drawn. The clover-leaf structure of tRNA^{Ser} shown is in Fig. 25 (Zachau et al., 1966d). Arrows indicate the sites of rupture by RNases, and the length of the arrow corresponds to the lability of the bond. It is interesting to note that bonds which were mainly ruptured on partial hydrolysis by pyrimidyl RNase are located at approximately the same positions in the molecule as bonds ruptured by T_1-RNase under unfavorable conditions; this indicates that the same basic principles of structure of the molecule play a role in partial hydrolysis by both these enzymes. The most labile bonds in tRNA^{Ser} with respect to T_1-RNase were those in the trinucleotide I-G-A and the tetranucleotide G-C-C-A of the acceptor end of the molecule. Cleavage in the trinucleotide I-G-A led to the formation of half-molecules; the nonacceptor halves, which are identical for both serine tRNAs, were further hydro-

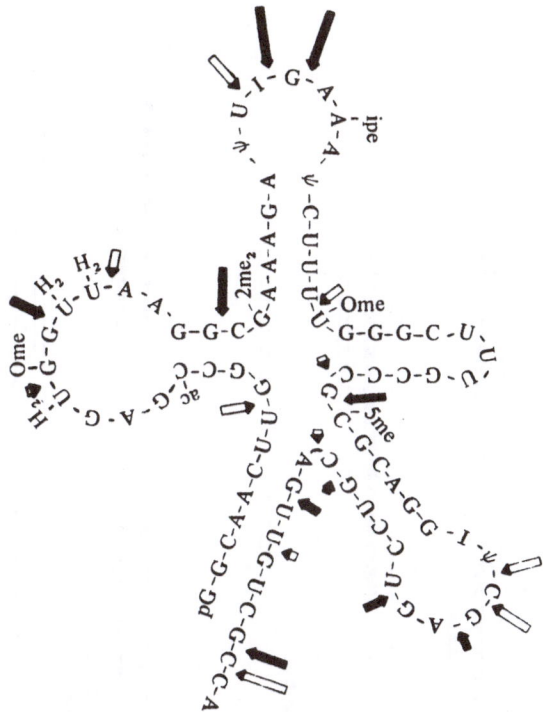

Fig. 25. tRNA^Ser in the clover-leaf configuration. Empty arrows denote hydrolysis with pyrimidyl RNase; filled arrows denote hydrolysis with guanyl RNase.

lyzed in the region of the H₂U-loop. During the further action of the enzyme, the 3′-halves of tRNA$_1^{Ser}$ and tRNA$_2^{Ser}$ behaved differently because of their different nucleotide composition.

The metamers obtained by partial hydrolysis of tRNA^Ser are shown in Fig. 26 (Zachau et al., 1966b). The T-metamers were chiefly analyzed by exhaustive hydrolysis by T₁-RNase, and the P-metamers by hydrolysis with pancreatic RNase.

Tyrosine tRNA from Bakers' Yeast (Madison et al., 1966a)

The molecule of tRNA^Tyr of bakers' yeast consists of 78 nucleotides, including 16 residues of nine different minor components. A tabulation of the nucleotide composition of tRNA^Tyr, calculated on the basis of the assortment of nucleotides in two RNase hydrolysates, appears on page 114.

Tyrosine tRNA is distinguished by its low content of U and its high content of H₂U; if the view regarding their mutual interconvertibility is held,

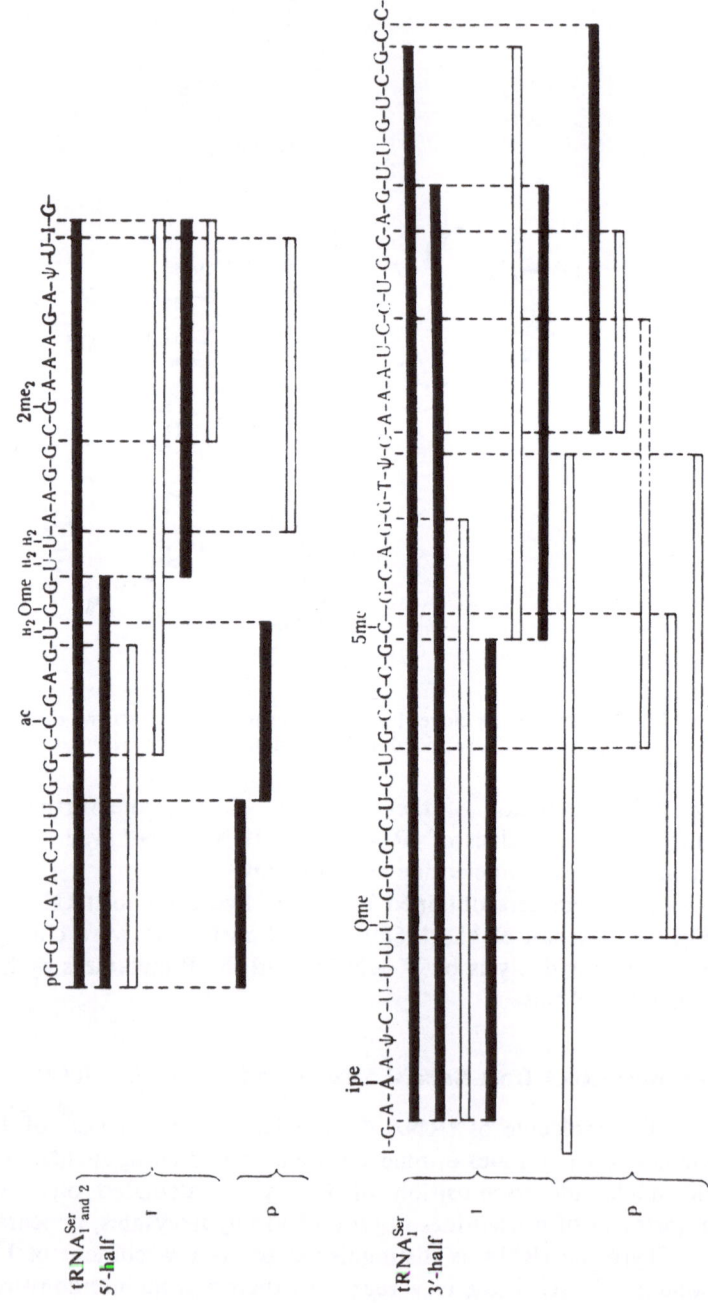

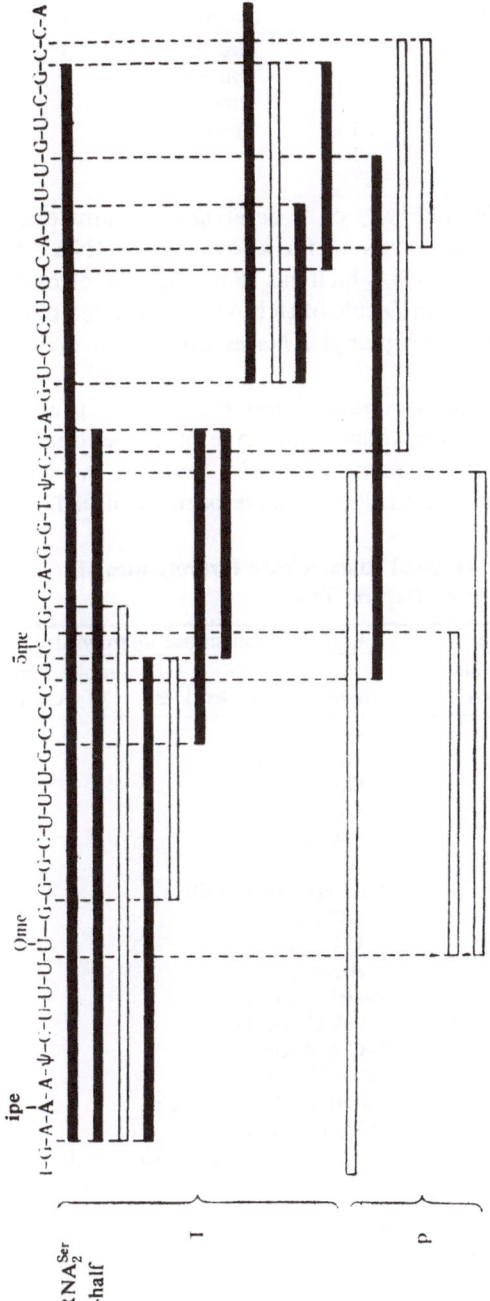

Fig. 26. Structural formulas of tRNA$_1^{Ser}$ and tRNA$_2^{Ser}$ and metamers used for their reconstruction. T denotes metamers obtained by the action of guanyl RNase; P denotes metamers obtained by the action of pyrimidyl RNase. Metamers of decisive importance in the reconstruction are shaded black.

Nucleotides, principal	Moles/mole of tRNA	Nucleotides, minor	Moles/mole of tRNA	Nucleotides, minor	Moles/mole of tRNA
A	15	ψ	3	2me$_2$G	1
G	20	H$_2$U	6	OmeG	1
U	7	T	1	1meA	1
C	20	5meC	1	ipeA	1
		2meG	1		

this is perhaps not fortuitous. The tRNA^{Tyr} of *S. cerevisiae* contains one of the largest numbers of minor components so far known in a tRNA; it is exceeded only by the tRNA^{Tyr} of *T. utilis*, which has 18 modified nucleotides.

Oligonucleotides into which the molecule of tRNA^{Tyr} is degraded during exhaustive hydrolysis by pyrimidyl and guanyl RNases are shown in Table 8.

In the pyrimidyl-ribonuclease hydrolysate 22 mono-, four di-, four tri-, three tetra-, three penta-, and one heptanucleotide have been identified. In the guanylribonuclease hydrolysate six mono-, four di-, two tri-, five tetra-, two penta-, three hexa-, and one nonanucleotide have been identified.

TABLE 8. Oligonucleotides of Total Ribonuclease Hydrolysates of tRNA^{Tyr} from Bakers' Yeast

Pyrimidyl-ribonuclease hydrolysate		Guanly-ribonuclease hydrolysate	
Oligo(mono)nucleotides	moles/mole of tRNA	Oligo(mono)nucleotides	moles/mole of tRNA
C$_{OH}$	1	Gp	6
Cp	9	C-2me$_2$G > p	1
Up	6	C-Gp	1
H$_2$Up	3	A-Gp	2
5meCp	1	A-C-C$_{OH}$	1
ψp	1	U-A-2meGp	1
pCp	1	H$_2$U-H$_2$U-OmeG-Gp	1
G-Tp	1	T-ψ-C-Gp	1
G-Cp	1	A-C-U-Gp	1
2me$_2$G-Cp	1	C-A-A-Gp	1
G-ψp	1	A-H$_2$U-meC-Gp	1
G-1meA-Cp	1	1meA-C-U-C-Gp	1
A-2meG-Cp	1	C-C-A-A-Gp	1
G-G-Up	1	C-C-C-C-C-Gp	1
OmeG-G-H$_2$Up	1	H$_2$U-H$_2$U-H$_2$U-A-A-Gp	1
A-ipeA-A-ψp	1	pC-U-C-U-C-Gp	1
A-A-G-H$_2$Up	1	ψ-A-ipeA-A-ψ-C-U-U-Gp	1
G-G-G-Cp	1		
A-A-G-A-Cp	1		
A-A-G-G-Cp	1		
G-A-G-A-H$_2$Up	1		
G-G-G-A-G-A-Cp	1		

Tyrosine tRNA contains a long polycytidylic sequence, including five C (in the oligonucleotide C-C-C-C-C-Gp), and a polydihydrouridylic sequence consisting of three H_2U residues (in the oligonucleotide H_2U-H_2U-H_2U-A-A-Gp). Another distinguishing feature of tRNA^{Tyr} is that its 5'-end, unlike that of most other tRNAs, is formed by cytidylic and not by guanylic acid. Like other tRNAs obtained by the countercurrent method, tRNA^{Tyr} has no acceptor adenosine. On hydrolysis of tRNA^{Tyr} into meta-mers by T_1-RNase, it was more readily attacked than tRNA^{Ala} and much more readily than tRNA^{Ser}. The specificity of this hydrolysis is interesting. All oligonucleotides of tRNA^{Ala} except C-G are found in the metamers, whereas in the case of tRNA^{Tyr}, if the hydrolysis took place under unfavorable conditions, the oligonucleotides C-C-A-A-G, H_2U-H_2U-OmeG-G, H_2U-H_2U-H_2U-A-A-G, A-C-U-G, and ψ-A-IpeA-A-ψ-C-U-U-G did not remain in the longer fragments. The impossibility of obtaining metamers from the 5'-half during hydrolysis of the whole tRNA^{Tyr} molecule by T_1-RNase compelled the workers cited to conduct the hydrolysis initially to halves, and then, by partial hydrolysis of the 5'-half, to obtain a sufficient number of overlaps for it. The most labile bond by rupture of which halves could be formed was the phosphodiester bond between G and ψ in the trinucleotide G-ψ-A, as was shown by the appearance of an extra G and by the absence of G-ψ in the total pancreatic hydrolysate of the 5'-half of tRNA^{Tyr}.

The halves and large fragments obtained by partial hydrolysis of tRNA^{Tyr} and used for reconstructing the sequence of the complete molecule are given below. The metamers were analyzed not only by total hydrolysis with T_1-RNase, but also with pyrimidyl RNase, which provided additional useful information (5).

A clover-leaf configuration can easily be given to the polynucleotide chain thus established (Fig. 27); in this way the largest possible number of hydrogen bonds is formed, and the hypothetical anticodon, at which the

2me H_2H_2 Ome $H_2H_2H_2$ 2me₂

pC-U-C-U-C-G-G-U-A-Ġ-C-C-A-A-G-U-U-G-G-U-U-U-A-A-G-G-C-Ġ-C-A-A-G-A-C-U-G-

2me $H_2H_2H_2$ 2me₂

pC-U-C-U-C-G-G-U-A-Ġ-C-C-A-A-G-   U-U-U-A-A-G-G-C-Ġ-C-A-A-G-A-C-U-G-

2me   2me₂

pC-U-C-U-C-G-G-U-A-Ġ-   C-Ġ-C-A-A-G

5'-half

ipe $H_2$5me   1me

ψ-A-A-A-ψ-C-U-U-G-A-G-A-U-Ċ-G-G-G-C-G-T-ψ-C-G-A-C-U-C-G-C-C-C-C-C-G-G-G-A-G-A-C-C-A$_{OH}$

1me

A-C-U-C-G-C-C-C-C-C-G-G-G-A-G-A-C-C$_{OH}$

H_2 5me

A-U-Ċ-G-G-G-C-G-T-ψ-C-G-   C-C-C-C-C-G-G-G-A-G-A-C-C$_{OH}$

3'-half

5

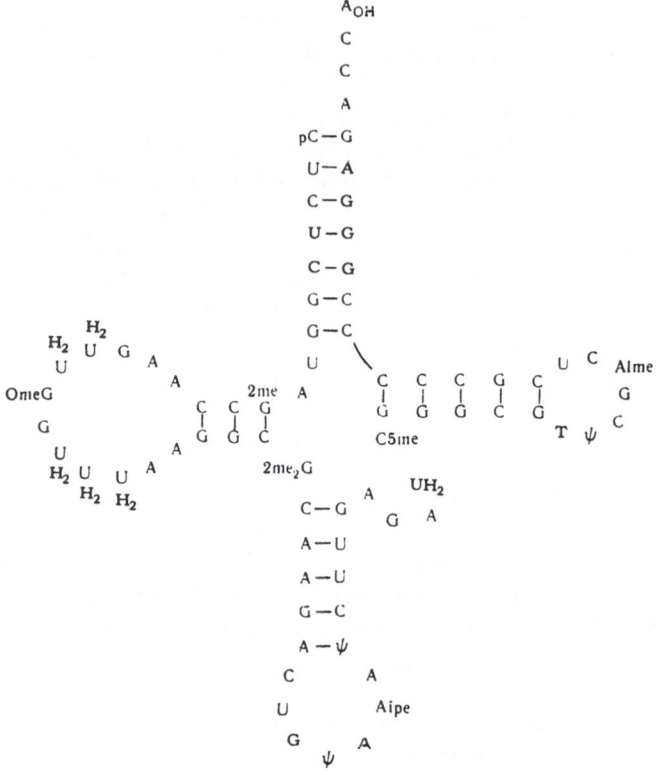

Fig. 27. tRNA^Tyr of yeast in the clover-leaf configuration.

first rupture takes place, lies in the same position as in other tRNAs. The presence of ψ in the anticodon was demonstrated for the first time; there is no question about this discovery, for with its spare N–H group, ψ can form two hydrogen bonds with A as in the case of U (Pochon et al., 1964).

Phenylalanine tRNA from Bakers' Yeast (RajBhandary et al., 1967)

The molecule of tRNA^Phe contains 76 nucleotides, including 14 minor components:

Nucleotides, principal	Moles/mole of tRNA	Nucleotides, minor	Moles/mole of tRNA	Nucleotides, minor	Moles/mole of tRNA
A	17	ψ	2	OmeG	1
G	18	H_2U	2	2meG	1
U	12	T	1	$2me_2G$	1
C	15	5meC	2	7meG	1
		OmeC	1	1meA	1
				Y	1

Phenylalanine tRNA is comparatively poor in ψ and H_2U (two residues of each), but is rich in methylated derivatives of C (three compared with their total absence in tRNA^{Ala} and only one residue in most other tRNAs) and, in particular, in methylated derivatives of G, of which there are four residues (2meG, 2me$_2$G, 7meG, OmeG). The nucleoside Y, neighboring the anticodon on the 3'-side, possesses unusual properties. It has been identified only recently (Nakanishi et al., 1970); it is the most complex modification of all the minor components so far known (see page 238). The strong greenish-blue fluorescence of this compound is transmitted not only to the corresponding oligonucleotides (OmeG-A-A-Y-A-ψ and A-OmeC-U-OmeG-A-A-Y-A-ψ-5meC-U-G), but also to the whole tRNA^{Phe} molecule. The fluorescence disappears after fractionation in 7 M urea at acid pH values, and this process is accompanied by changes in the spectrum, chromatographic mobility, and susceptibility to enzymic attack of the corresponding oligonucleotides. The phosphodiester bond formed by Y is ruptured by pancreatic RNase only after modification in this way. Analogous fluorescence was found during a study of the structure of tRNA^{Phe} from wheat germ (Dudock et al., 1968), bovine liver (Yoshikami and Keller, 1969), and rat liver (Fink et al., 1968,

TABLE 9. Oligonucleotides of Total Ribonuclease Hydrolysates of tRNA^{Phe} from Bakers' Yeast

Pyrimidyl-ribonuclease hydrolysate		Guanyl-ribonuclease hydrolysate	
Oligo(mono)nucleotides	moles/mole of tRNA	Oligo(mono)nucleotides	moles/mole of tRNA
C$_{OH}$	1	Gp	4
H$_2$U > p	1	pGp	1
Cp	7	C-Gp	1
5meCp	2	U-Gp	1
Up	6	A-Gp	3
ψp	1	C-2me$_2$G > p	1
A-Cp	2	H$_2$U-H$_2$U-Gp	1
2me$_2$G-Cp	1	C-A-C-C$_{OH}$	1
G-Cp	1	T-ψ-C-Gp	1
G-Tp	1	C-C-A-Gp	1
G-Up	1	C-U-C-A-Gp	1
pG-Cp	1	A-U-U-U-A-2meG > p	1
A-2meG-Cp	1	7meG-U-C-5meC-U-Gp	1
G-1meA-Up	1	A-A-U-U-C-Gp	1
A-G-H$_2$Up	1	1meA-U-C-C-A-C-A-Gp	1
G-G-A-Up	1	A-OmeC-U-OmeG-A-A-	
A-G-A-A-Up	1	Y-A-ψ-5meC-U-Gp	1
A-G-A-OmeC-Up	1		
OmeG-A-A-Y-A-ψp	1		
G-G-A-G-7meG-Up	1		
G-G-G-A-G-A-G-Cp	1		

1969). Because the tRNAs of other specificity evidently do not possess this fluorescence, and since its intensity is high, a specific and simple method of quantitative determination of phenylalanine tRNA has been suggested (Yoshikami et al., 1968).

The oligonucleotides of exhaustive pyrimidyl- and guanyl-RNase hydrolysates of tRNA[Phe] are listed in Table 9.

Under the influence of pyrimidyl RNase the molecule is split into 18 mono-, seven di-, three tri-, one tetra-, two penta-, two hexa-, and one octanucleotide; five mono-, six di-, one tri-, three tetra-, one penta-, three hexa-, one octa-, and one dodecanucleotide have been found in the guanyl-ribonuclease hydrolysate. The 5′-terminal nucleotide of tRNA[Phe] is Gp, as is shown by the presence of pG-Cp and pGp in the pyrimidyl- and guanyl-ribonuclease hydrolysates respectively. The only oligonucleotide of the guanyl-ribonuclease hydrolysate not possessing a 3′-terminal G was C-A-C-C$_{OH}$, which was therefore taken to be the acceptor end of the molecule, minus its adenosine. Methylated derivatives occupy the 3′-terminal position

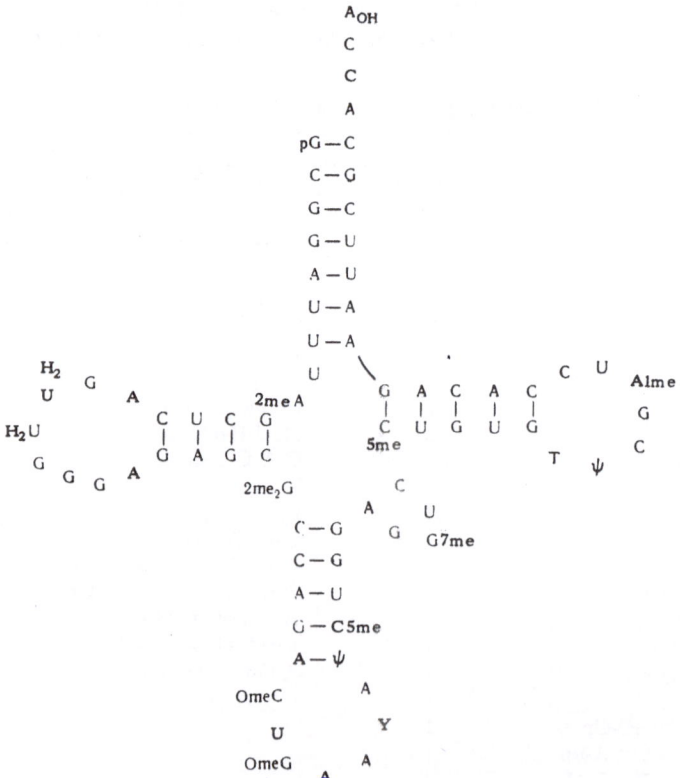

Fig. 28. tRNA[Phe] from yeast in the "clover-leaf" configuration.

in two of the T-oligonucleotides, but they are mainly left as cyclic phosphates (C-2me$_2$G > p and A-U-U-U-A-2meG > p); 7meGp and OmeGp are located in the inner part of the oligonucleotides, so that the internucleotide bonds formed by them are not ruptured by T$_1$-RNase.

In the final stages of the analysis to reconstruct the whole polynucleotide chain of tRNA^{Phe} partial hydrolysis by pancreatic and guanyl RNases followed by exhaustive hydrolysis of the metamers by the same two enzymes was used. Small quantities of pancreatic RNase ruptured only one phosphodiester bond, that neighboring the anticodon on the 5'-side (Chang and RajBhandary, 1968). Fractionation and analysis of the resulting fragments showed that the acceptor and nonacceptor halves of the molecule are formed under these conditions. After analysis of the tRNA^{Phe} halves obtained by the action of pyrimidyl RNase, two alternative structures remained; the choice between them was made by analysis of the partial guanyl-RNase hydrolysate of the molecule. As was mentioned above, T$_1$-RNase cannot split tRNA^{Phe} into halves since its anticodon contains a guanylic acid residue methylated in the ribose group. However, smaller hydrolysis products were obtained by means of guanyl RNase. The structure of tRNA^{Phe} and of its fragments formed by partial hydrolysis with T$_1$-RNase which played a decisive role in the reconstruction of the molecule (6) is shown facing page 120 (RajBhandary and Chang, 1968).

The structure of tRNA^{Phe} in the form of a clover-leaf model is shown in Fig. 28.

Valine tRNA from Bakers' Yeast (Baev et al., 1967a)*

The molecule of tRNA$_1^{Val}$ of bakers' yeast (Fig. 29) consists of 77 nucleotides, of which 63 are principal and 14 minor components. The nucleotide composition of tRNA$_1^{Val}$ of *S. cerevisiae,* determined by alkaline hydrolysis, and then compared with the results of ribonuclease hydrolysis, is given below:

Nucleotides, principal	Moles/mole of tRNA	Nucleotides, minor	Moles/mole of tRNA	Nucleotides, minor	Moles/mole of tRNA
A	14	ψ	4	5meC	1
G	18	H$_2$U	4	1meG	1
U	11	T	1	1meA	1
C	20	I	1	7meG	1

Unlike other yeast tRNAs, tRNA$_1^{Val}$ contains no 2-dimethylguanosine. Like tRNA^{Ala}, tRNA^{Val} contains no nucleotides methylated in the ribose

---

*The data for tRNA$_1^{Val}$ of bakers' yeast have been corrected in accordance with the results of a combined investigation undertaken by the laboratories of Ebel and Baev (Ebel and Baev, unpublished data, see page 125).

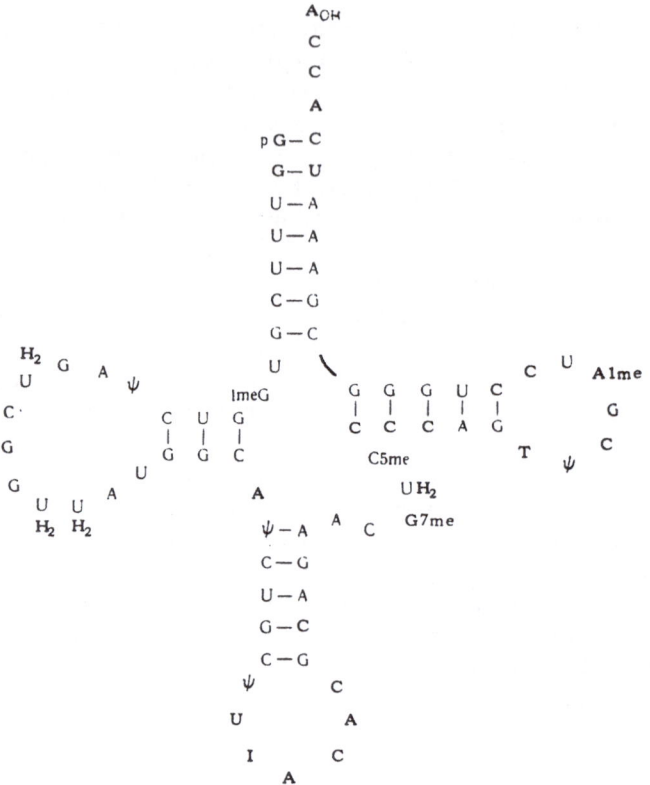

Fig. 29. tRNA$_1^{Val}$ of bakers' yeast in the clover-leaf configuration.

group, the presence of which could be shown by the appearance of dinucleotides in the alkaline hydrolysate or by the resistance to nuclease of bonds
located in the 3′-position relative to the nucleotide methylated in the 2-OH
group of the ribose. Six of the eight minor components are present in
tRNA$_1^{Val}$ in the proportion of 1 mole per mole of tRNA; ψ and H$_2$U are
present in the proportion of 4 moles per mole tRNA.

Oligonucleotides obtained by exhaustive hydrolysis of tRNA$_1^{Val}$ by
pyrimidyl RNase are given on the left-hand side of Table 10. Besides the
pyrimidine mononucleotides removed from polypyrimidine sequences, and
the 3′-terminal adenosine, the hydrolysate also contains 17 different oligonucleotides, including eight di-, seven tri-, one tetra-, and two pentanucleotides.
Only two dinucleotides, A-Cp and G-Cp, are repeated twice in the molecule,
and all other oligonucleotides are present in the proportion of 1 mole per
mole of tRNA$_1^{Val}$. The 5′-end is represented by the trinucleotide pG-G-Up,

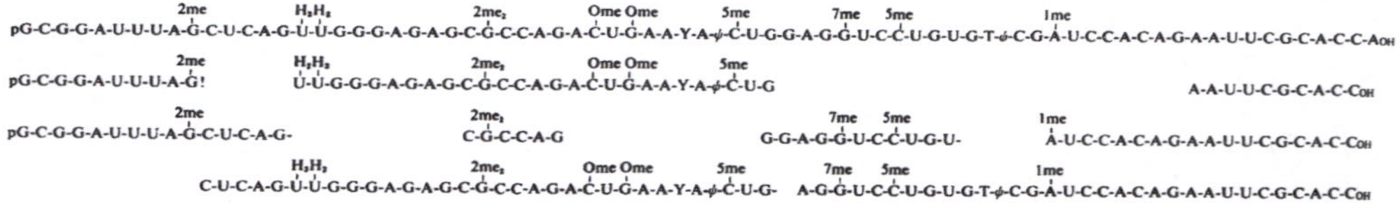

6

Yeast tRNA^Phe

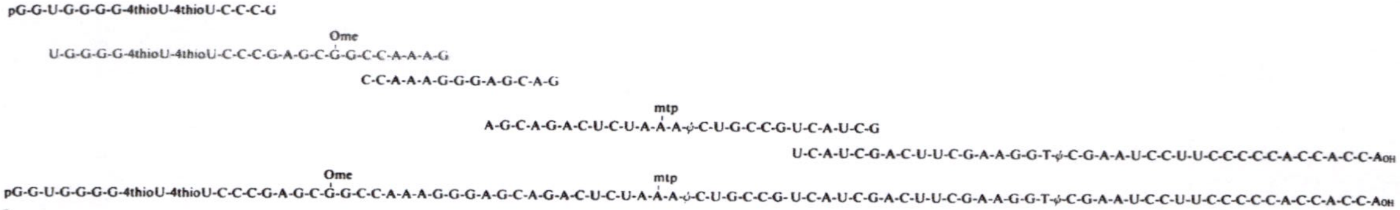

7

E. coli tRNA^Tyr Su_III+

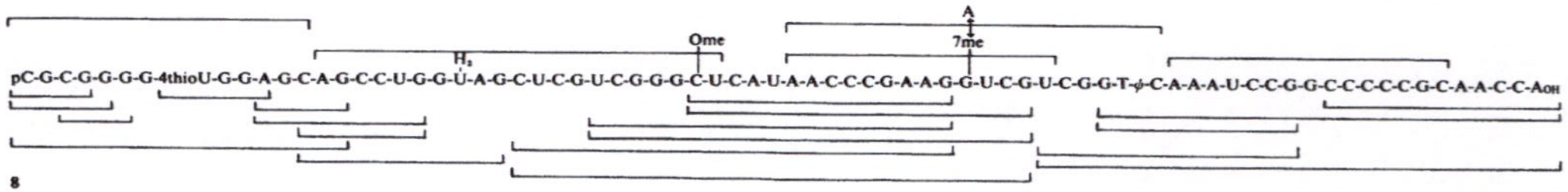

8

E. coli tRNA_f^Met

and the acceptor end by adenosine alone, for the procedure used to isolate tRNA$_1^{Val}$ rules out the possibility of regeneration of the terminal C-C-A (Mirzabekov et al., 1965b).

Oligonucleotides of the total pyrimidyl-ribonuclease hydrolysate of tRNA$_1^{Val}$ are short in length, attaining five monomers in only two cases.

Oligonucleotides identified in the guanyl-ribonuclease hydrolysate of tRNA$_1^{Val}$ are shown on the right-hand side of Table 10.

Among these oligonucleotides, two di-, two tri-, two tetra-, four penta-, two hexa-, one nona-, and one undecanucleotide have been identified. The 5'-end is represented by the nucleoside diphosphate pGp, and the 3'-end by the nonanucleotide A-A-A-U-C-A-C-C-A$_{OH}$, containing the acceptor adenosine. Corresponding with the specificity of guanyl RNase, all the oligonucleotides except the acceptor oligonucleotide have Gp at the 3'-end; the 3'-terminal position of 1meGp (the dinucleotide U-1meGp) and Ip (the

TABLE 10. Oligonucleotides of Total Ribonuclease Hydrolysates
of tRNA$_1^{Val}$ from Bakers' Yeast

Pyrimidyl-ribonuclease hydrolysate		Guanyl-ribonuclease hydrolysate	
Oligo(mono)nucleotides	moles/mole of tRNA	Oligo(mono)nucleotides	moles/mole of tRNA
Cp	12	Gp	6
Up	5	C-Gp	1
H$_2$Up	1	A-Gp^a	—
ψp	3	U-1meGp	1
5meCp	1	U-Gp^a	—
7mG-H$_2$Up	1	H$_2$U-C-Gp	1
A-Cp	2	C-A-Gp	1
G-Cp	2	T-ψ-C-Gp	1
A-ψp	1	C-ψ-U-Ip	1
A-Up	1	H$_2$U-H$_2$U-A-U-Gp	1
G-Up	1	U-C-ψ-A-Gp	1
G-1meA-Up	1	U-U-U-C-Gp	1
I-A-Cp	1	A-C-A-C-Gp	1
G-G-Cp	1	C-A-ψ-C-U-Gp	1
A-G-H$_2$Up	1	1meA-U-C-C-U-Gp	1
A-G-Tp	1	A-A-C-7mG-H$_2$U-5meC-C-	
1meG-G-Up	1	C-C-A-Gp	1
G-G-H$_2$Up	1	pGp (5'-end)	1
G-G-G-Cp	1	A-A-A-U-C-A-C-C-A$_{OH}$	
A-G-A-A-Cp	1	(3'-end)	1
G-A-A-A-Up	1		
pG-G-Up (5'-end)	1		
A (3'-end)	1		

^{a}A-Gp and U-Gp were products of nonspecific hydrolysis, and in later experiments they were not found among the products of guanyl-RNase hydrolysis.

tetranucleotide C-ψ-U-Ip) agrees with the specificity of guanyl RNase from actinomycetes. All the oligonucleotides are present in the molecule of tRNA$_1^{Val}$ in the proportion of 1 mole per mole of tRNA.

As was mentioned above, splitting the molecule into two halves played an important role in the study of tRNA$_1^{Val}$ structure. At a low temperature and in the presence of Mg^{2+}, guanyl RNase ruptures one phosphodiester bond in the tRNA$_1^{Val}$ molecule, namely that between the I and A residues within the I-A-C anticodon. After clear separation of the half-molecules on DEAE-cellulose and their isolation in preparative quantities, it was possible to continue the investigation of their structure and the work of reconstructing the tRNA$_1^{Val}$ molecule was done on its halves (Baev et al., 1967b,c). Practically pure halves of tRNA$_1^{Val}$ were subjected to exhaustive hydrolysis by guanyl, and then by pyrimidyl RNases. The hydrolysates were fractionated by previously perfected methods of ion-exchange and paper chromatography and the results were compared with the well known distribution pattern of the oligonucleotides of the whole molecule on the chromatographic profile and on paper. The distribution of the oligonucleotides between the two halves of tRNA$_1^{Val}$ could thus easily be established:

3'-fragment (acceptor)		5'-fragment (nonacceptor)	
A-Cp (2 moles)	C-Gp	G-Cp	pG-G-Up
G-Cp	C-A-Gp	A-ψp	U-1meGp
G-1meA-Up	T-ψ-C-Gp	A-Up	H$_2$U-C-Gp
A-Cp	A-C-A-C-Gp	G-Up	C-ψ-U-Ip
A-G-Tp	1meA-U-C-C-U-Gp	Ip	H$_2$U-H$_2$U-A-U-Gp
A-G-A-A-Cp	A-A-C-7meG-H$_2$U-	G-G-Cp	U-C-ψ-A-Gp
G-A-A-A-Up	5meC-C-C-C-A-Gp	A-G-H$_2$Up	U-U-U-C-Gp
G-G-G-Cp	A-A-A-U-C-A-C-	1meG-G-Up	C-A-ψ-C-U-Gp
A$_{OH}$	C-A$_{OH}$	G-G-H$_2$Up	pGp
7meG-H$_2$Up			

Instead of the trinucleotide I-A-C, its fragments naturally were found in the pyrimidyl-RNase hydrolysate: A-Cp in the 3'-half and Ip in the 5'-half of the molecule. The two-dimensional chromatograms of guanyl-RNase hydrolysates of tRNA$_1^{Val}$ and of its two halves respectively are compared in Fig. 9.

The distribution of oligonucleotides of the ribonuclease hydrolysates between the 3'- and 5'-halves of the tRNA$_1^{Val}$ molecule considerably reduced the number of possible versions of the structure, but an unequivocal solution was still impossible. However, an attempt was in fact made at this stage to reconstruct the primary structure of tRNA$_1^{Val}$ (Baev et al., 1966b,c).

The last stage of the analysis used to obtain missing information regarding the structure of tRNA$_1^{Val}$ and introducing the necessary precision into the probable structural formula consisted of partial hydrolysis of the halves with guanyl-RNase, with the resulting formation of metamers, i.e., of fragments consisting of several oligonucleotides of the exhaustive hydrolysates. The metamers were analyzed in the same way as the halves, i.e., by total hydrolysis with guanyl RNase. As a result, oligonucleotides of the exhaustive guanyl-RNase hydrolysate were obtained. The use of metamers of tRNA$_1^{Val}$ halves, and not of the whole molecule, as was done in the investigation of most other tRNAs, proved to be greatly advantageous. In the first place, it facilitated separation of the metamers and their purification.

Second, the formation of metamers from each half separately naturally simplifies the procedure of identification of the component oligonucleotides, for in this case the choice must be made not among 15–16, but among only 7–8 oligonucleotides. Analysis of the oligonucleotide composition of the metamers showed that on partial hydrolysis of the half-molecules of tRNA$_1^{Val}$, overlapping and neighboring fragments were obtained. Comparison of these fragments revealed the sequence of the oligonucleotides in the 3'- and 5'-halves of the molecule.

As an example, the fractionation of metamers from the 3'-half of tRNA$_1^{Val}$ is shown in Fig. 30A (Mirzabekov et al., 1970). Because the metamers were obtained from the two halves of the molecule separately, and also because of the high resolving power of the system used for their fractionation, all the metamers except one were obtained in the individual state after the first cycle of fractionation. Only metamer 3'-12 had to be fractionated twice. The criteria of homogeneity of the metamers were, first, correlation between their size and their position on the chromatographic profile and, second, the equimolar proportions of the oligonucleotides composing them.

All the fractions present in sufficient quantity for analysis were then subjected to exhaustive hydrolysis by guanyl RNase. The resulting hydrolysates were fractionated on DEAE-cellulose in 7 M urea at pH 8 in a 0–0.4 M NaCl gradient. With strict standardization of the fractionation procedure (the same batch of DEAE-celluslose, standard columns and eluting solutions) identical oligonucleotides from the total hydrolysate of the half-molecule (Fig. 30B) and hydrolysates of the various metamers (Fig. 30C) occupy identical places on the chromatographic profile. Once a reliable identification of the profile of the total hydrolysate of the half-molecules has been carried out it is therefore possible to identify oligonucleotides composing the metamers from their position on the chromatographic profile and from the eluting volume with resort to additional criteria (spectra, H$_2$U content, etc.) only in individual cases. Besides the economy in time, this also gives considerable economy in material, and for a long time yet this will be an important factor during work with individual tRNAs.

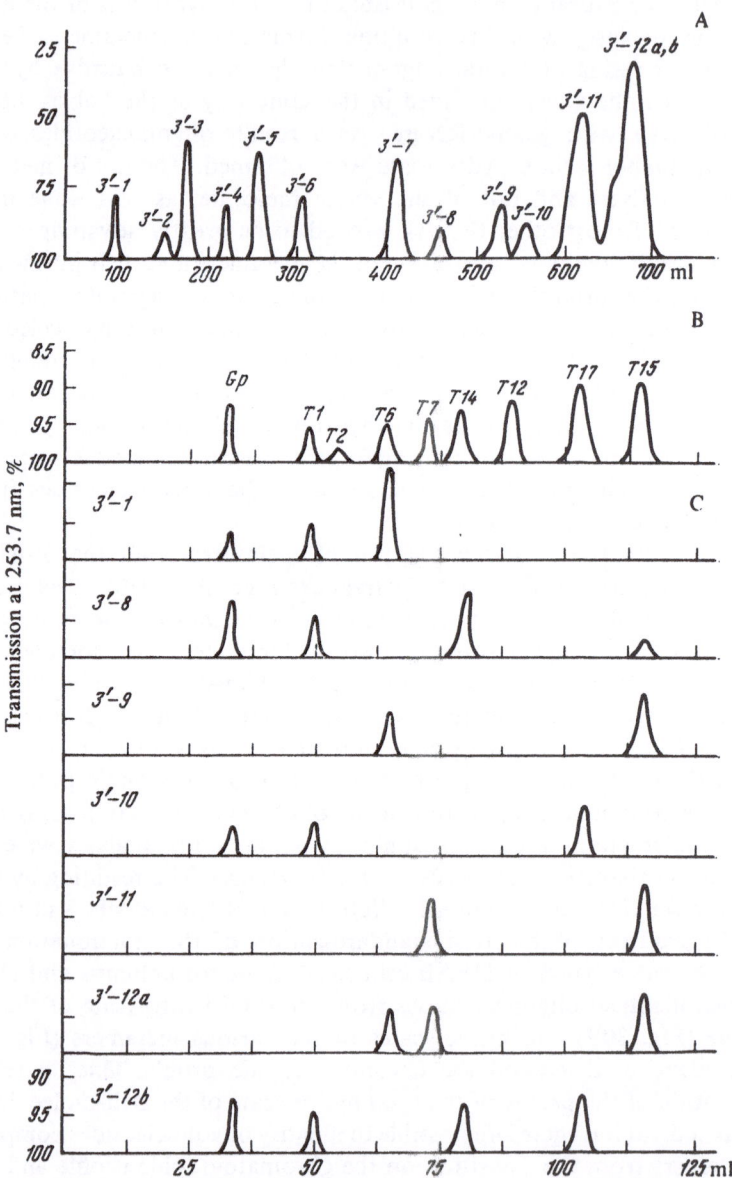

Fig. 30. Fractionation on DEAE-cellulose of metamers obtained by partial hydrolysis of the
3'-half of tRNA$_1^{Val}$ by guanyl RNase (A) and of oligonucleotides obtained by total guanyl-
RNase hydrolysis of the 3'-half of tRNA$_1^{Val}$ (B) and its metamers (C) from 3'-1 to 3'-12
(a, b): metamers; (T$_1$, T$_2$, ...): oligonucleotides.

By analysis of metamers in this way, only their oligonucleotide composition can be determined; the sequence of the oligonucleotides in the metamers and the mutual arrangement of the metamers in the halves of $tRNA_1^{Val}$ were established by comparing the composition, overlapping, and abutting of the metamers. Reconstruction of the 3'-half of $tRNA_1^{Val}$ from its component metamers is shown schematically in Fig. 31.

All the analytical data thus obtained were used for the final reconstruction of the $tRNA_1^{Val}$ molecule: overlapping of the oligonucleotides of the total pyrimidyl- and guanyl-RNase hydrolysates, the distribution of the oligonucleotides among the halves and, finally, data for the composition and overlapping of the metamers (Baev et al., 1967a).

At a conference on tRNA held in Göttingen in 1971, Ebel and Dirheimer reported having determined the primary structure of $tRNA_2^{Val}$ from brewers' yeast. The results, which these workers also published in *FEBS Letters* (Bonnet et al., 1971),* compelled a fresh look at the structure of $tRNA_1^{Val}$ from bakers' yeast for the following reasons: the $tRNA_2^{Val}$ from brewers' yeast investigated by the French workers corresponds in its countercurrent distribution in the phosphate–isopropanol—formaldehyde system of Holley et al. (1963) to the $tRNA_1^{Val}$ from bakers' yeast studied by Baev et al. (1967a); nevertheless, differences were found between these two structures, in two sequences, as follows:

tRNA^{Val} from brewers' yeast: A-A-C-7meG-H$_2$U-C-5meC-C-C-A-G and
   G-G-G-C;
tRNA^{Val} from bakers' yeast: A-A-C-H$_2$U-5meC-C-C-C-A-G and
   G-G-G-G-C.

Further tests carried out jointly in Ebel's and Baev's laboratories (Ebel and Baev, unpublished data) led to the conclusion that the following modifications must be introduced into the structural formula of $tRNA^{Val}$ from bakers' yeast: the structure of the decanucleotide must be changed to A-A-C-7meG-H$_2$U-5meC-C-C-C-A-G, because the dinucleotide 7meG-H$_2$U was found in its pyrimidyl-RNase hydrolysate. The sequence G-G-G-G-C must be changed to G-G-G-C for the following reason: when the composition of this oligonucleotide was determined originally the ratio G : C was always slightly higher than 3.5 : 1, and by the rules of arithmetic it was rounded up to 4 : 1 (Baev et al., 1967a). Results obtained by Vasilenko et al. (1969) and also in Ebel's laboratory, plus the fact that the existence of the fourth G infringes on some of the general rules governing tRNA structure, to which no other exception has yet been found, compels the removal of one G. The corrected structural formula of $tRNA_1^{Val}$ from bakers' yeast is

*The authors kindly provided an advance copy of their paper.

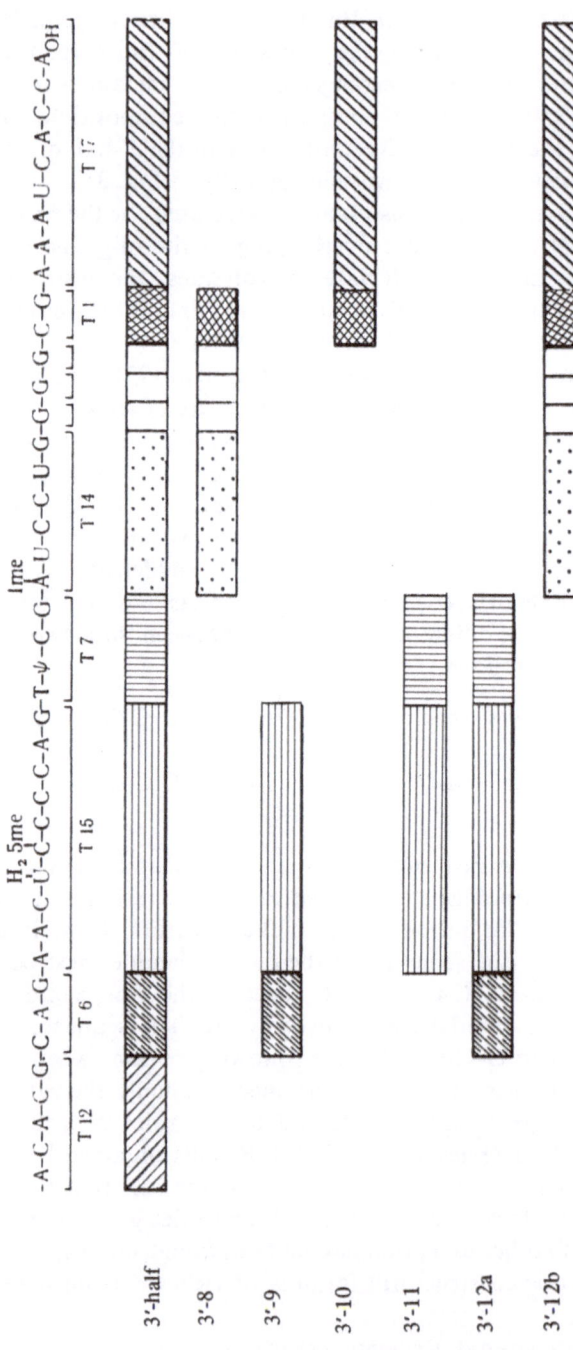

Fig. 31. Scheme of determination of nucleotide sequence in the 3′-half of tRNA$_1^{Val}$. For legend, see Fig. 30.

accordingly shown in Fig. 29. The corrections have also been made to this structure elsewhere in the book.

The other difference discovered between $tRNA_2^{Val}$ of brewers' and $tRNA_1^{Val}$ of bakers' yeast, namely the position of the 5meC residue, has yet to be investigated.

The structures of the transfer RNAs from *E. coli* described below were determined in the Laboratory of Molecular Biology in Cambridge. The structures of these tRNAs were studied basically by the same unit method as was used in all the other cases. The only difference was that the *E. coli* cells were grown on medium containing ^{32}P, so that the samples of individual tRNAs isolated from the cells contained uniformly distributed label. This meant that the structure of these tRNAs could be elucidated by the use of the system of methods devised by Sanger and co-workers (Sanger et al., 1965; Brownlee and Sanger, 1967, 1969; Sanger and Brownlee, 1967; Brownlee et al., 1968), in which the entire analysis is conducted at the radiochemical level. The products of enzymic analysis (total and partial) of the tRNA were fractionated by the two-dimensional method and the nature and number of the nucleic-acid derivatives were obtained exclusively on the basis of their radioactivity. Subsequent structural studies of the hydrolysis products of tRNA also were undertaken by the ordinary methods of enzymic analysis, in which the compounds were identified by the position of the label and were estimated quantitatively by the intensity of its activity.

Tyrosine tRNA from *E. coli* (Goodman et al., 1968, 1970)

Earlier work showed that cells of amber-suppressor mutants of *E. coli* Su_{III}^+ contain modified tRNA (Smith et al., 1966). There were grounds for supposing that the triplet coding for chain termination in the bacteria is read in these mutants as the codon for tyrosine. In their attempt to determine the nature of this phenomenon, Goodman and his co-workers showed that the amber-suppressor gene (Su^+) is the structural gene for one of the tyrosine tRNAs of *E. coli*, and the $Su_{III}^- \longrightarrow Su_{III}^+$ mutation leads to the replacement of one nucleotide in the anticodon of this tRNA. The character of the mutation was elucidated by determining the complete nucleotide sequence of $tRNA^{Tyr}$ Su_{III}^+ and $tRNA^{Tyr}$ Su_{III}^-. There were, in fact, two of the latter forms, so that as a result of the work of Goodman and co-workers the structure of essentially three different transfer RNAs was determined. Since Su_{III}^+ suppressor tRNA constitutes only a very small proportion of the total $tRNA^{Tyr}$ of *E. coli*, a method of accumulating the product of action of the Su_{III}^+ gene was developed by increasing the number of copies taken from it. At the same time, this method allowed the newly synthesized tRNA to be labeled.

Preliminary purification of tRNA^{Tyr} from the phage-infected cells was carried out on DEAE-Sephadex by the method described by Landy et al. (1967). The tRNA^{Tyr} was then aminoacylated and separated from the other tRNAs oxidized with sodium periodate either by electrophoresis on poly-acrylamide gel containing polyacrylic acid hydrazide (Marcker et al., 1967) or by polymerization of tyrosyl-tRNA^{Tyr} with β-benzyl-L-aspartate and isolation of the reaction product by the method of Katchalsky et al. (1966). The tRNA$_1^{Tyr}$ and tRNA$_2^{Tyr}$ from uninfected cells were isolated on DEAE-Sephadex (Nishimura et al., 1967b), and subsequently purified on DEAE-cellulose (Gillam et al., 1967) and by reverse-phase chromatography, a modi-fication of the procedures of Kelmers et al. (1965) and of Gefter and Russell (1969).

Total and partial hydrolysis by nucleases, two-dimensional electropho-resis, and paper chromatography were carried out by Sanger's methods (Sanger et al., 1965; Brownlee and Sanger, 1967).

The nucleotide composition of the two forms tRNA^{Tyr} Su$^-_{III\ 1\ and\ 2}$ and of tRNA^{Tyr} Su$^+_{III}$ is given below:

Nucleotides	Moles/mole of tRNA		
	tRNA$_1^{Tyr}$ Su$^-_{III}$	tRNA$_2^{Tyr}$ Su$^-_{III}$	tRNA^{Tyr} Su$^+_{III}$
principal			
A	18	19	18
G	21	21	21
U	11	10	11
C	27	27	28
minor			
ψ	2	2	2
T	1	1	1
OmeG	1	1	1
G*	1	1	—
thioU	2	2	2
mtpA	1	1	1

The molecules of all three forms of tRNA^{Tyr} from *E. coli* each contain 95 nucleotides, including only seven (Su$^+_{III}$) or eight (Su$^-_{III\ 1\ and\ 2}$) minor compo-nents. A characteristic feature of the tRNA^{Tyr} of *E. coli* is the total absence of H$_2$U in its molecule, and the presence of thiouridine and of a deriva-tive of A, which has recently been identified in tRNA$_2^{Tyr}$ from *E. coli* as 2-methylthio-N^6-($\varDelta^2$-isopentenyl)adenosine (Harada et al., 1968).

The G* in the anticodon of tRNA^{Tyr} Su$^-_{III}$ is similar in its spectral properties to unmodified G, and is indistinguishable from it also in its coding properties. However, the phosphodiester bond formed by G* is not ruptured by guanyl RNase. The modification converting G into G* and consisting, evidently, of the introduction of a substituent of basic character (pK_a between 5 and 6) into the imidazole part of the purine ring, was found not to take place under some experimental conditions (i.e., *E. coli* starvation), as a result of which the guanyl RNase is able to split the tRNA^{Tyr} Su$\overline{\text{III}}$ molecule at the G of the anticodon.

The minor components A* (mtpA) and OmeG likewise are not formed in cells starved before infection. The U* is evidently 4thioU. According to Lipsett and Doctor (1967), tRNA$_2^{Tyr}$ from *E. coli* B contains two 4thioU residues, in positions 8 and 9. Smith et al. (1970) do not rule out the possibility that the 4thioU is partially oxidized in the course of isolation of the tRNA, and that their specimens also contain two 4thioU residues; corresponding change was made in their structural formula.

Products of total guanyl- and pyrimidyl-RNase hydrolysates of tRNA^{Tyr} Su$_{\text{III}}^+$ are listed in Table 11. Only two important reconstructions were possible from the comparison of these products, and as a result fragments containing nine and 11 monomers (G-A-A-G-G-T-ψ-C-G and C-C-A-A-A-G-G-G-A-G-C) were obtained. The longest oligonucleotide of the exhaustive T$_1$-hydrolysate contains 19 mononucleotides, the largest oligonucleotide found in

TABLE 11. Oligonucleotides of Total Ribonuclease Hydrolysates
of tRNA^{Tyr} from *E. coli* Su$_{\text{III}}^+$

Pyrimidyl-ribonuclease hydrolysate		Guanyl-ribonuclease hydrolysate	
Oligo(mono)nucleotides	moles/mole of tRNA	Oligo(mono)nucleotides	moles/mole of tRNA
U*	1	G	8
C	20	A-G	2
U	7	U-G	1
ψ	1	C-C-G	1
A$_{OH}$	1	C-A-G	1
A-C	2	A-A-G	1
G-C	1	C-G-OmeG	1
A-U	1	C-C-A-A-A-G	1
G-U	1	T-ψ-C-G	1
G-A-C	1	U*-U*-C-C-C-G	1
A-G-A-C	1	U-C-A-U-C-G	1
G-OmeG-C	1	A-C-U-U-C-G	1
A-mtpA-A-ψ	1	A-A-U-C-C-U-U-C-C-	
G-A-G-C	1	C-C-A-C-C-A-C-C-A$_{OH}$	1
G-A-A-U	1	A-C-U-C-U-A-mtpA-A-ψ-	
A-A-A-G-G-G-A-G-C	1	C-U-G-	1
G-A-A-G-G-T	1		
G-G-G-G-U*	1		
pG-G-U	1		

*4thioU.

any tRNA so far studied. tRNA^{Tyr} Su$_{III}^+$ contains a polypurine sequence of eight monomers and a polypyrimidine sequence of 10 monomers.

To reconstruct the molecule as a whole, partial hydrolysis was carried out with a series of enzymes, yielding more than a hundred different and partially overlapping fragments, the subsequent analysis of which by total hydrolysis with guanyl and pyrimidyl RNases enabled the structure of the molecule to be reconstituted. Five metamers (7) obtained by partial hydrolysis of tRNA^{Tyr} Su$_{III}^+$ by guanyl RNase, which served as the basis for reconstitution of its structure, are shown facing page 120.

The formulae of the tyrosine tRNAs from *E. coli* Su$_{III}^-$ and Su$_{III}^+$ are shown in Fig. 32 in the clover-leaf form. The letters outside the clover leaf (Fig. 32A) show the structural differences between the two tyrosine tRNAs from *E. coli* Su$_{III}^-$: one of them differs from tRNA^{Tyr} Su$_{III}^+$ only by the replacement of C in the first position of the anticodon by G*; in the second, in addition, there are two differences in the extra loop (U→C and C→A). The anticodon of tRNA$_1^{Tyr}$ and tRNA$_2^{Tyr}$ of *E. coli* Su$_{III}^-$ is formed by the triplet G*-U-A which, in accordance with Crick's wobble hypothesis, can pair with the codons U-A-U and U-A-C established for tyrosine. The anticodon of tRNA^{Tyr} Su$_{III}^+$ consists of the triplet C-U-A, by which this tRNA recognizes the chain-termination codon U-A-G.

The primary structure of tRNA^{Tyr} from a number of mutants of *E. coli* Su$_{III}^+$ which are weak suppressors has been studied (Smith et al., 1969; Abelson et al., 1970). The studies have shown that some of them contain mutant types of tRNA^{Tyr} which differ from tRNA^{Tyr} Su$_{III}^+$ in only one nucleotide in various parts of the polynucleotide chain. For example, in tRNA^{Tyr} of the mutant Su$_{15}^-$ the 15th nucleotide G is replaced by A; in Su$_{24}^-$ the OmeG in position 17 is replaced by A; in Su$_{12}^-$ the G in position 31 is replaced by A. Second site revertants were then obtained from some of the mutants with respect to the suppressor genes, in which the suppressor activity was restored (Smith et al., 1970). It was found that one further base is substituted in these complete revertants, restoring the clover-leaf pairing disturbed by the first mutation. For example, a mutant exists in which G2 is replaced by A2 and G2–C80 pairing is impossible (the A2 is opposite C80); in the revertant, a second substitution C80→U80 takes place, as the result of which the second pair in the double-helical segment linking the 3'- and 5'-ends of the molecule is restored, but by the formation of a different base pair (A2–U80).

A study of the functional properties of the mutants with precisely determined changes in the polynucleotide chain of the tRNA^{Tyr} offers extraordinarily wide prospects for the study of functional centers and the conformation of the molecule (Smith et al., 1971). The structure of tRNA^{Tyr} Su$_{III}^+$ and the nucleotide substitutions in the mutant tRNAs are shown in Fig. 33.

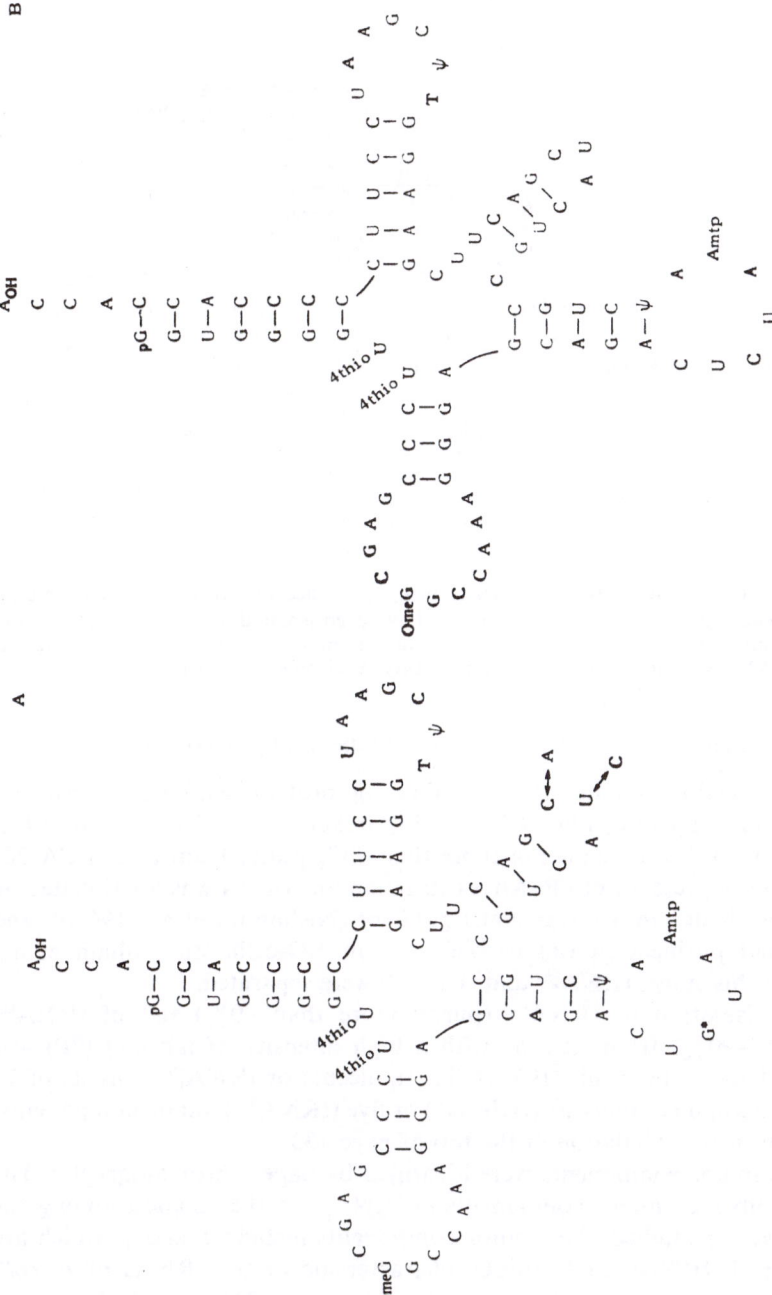

Fig. 32. tRNA^{Tyr} from *E. coli* in the clover-leaf configuration. (A) tRNA^{Tyr} Su$_{\mathrm{III}}^{-}$; (B) tRNA^{Tyr} Su$_{\mathrm{III}}^{+}$. Arrows indicate two nucleotides which distinguish tRNA$_2^{\mathrm{Tyr}}$ from tRNA$_1^{\mathrm{Tyr}}$ of *E. coli* Su$_{\mathrm{III}}^{-}$.

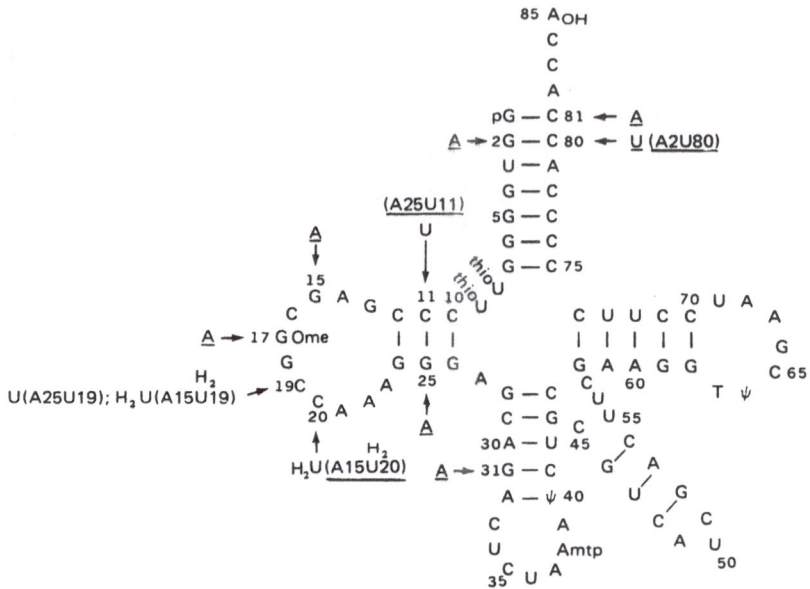

Fig. 33. Sequence of Su$_{III}^+$ tyrosine tRNA showing the nucleotide substitutions in the mutant tRNAs. Nucleotide substitutions underlined have been isolated as single mutants. Those not underlined have only been isolated as double mutants with the sequence changes in brackets. A15, A17, and A31 have been described (Abelson et al., 1970).

N-Formylmethionine tRNA from *E. coli* (Dube et al., 1968)

N-Formylmethionine tRNA, initiating protein synthesis in bacteria (Adams and Capecchi, 1966; Clark and Marcker, 1966; Webster et al., 1966) was obtained as a specimen of more than 90% purity from *E. coli* CA 265 grown in the presence of ^{32}P. After extraction, the tRNA was fractionated on DEAE-Sephadex in a linear NaCl gradient (Nishimura et al., 1967b), and subsequent purification was carried out on BD-cellulose (Gillam et al., 1967); at this stage, tRNA$_f^{Met}$ and tRNA^{Met} were separated.

The isolation of tRNA$_f^{Met}$ (purity more than 90%) and of tRNA^{Met} (purity 80–90%) from *E. coli*, with a high intensity of labeling (^{32}P) was described by Dube et al. (1969a). The molecule of tRNA$_f^{Met}$ consists of 77 nucleotides and contains six (tRNA$_{f1}^{Met}$) or five (tRNA$_{f2}^{Met}$) minor components, as shown in the tabulation at the top of page 133.

The minor components were identified by paper chromatography. The total number of minor components in tRNA$_f^{Met}$ is the smallest among the tRNAs so far studied. The minor components include T and ψ, which are found in all tRNAs, and 4thioU, characteristic of the tRNAs of *E. coli*.

The products of total pyrimidyl- and guanyl-RNase hydrolysis were

Nucleotides, principal	Moles/mole tRNA		Nucleotides, minor	Moles/mole tRNA	
	$tRNA_{f1}^{Met}$	$tRNA_{f2}^{Met}$		$tRNA_{f1}^{Met}$	$tRNA_{f2}^{Met}$
A	14	15	ψ	1	1
G	24	24	H_2U	1	1
U	8	8	T	1	1
C	25	25	OmeC	1	1
			7meG	1	—
			U*	1	1

fractionated by the two-dimensional method on DEAE-paper in 7% HCOOH and on cellulose acetate at pH 3.5 (Sanger et al., 1965); their structure was determined by hydrolysis with alkali and with pancreatic and T_1-RNases, and also by total and partial hydrolysis with spleen and snake-venom PDEases (Dube et al., 1969a). The oligonucleotides of total RNase hydrolysates of $tRNA_f^{Met}$ from *E. coli* are listed in Table 12.

The large fragments were separated by the two-dimensional method (homochromatography) (Brownlee et al., 1968), enabling oligonucleotides

TABLE 12. Oligonucleotides of Total Ribonuclease Hydrolysates of
N-Formylmethionine tRNA from *E. coli*

Pyrimidyl-ribonuclease hydrolysate		Guanyl-ribonuclease hydrolysate	
Oligo(mono)nucleotides	moles/mole G-G-Cp*	Oligo(mono)nucleotides	moles/mole C-C-C-C-Gp*
pCp	0.8	C-Gp	1.4
G-Cp	1.9	A-Gp	1.1
A-Up	0.9	U**-Gp ⎫	
G-Up	2.1	U-Gp ⎬	0.8
A-A-Cp	2.1	A-A-Gp	0.9
A-G-Cp	2.3	C-A-Gp	0.9
G-G-Cp	1.00	U-C-Gp	2.1
G-G-Tp ⎫		H_2U-A-Gp	0.9
G-G-H_2Up ⎬	1.9	pC-Gp	0.8
A-A-A-Up	1.0	7meG-U-C-Gp	0.75
C-G-G-OmeC-Up	0.9	A-U-C-Gp	0.25
G-G-A-G-Cp	0.8	C-C-U-Gp ⎫	
G-G-G-G-Up** ⎫		C-U-C-Gp ⎬	1.9
G-G-G-G-Up ⎬	0.7	C-C-C-C-C-Gp	1.00
G-A-A-G-7meG-Up	0.75	C-A-A-C-C-A_OH	0.9
G-A-A-G-A-Up	0.25	T-ψ-C-A-A-A-U-C-G-Gp	1.0
ψp		OmeC-U-C-A-U-A-A-C-C-C-Gp	0.9
Up		Gp	9.0
Cp			
A_OH			

*Results obtained by measuring radioactivity in the part of the paper containing the oligonucleotide.
**4thioU.

containing 25 monomers each to be separated. Smaller partial hydrolysis products also were fractionated by the two-dimensional method, using 7% HCOOH in one direction (DEAE-paper) and electrophoresis on cellulose acetate in buffer at pH 3.5 in the other direction (Sanger et al., 1965). During the study of products of partial guanyl-RNase hydrolysis, a fragment including the whole anticodon loop U-C-G-G-G-OmeC-U-C-A-U-A-A-C-C-C-G-A-A-G was isolated and its coding properties were studied (Clark et al., 1968b). Analysis of the partial hydrolysis products was carried out chiefly by exhaustive hydrolysis with guanyl and pancreatic RNase, the hydrolysis products being detected in all cases by their radioactivity.

The products of partial hydrolysis of tRNA$_f^{Met}$ (8) are shown facing page 120 (Dube and Marcker, 1969). The lines above the formula show products obtained by partial hydrolysis with pancreatic RNase, the lines below the formula show the corresponding products of partial hydrolysis by T$_1$-RNase, except the fragment U*GGA, which was obtained by partial hydrolysis with spleen RNase.

The preparation studied evidently consisted of two components present in the ratio of 3 : 1, for in 25% of cases A was identified instead of 7meG (Fig. 34); it is interesting to note that this substitution also was localized in the extra loop of the clover leaf (see tRNA^{Ser} from yeast and tRNA^{Tyr} from E. coli).

There are two essential features of the primary structure of tRNA$_f^{Met}$ which distinguish it from most other transfer RNAs. The "universal" oligonucleotide consists of the sequence G-T-ψ-C-A, as in one of the serine tRNAs, and not of the sequence G-T-ψ-C-G, Furthermore, to the right of the anticodon, unlike in most tRNAs there is a residue of adenosine and not of one of its derivatives.

The polynucleotide chain of tRNA$_f^{Met}$ can be arranged as a clover leaf (Fig. 34) with the standard characteristics, but differing in some details from the models constructed earlier. In particular, base pairing starts from the sixth, and not the fifth base of the acceptor end, and from the second and not the first base of the 5'-end of the molecule; at the 3'-end five nucleotides thus remain free, along with the terminal pCp at the 5'-end. The number of pairs in this part is one less than in other tRNAs. Some of these differences might be responsible for the unusual functional properties of tRNA$_f^{Met}$.

The view is now increasingly held that the two features described in the structure of tRNA$_f^{Met}$ from E. coli—the unusual position of the hydrogen bonds in the stem linking the 3'- and 5'-ends of the molecule and the absence of modification of the nucleotide located on the 3'-side of the anticodon— cannot provide a basis for the recognition of this molecule by the formylating enzyme or for the characteristics of its coding properties. A comparison made by RajBhandary and co-workers (RajBhandary and Ghosh, 1969; RajBhan-

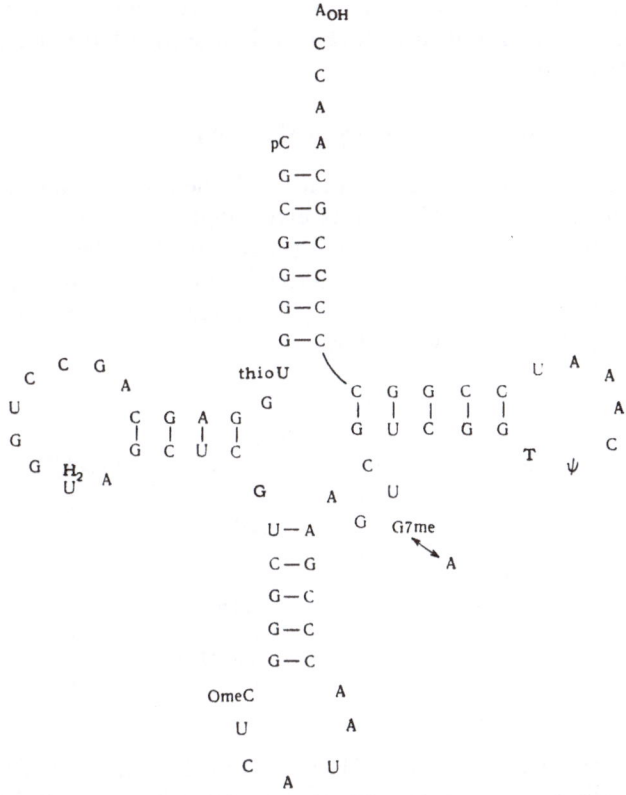

Fig. 34. tRNA$_f^{Met}$ from *E. coli* in the clover-leaf configuration.

dary and Kumar, 1970) showed that in the arrangement of the hydrogen bonds linking the 3'- and 5'-ends, and in the presence of a modification at the A residue on the 3'-side of the anticodon, yeast tRNA$_f^{Met}$ (A) resembles the ordinary tRNAs more closely than tRNA$_f^{Met}$ from *E. coli* (B):

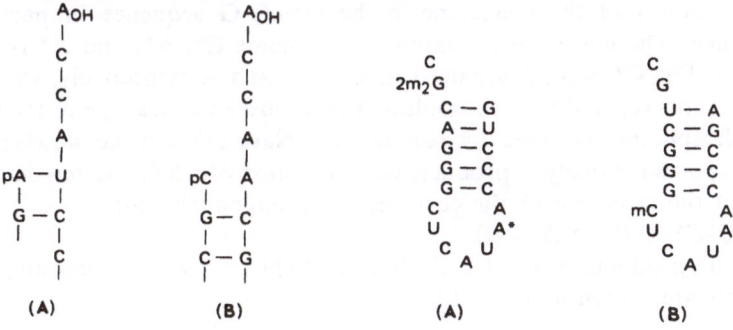

Nevertheless, yeast tRNA$_f^{Met}$ is formylated, it recognizes the codons AUG and GUG, and it cannot introduce methionine into the inner parts of the polypeptide chain.

Methionine tRNA from *E. coli* (Cory et al., 1968)

By means of Sanger's method and with the use of a highly radioactive label, the structure of tRNA^{Met} was elucidated. Like the ordinary tRNA, this type incorporates an amino acid (methionine) in the inner part of the polypeptide chain. Details of the determination of the structure of tRNA^{Met} are described by Cory and Marcker (1970). The nucleotide composition of tRNA^{Met} from *E. coli* is given below:

Nucleotides, principal	Moles/mole of tRNA		Nucleotides, minor	Moles/mole of tRNA	
	tRNA$_1^{Met}$	tRNA$_2^{Met}$		tRNA$_1^{Met}$	tRNA$_2^{Met}$
A	18	18	ψ	2	2
G	18	18	H$_2$U	3	4
U	10	9	T	1	1
C	19	19	7meG	1	1
			OmeG	1	1
			C*	1	1
			4thioU	1	1
			A*	1	1
			X*	1	1

Unlike the tRNA^{Tyr} and tRNA$_f^{Met}$ from *E. coli,* tRNA^{Met} is rich in minor components; of the 77 monomers forming its molecule 12 (or 13) are modified nucleotides. Just as in the tRNA^{Ala} from yeast, one uridyl residue is partially replaced by a dihydrouridyl residue. It is also not clear in this case whether two different tRNAs are present here or whether the substitution of H$_2$U for U reflects different stages in the modification of the same tRNA. A more definite sign of incomplete modification under these conditions of growth of *E. coli* is the absence of a methyl group attached to the ribose moiety of the guanosine in the OmeG-G sequence in part of the specimen. The nature of the minor components C*, A*, and X* is not yet known. The C* and A* are derivatives of C and A respectively, with bulky substituents (probably at the amino groups), because cleavage of the preceding phosphodiester bond by pancreatic RNase takes place slowly; as the result, partial hydrolysis products were obtained which facilitated determination of the structure of the corresponding oligonucleotide (C-A-C-A-U-C-A-C-U-C*-A-U-A*-A-ψ-G).

The products of total hydrolysis of tRNA^{Met} by pyrimidyl and guanyl RNases are shown in Table 12a.

Table 12a. Oligonucleotides of Total Ribonuclease Digests of tRNA^{Met} from *E.coli*

Pyrimidyl-RNase digest			Guanyl-RNase digest		
	moles/mole			moles/mole	
Oligo(mono)-nucleotides	Found	Calculated from structure	Oligo(mono)-nucleotides	Found	Calculated from structure
U, ψ, T, H$_2$U, C	—	7.0	G	7.3	5.0
C	6.0	9.0	A-G	1.7	1.0
A-C	4.9	5.0	C-C-A-C-C-A$_{OH}$	1.1	1.0
A-U	2.0	2.0	7meG-X-C-A-C-A-G	0.7	1.0
A-G-C	2.5	2.0	pGp	0.7	1.0
G-U	2.5	3.0	U-C-G	1.2	1.0
G-A-U	1.1	1.0	U-A-G	1.3	2.0
A-G-U	1.2	1.0	A-U-G	0.9	1.0
A*-A-ψ[a]	1.0	1.0	C-U-A-C-G	1.0	1.0
G-A-A-U	1.1	1.0	C-U-C-A-G	1.4	1.0
A-G-A-G-C	1.0	1.0	A-A-U-C-C-C-G	1.0	1.0
pG-G-C	0.4	1.0	T-ψ-C-G	1.4	1.0
G-G-H$_2$U	1.2 ⎫		H$_2$U-H$_2$U-A-G	1.2	1.0
OmeG-G-H$_2$U	0.5 ⎭ 1.0		H$_2$U-H$_2$U-G ⎫[b]		
A-G-G-T	1.0	1.0	U-H$_2$U-G ⎭	1.0 ⎫	
G-G-G-7meG-X-C	0.6	1.0	H$_2$U-H$_2$U-OmeG-G ⎫[b]	⎬	1.0
			U-H$_2$U-OmeG-G ⎭	0.14 ⎭	
			C-A-C-A-U-C-A-C-U-		
			C*-A-U-A*-A-ψ-G[a]		1.0

[a]C*, A* are unidentified nucleotides.
[b]Both forms are found.

To determine the nucleotide composition, the oligonucleotides were hydrolyzed not only with alkali, but also with either pancreatic RNase and spleen phosphodiesterase, or by ribonuclease T$_2$. The structure of the oligonucleotides was determined by using RNase U$_2$ (Arima et al., 1968), in addition to the methods described previously, as well as modification by carbodiimide and subsequent degradation by pancreatic RNase.

As in all the other cases, the overlaps detected by comparing the oligonucleotides of the two total RNase hydrolysates were insufficient to allow reconstruction of the molecule. For this purpose, partial hydrolysis by both RNases was used: the partial hydrolysis products were fractionated by two-dimensional ionophoresis and also by homochromatography (Brownlee and Sanger, 1967, 1969). The partial hydrolysis products used to reconstruct the structure of tRNA^{Met} are illustrated in Fig. 34a (facing page 138).

As in all the other cases, a clover-leaf configuration can be ascribed to tRNA^{Met} (Fig. 35), but the paired segment of the T-ψ loop contains only four and not five hydrogen-bonded base pairs. The segment of the double helix connecting the 3'- and 5'-ends of the molecule, unlike in tRNA$_f^{Met}$, has

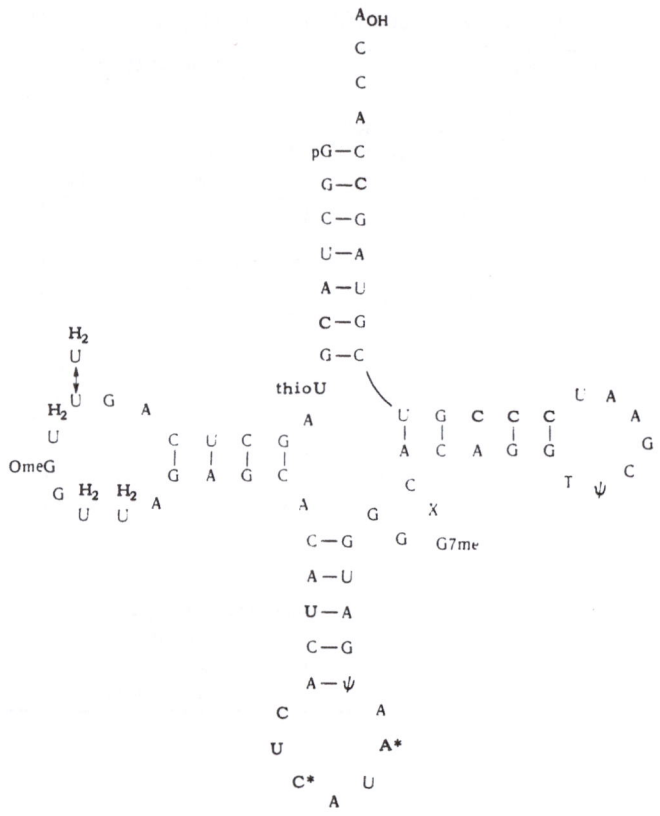

Fig. 35. tRNA^Met from *E. coli* in the clover-leaf configuration.

the same structure as in all other tRNAs. A modified A residue lies to the right of the anticodon.

Purified methionyl-tRNA-synthetase reacts equally with tRNA^Met and tRNA_f^Met (Bruton and Hartley, 1968; Heinrikson and Hartley, 1968); this indicates that its interaction with tRNA is due to certain structural features which are common to these two tRNAs. Meanwhile transformylase possesses absolute specificity for met-tRNA_f^Met and does not transfer the formyl residue from N-10-formyltetrahydrofolic acid to any other tRNA, including met-tRNA^Met; it evidently recognizes certain structural features characteristic of tRNA_f^Met only.

The coding properties of tRNA^Met and tRNA_f^Met also are different (Clark and Marcker, 1966; Ghosh et al., 1967); both tRNAs respond to A-U-G but, in addition, tRNA_f^Met also responds to G-U-G. The reason for this conceivably could be the absence of modification to the A residue neighboring the anticodon of tRNA_f^Met on the 3'-side.

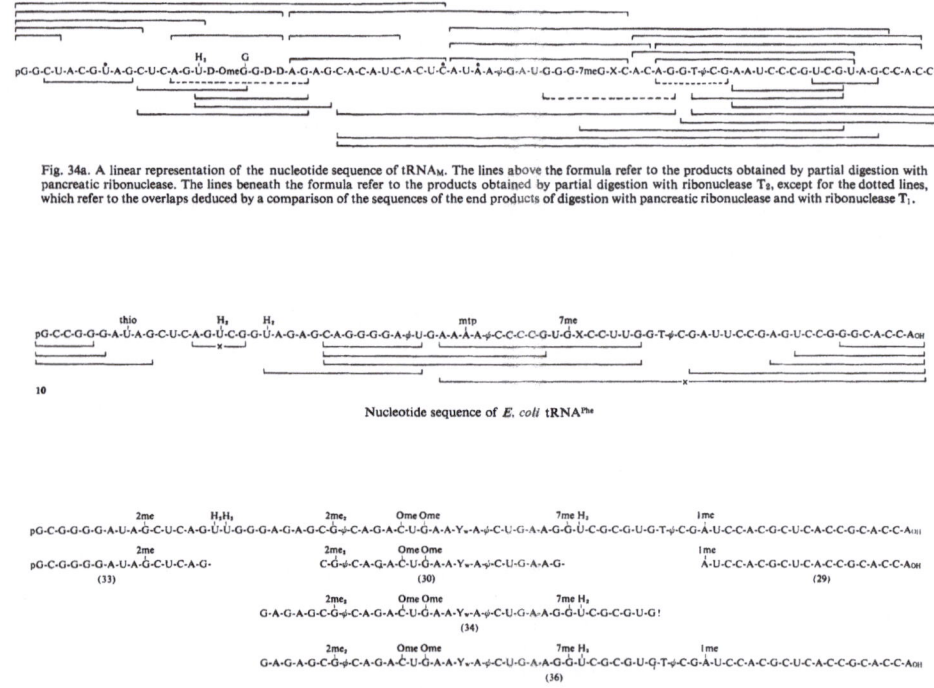

Fig. 34a. A linear representation of the nucleotide sequence of tRNA_M. The lines above the formula refer to the products obtained by partial digestion with pancreatic ribonuclease. The lines beneath the formula refer to the products obtained by partial digestion with ribonuclease T_B, except for the dotted lines, which refer to the overlaps deduced by a comparison of the sequences of the end products of digestion with pancreatic ribonuclease and with ribonuclease T_1.

Nucleotide sequence of *E. coli* tRNA^Phe

Fig. 43. Nucleotide sequence of wheat-germ tRNA^Phe (top) and large oligonucleotide crucial for the elucidation of the structure (bottom). Fragments 29 to 36 were obtained from partial RNase T_1 digestion, fragment *a* was obtained by partial pancreatic RNase digestion.

The formulas (9) of tRNA$_f^{Met}$ (top line) and tRNA^{Met} (bottom line) are compared below; the boxes enclose identical nucleotides, disregarding modifications.

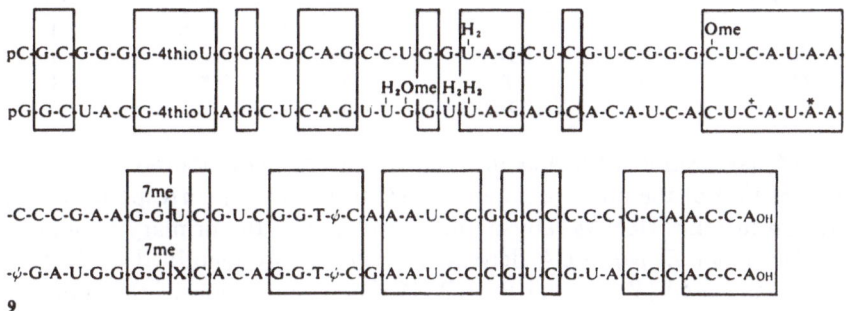

9

Comparison shows that despite some features of similarity, these two tRNAs differ sharply from each other. The similarity concerns principally those sequences which are common to all tRNAs, and the authors cited conclude that the degree of homology between tRNA^{Met} and tRNA$_f^{Met}$ is lower that between tRNAs of different amino-acid specificity isolated from the same source: for example, tRNA^{Met} and tRNA^{Val} of *E. coli* have 51 nucleotides in common, whereas tRNA^{Met} and tRNA$_f^{Met}$ have only 41.

Elucidation of the primary structure of tRNA^{Met} was awaited with great interest because it was thought that the comparison of tRNA^{Met} and tRNA$_f^{Met}$ would shed light on the structural basis of their functional differences. However, in this case also the primary structure itself still does not tell which part of the molecule is responsible for which function. To tackle these problems, functional investigations of large fragments of the molecule are being undertaken (Dube et al., 1969b).

Valine tRNA from *E. coli* (Yaniv and Barrell, 1969)

A tRNA$_1^{Val}$, accounting for about 80% of the total valine-acceptor activity, was isolated from *E. coli* cells grown on medium containing ^{32}P. The tRNA$_1^{Val}$ was separated from tRNA$_2^{Val}$ on benzoylated DEAE-cellulose (Gillam et al., 1967); the fraction containing tRNA$_1^{Val}$ was loaded with valine and then phenoxyacetylated by the method of Gillam et al. (1968) before being applied to the second column with BD-cellulose. Elution was carried out in an alcohol–salt gradient. The purity of the tRNA$_1^{Val}$ isolated in this manner was not less than 95%. Its primary structure was investigated by the unit method, using the radiochemical technique of Sanger and co-workers (Sanger et al., 1965; Brownlee and Sanger, 1967; Brownlee et al., 1968).

The molecule of tRNA$_1^{Val}$ from *E. coli* consists of 76 nucleotides, including seven minor components:

Nucleotides, principal	Moles/mole of tRNA	Nucleotides, minor	Moles/mole of tRNA	Nucleotides, minor	Moles/mole of tRNA
A	14	ψ	1	4thioU	1
G	23	H₂U	1	6meA	1
U	9	T	1	X*	1
C·	23	7meG	1		

X* has been identified as uridine 5-oxyacetic acid (Murao et al., 1970). The position of the minor components (4thioU, 7meG) corresponds to their position in other tRNAs. During determination of the primary structure of tRNA$_1^{Val}$ from *E. coli* an A derivative preliminarily identified as 1meA or 6meA was found (Yaniv and Barrell, 1969) on the 3′-side of the anticodon. Japanese workers (Harada et al., 1969; Saneyoshi et al., 1969), having taken measures to prevent the conversion 1meA→6meA, showed that the minor component in tRNA$_1^{Val}$ (and also, probably, in tRNA$_2^{Val}$) is 6meA; tRNA$_1^{Val}$ is shown in Fig. 36 in the clover-leaf configuration.

The structure of the tRNA$_2^{Val}$, responding to the codons GUU and GUC, has also been determined; it consists of two fractions differing in three pairs of nucleotides (see page 266 and isoacceptor tRNAs) (Yaniv and Barrell, 1971). The tRNA$_2^{Val}$ molecule contains 77 nucleotides, including nine minor components.

Phenylalanine tRNA from *E. coli* (Barrell and Sanger, 1969)

Initially ³²P-labeled tRNA^{Phe} was isolated from *E. coli* cells grown on medium containing ³²P, by using its property of specific binding with ribosomes in the presence of polyU (Nirenberg and Leder, 1964). However, the yield by the use of this method is low, and purification was subsequently carried out by reverse-phase partition chromatography (Kelmers et al., 1965).

The nucleotide composition of tRNA^{Phe} from *E. coli* is given below:

Nucleotides, principal	Moles/mole of tRNA	Nucleotides, minor	Moles/mole of tRNA	Nucleotides, minor	Moles/mole of tRNA
A	14	ψ	3	7meG	1
G	23	H₂U	2	4thioU	1
U	8	T	1	X*	1
C	21	mtpA	1		

X* has not yet been identified; hypothetically it is the same unstable nucleotide as is present in tRNA^{Met} from *E. coli*. To the right of the anticodon there is a minor component which is evidently mtpA, just as in tRNA^{Tyr} of *E. coli*.

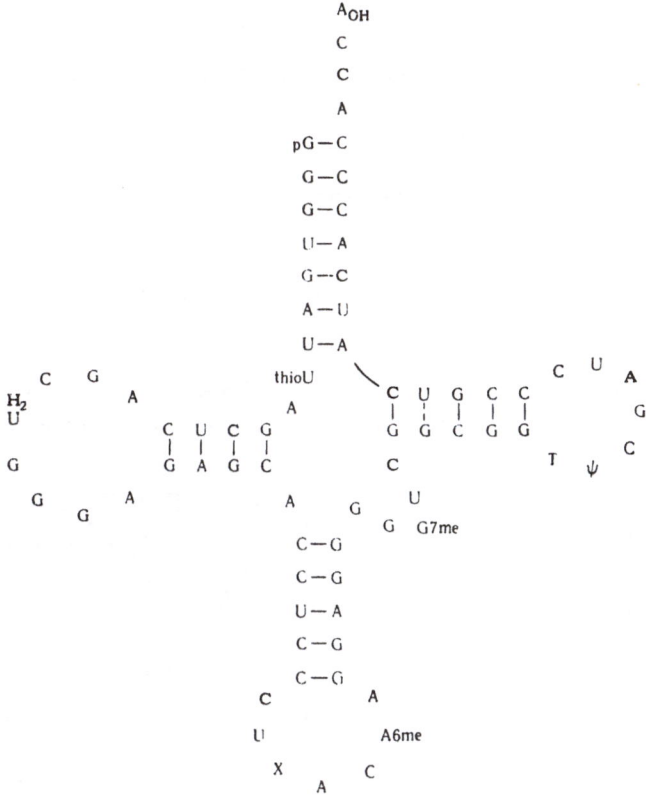

Fig. 36. tRNA$_1^{Val}$ from *E. coli* in the clover-leaf configuration.

Total RNase hydrolysates of tRNA^{Phe} were fractionated by Sanger's method (Sanger et al., 1965). Fractionation of a pyrimidine-RNase digest of tRNA^{Phe} from *E. coli* is illustrated in Fig. 37. Pancreatic and T$_1$-RNases and also RNase from *Bacillus subtilis* were used for partial hydrolysis; large fragments were fractionated by homochromatography. The structural formula of tRNA^{Phe} from *E. coli* and the partial hydrolysis products used for its reconstruction (10) are shown facing page 138. Fragments isolated only once are marked by a cross.

Phenylalanine tRNA from *E. coli* is shown in Fig. 38 in the clover-leaf form. tRNA^{Phe} from *E. coli* has also been studied by Gassen and Uziel (1969). Besides the ordinary methods, they also used stepwise chemical degration and reduction with ^{3}H-NaBH$_4$. The structural formula (11) which they suggested differs considerably from the formula established by Sanger and Barrell, and it is given on page 142.

H₃   H₃

pG-C-C-C-G-C-U-C-A-G-C-G-G-U̇-C-G-A-S-C-(C)-A-G-U̇-A-G-G-4thioU-A-G-A-

7me

-G-C-A-G-G-G-G-G-A-ψ-U-G-G-A-A-A-ψ-C-C-C-C-G-U-Ġ-C-C-U-U-G-G-T-ψ-C-

-G-A-U-U-C-C-G-A-G-U-C-C-G-G-G-C-A-C-C-A_{OH}

11

*   *   *

The primary structure of three tRNAs from *T. utilis*—valine, isoleucine, and tyrosine—was established by Takemura and co-workers. For

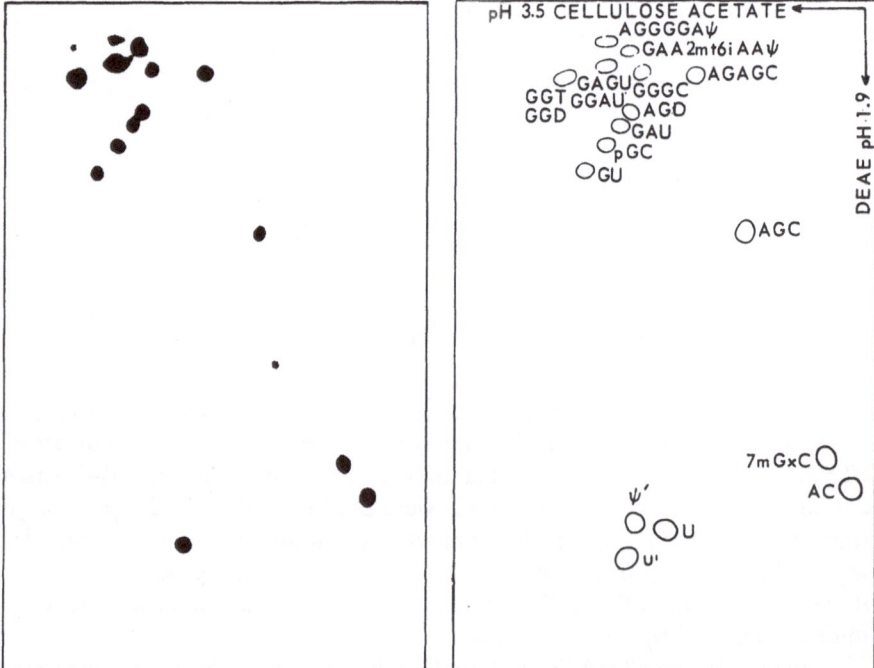

Fig. 37. A two-dimensional fractionation of a pancreatic ribonuclease digest of tRNA^{Phe}. Radioautograph and diagram showing the sequences of the nucleotides. C and C> are not included in this fingerprint. A-G-G-G-G-A-ψ is always obtained in low yield and some of the minor spots on the fingerprint are due to breakdown of this nucleotide. GGAU is also obtained in low yield and is formed from G-G-A-4thioU after electrophoresis at acid pH.

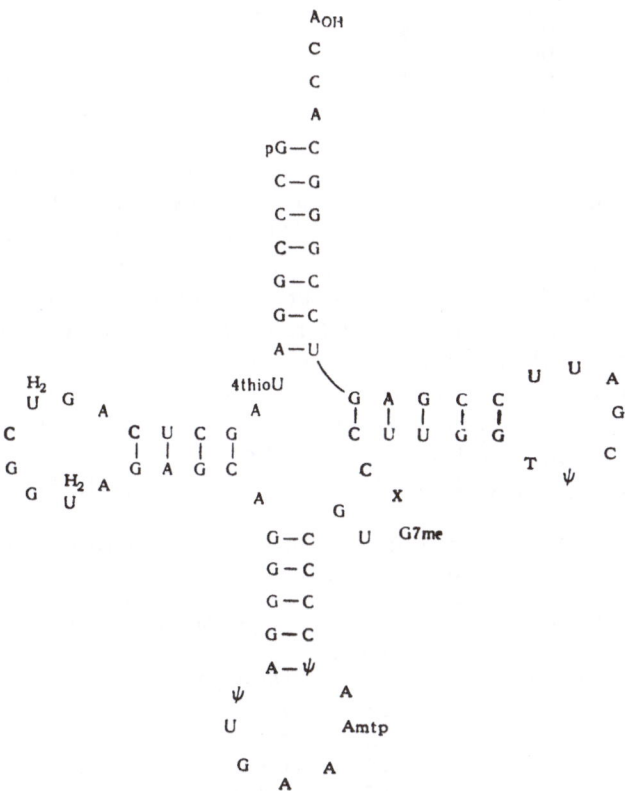

Fig. 38. tRNA^Phe from *E. coli* in the clover-leaf configuration.

their analysis they used the standard methods of total and partial hydrolysis by pancreatic and T_1-RNases, followed by fractionation of the resulting fragments on DEAE-cellulose in $7M$ urea. To study the structure of the oligonucleotides, besides the usual methods of total and partial hydrolysis with snake-venom PDEase and hydrolysis by RNase of an additional specificity, they also used T_2- and U_2-RNases (Arima et al., 1968).

Valine tRNA₁ from *T. utilis* (Takemura et al., 1968a)

The purity of the tRNA₁^Val sample isolated for structural investigations by chromatography on DEAE-Sephadex, judging from its acceptor activity and chemical analysis, was 85%. The tRNA₁^Val molecule consists of 75 nucleotides, including 12 minor components:

Nucleotides, principal	Moles/mole of tRNA	Nucleotides minor	Moles/mole of tRNA	Nucleotides, minor	Moles/mole of tRNA
A	14	ψ	4	1meA	1
G	18	H_2U	3	1meG	1
U	11	T	1	5meC	1
C	20	I	1		

The mono- and oligonucleotides formed from the tRNA$_1^{Val}$ molecule by total hydrolysis by the two RNases are shown in Table 13 (Takemura et al., 1968b). As in the case of tRNA$_1^{Val}$ from *S. cerevisiae,* the anticodon of tRNA$_1^{Val}$ from *T. utilis* is the triplet I-A-C. Complete agreement was established between the two hydrolysates and a number of reconstructions could be carried out, but before the structure of the whole molecule could be reconstituted partial hydrolysis was required, and T$_1$-RNase was used for this purpose (Mizutani et al., 1968). The results of partial hydrolysis of tRNA$_1^{Val}$ suggest that its molecule contains segments more easily attacked by T$_1$-RNase. Just as in the other cases, in tRNA$_1^{Val}$ from *T. utilis* they correspond to unpaired segments of the clover-leaf model (Fig. 39). However, unlike

TABLE 13. Oligonucleotides of Total Ribonuclease Hydrolysates of tRNA$_1^{Val}$ from *Torulopsis utilis*

Pyrimidyl-ribonuclease hydrolysate		Guanyl-ribonuclease hydrolysate	
Oligo(mono)nucleotides	moles/mole A	Oligo(mono)nucleotides	moles/mole pGp
A	1.0	Gp	8.6
5meCp	1.1	U-1meGp	1.2
Cp	11.7	C-Gp	1.4
H₂Up	0.7	A-Gp	0.5
ψp	2.8	H₂U-H₂U-Gp	1.2
Up	4.8	C-A-Gp	1.2
A-Cp	1.9	pGp	1.0
G-Cp	1.9	C-ψ-U-Ip	1.2
A-ψp	0.9	T-ψ-C-Gp	1.2
A-Up	0.9	1meA-U-C-C-U-Gp	0.9
G-1meA-Up	0.8	H₂U-C-A-U-Gp	1.1
G-Up	1.3	U-C-ψ-A-Gp	1.1
I-A-Cp	0.9	U-U-U-C-Gp	1.1
G-G-Cp	0.9	C-A-ψ-C-U-Gp	0.8
A-G-H₂Up	0.8	A-C-A-C-Gp	0.8
A-G-Tp	0.9	A-A-A-U-C-A-C-C-A$_{OH}$	1.0
1meG-G-Up	1.1	A-A-C-5meC-C-C-C-A-Gp	1.1
G-G-H₂Up	1.0		
G-G-G-Cp	0.7		
A-G-A-A-Cp	0.8		
G-A-A-A-Up	0.8		
pG-G-Up	0.6		

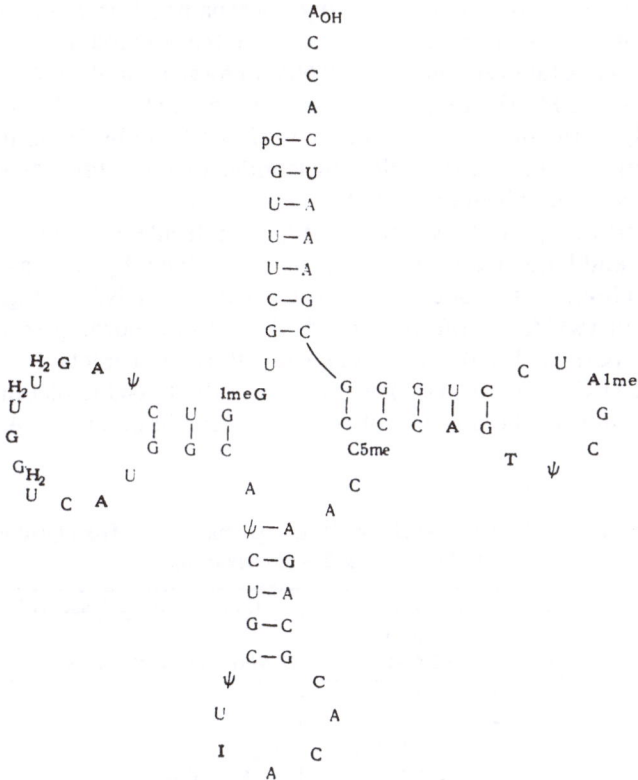

Fig. 39. tRNA$_1^{Val}$ from *T. utilis* in the clover-leaf configuration.

other tRNAs and, in particular, in tRNA$_1^{Val}$ from *S. cerevisiae,* it was impossible to split the molecule into halves at the anticodon, a result which these workers attribute to the slower action of T$_1$-RNase on the I-N bonds than on the G-N bonds.

Isoleucine tRNA from *T. utilis* (Takemura et al., 1969a)

tRNA^{Ile} of 95% purity was isolated by chromatography on DEAE-Sephadex. As the results given below show, the molecule consists of 77 nucleotides, 13 of which are minor components:

Nucleotides, principal	Moles/mole of tRNA	Nucleotides, minor	Moles/mole of tRNA	Nucleotides, minor	Moles/mole of tRNA
A	14	ψ	2	1meA	1
G	21	H$_2$U	5	2me$_2$G	1
U	9	T	1	5meC	1
C	20	I	1	thrA	1

The minor component N-(purinyl-6-carbamoyl)threonine nucleoside, (thrA) found in tRNA^{Ile} on the 3′-side of the anticodon, was identified previously in a total preparation of tRNA (Schweizer et al., 1965), and it has now been found for the first time in an individual tRNA. Technical details of the study of the nucleotide sequence in tRNA^{Ile} can be found in the paper by Takemura et al., (1969b). Oligonucleotides of the total RNase hydrolysates of tRNA^{Ile} are given in Table 14.

Even before partial hydrolysis of the molecule by RNase had been carried out and large overlapping fragments obtained, by analogy with other tRNAs a clover-leaf model was constructed for tRNA^{Ile} (Fig. 40). The anticodon of tRNA^{Ile} is the triplet I-A-U, which, according to the wobble hypothesis (Crick, 1966), can recognize all three triplets for isoleucine (A-U-A, A-U-C, and A-U-U). The position of the I-A-U triplet in the clover leaf corresponds to the position of the anticodons in other tRNAs.

TABLE 14. Oligonucleotides of Total Ribonuclease Hydrolysates of
tRNA^{Ile} from *Torulopsis utilis*

Pyrimidyl-ribonuclease hydrolysate		Guanyl-ribonuclease hydrolysate	
Oligo(mono)nucleotides	moles/mole pG-G-Up[a]	Oligo(mono)nucleotides	moles/mole pGp[a]
A	0.9	Gp	7.9 (7)
5meCp	1.0	U-Gp	1.0
Cp	11.6 (11)	A-Gp	0.1 (0)
H$_2$Up	2.2	H$_2$U-H$_2$U-Gp	0.9
ψp	1.1	C-U-Ip	0.9
Up	5.7 (5)	C-A-Gp	1.0
2me$_2$G-ψp	0.9	pGp	1.0
G-1meA-Up	0.8	C-2me$_2$G-ψ-Gp	1.0
A-Cp	1.2	T-ψ-C-Gp	0.9
G-Cp	3.4 (3)	C-U-A-Gp	0.9
A-G-Cp	1.1	1meA-U-C-C-U-Gp	0.6 (1)
G-G-Cp[b]	1.0	A-C-C-A-C-C-A$_{OH}$	1.0
I-A-Up	1.0	H$_2$U-H$_2$U-A-A-Gp	1.0
A-G-H$_2$Up	1.0	C-C-C-A-Gp	1.0
A-G-Tp	1.0	C-C-A-A-Gp	0.9
G-G-H$_2$Up	1.0	A-H$_2$U-5meC-A-Gp	1.0
G-G-Up	1.1	6meA-U-C-C-U-Gp	0.3
N-A-Cp	1.0	U-C-C-C-U-U-Gp	1.0
A-A-G-G-Cp	1.0	A-U-N-A-C-Gp	1.0
A-A-G-A-H$_2$Up	1.0		
pG-G-Up	1.0		
A-G-G-G-A-Cp	1.0		

[a] Number of moles given in parentheses was determined from results obtained by hydrolysis with RNase of opposite specificity.
[b] This fragment contained a small quantity of a modified G, probably 2meG.

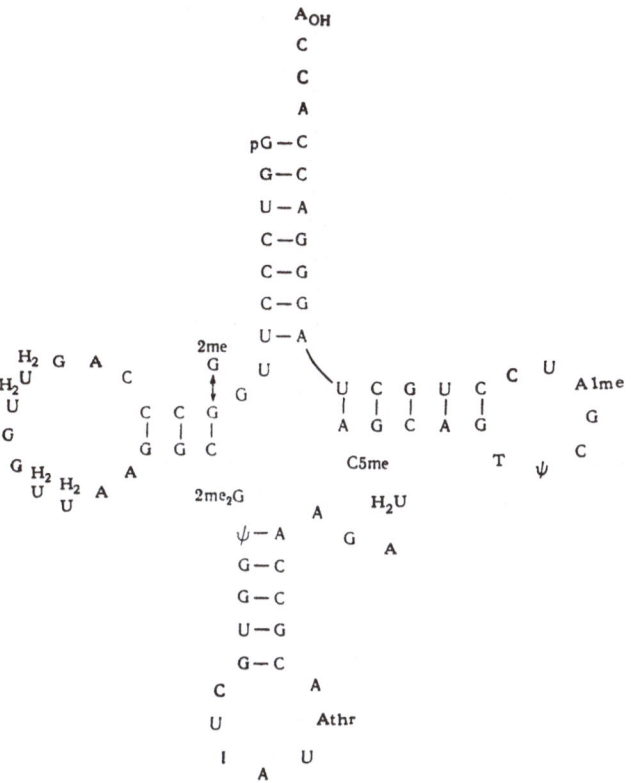

Fig. 40. tRNA^Ile from *T. utilis* in the clover-leaf configuration.

Tyrosine tRNA from *T. utilis* (Hashimoto et al., 1969)

The tRNA^Tyr was isolated from a total preparation of tRNA from *T. utilis* by chromatography on DEAE-Sephadex in an ammonium sulfate gradient (0.75–1.5 M) at pH 5.3 (Miyazaki et al., 1966) and by rechromatography of fractions possessing tyrosine activity on the same columns in 1 M phosphate containing dimethylformamide (KCl gradient) (Miyazaki and Takemura, 1966); in the third stage, where chromatography on BD-cellulose (Gillam et al., 1967) was used, the purity of the specimens reached 80%. The nucleotide composition of the tRNA^Tyr from *T. utilis* is given below:

Nucleotides, principal	Moles/mole of tRNA	Nucleotides, minor	Moles/mole of tRNA	Nucleotides, minor	Moles/mole of tRNA
A	16	ψ	4	1meG	1
G	18	H₂U	6	2meG	1
U	6	T	1	2me₂G	1
C	20	1meA	1	5meC	1
		ipeA	1	OmeG	1

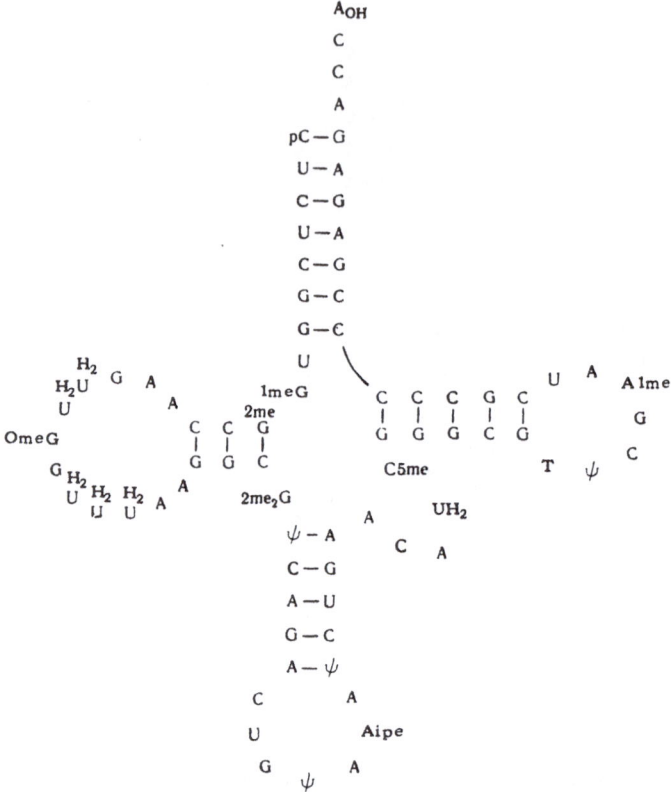

Fig. 41. tRNA^Tyr from *T. utilis* in the clover-leaf configuration.

The molecule consists of 78 nucleotides, including 18 minor components, the largest number found so far among the tRNAs. The arrangement of the minor components in the clover-leaf model constructed for tRNA^Tyr (Fig.41) corresponds to that expected by analogy with other structural formulas.

The hypothetical anticodon is G-ψ-A which, as in all tRNAs, is located at the tip of the anticodon branch, and recognizes both codons of tyrosine (U-A-U and U-A-C) (Doctor et al., 1966).

Serine tRNA from Rat Liver (Staehelin et al., 1968)

Serine tRNA was isolated from rat liver by partition chromatography (Muench and Berg, 1966a). This method, like chromatography on MAK columns, revealed several fractions possessing serine acceptor activity. The largest of the fractions were investigated.

The nucleotide composition of tRNA^{Ser} from rat liver is given below:

Nucleotides, principal	Moles/mole of tRNA	Nucleotides, minor	Moles/mole of tRNA	Nucleotides, minor	Moles/mole of tRNA
A	14	ψ	2	OmeU	1
G	25	H$_2$U	3	Omeψ	1
U	10	T	1	OmeG	1
C	19	I	1	2me$_2$G	1
		5meC	1	1meA	1
		3meC	2	ipeA	1
		acC	1		

A detailed comparison of the serine-specific tRNAs from yeast and rat liver will be given on page 199, and all that will be said here is that they both contain 85 monomers and differ in approximately one-quarter of their nucleotides. Liver tRNA^{Ser} contains 17 minor components, two of which (3meC and Omeψ) were identified for the first time in tRNA. The 1meA component occupies the same place in liver tRNA^{Ser} as in the tyrosine, phenylalanine, and valine tRNAs from yeast; 3meC is located where 7meG is found in many tRNAs, namely in the extra loop; this is interesting because both minor components carry a positive charge at neutral pH values.

The oligonucleotides of the total ribonuclease hydrolysates were fractionated on DEAE-cellulose columns and their structure was determined by hydrolysis with the complementary enzyme, micrococcal nuclease, and snake-venom PDEase, with subsequent fractionation of the reaction products by thin-layer chromatography. Partial hydrolysis of the molecule was carried out with T$_1$-RNase.

Serine tRNA from rat liver is shown in Fig. 42 as the clover-leaf model. It does not differ in principle from the corresponding models for tRNA from yeast and *E. coli*.

Staehelin et al. (1969) later reported that by the combined application of partition chromatography and reverse-phase chromatography they had succeeded in separating serine-specific tRNA from rat liver into five fractions. A preliminary comparison was made of the primary structures, coding properties, and interaction of these tRNAs with aminoacyl-tRNA-synthetase (see the section *Structure of Isoacceptor tRNAs* in this chapter).

Phenylalanine tRNA from Wheat Germ (Dudock et al., 1969)

A sample of tRNA^{Phe} of 90% purity was isolated from a total preparation obtained from wheat germ by the phenol method (Holley et al., 1963) in two stages: 10- to 15-fold purification of comparatively large quantities of tRNA was achieved on BD-cellulose (Gillam et al., 1967); the second stage,

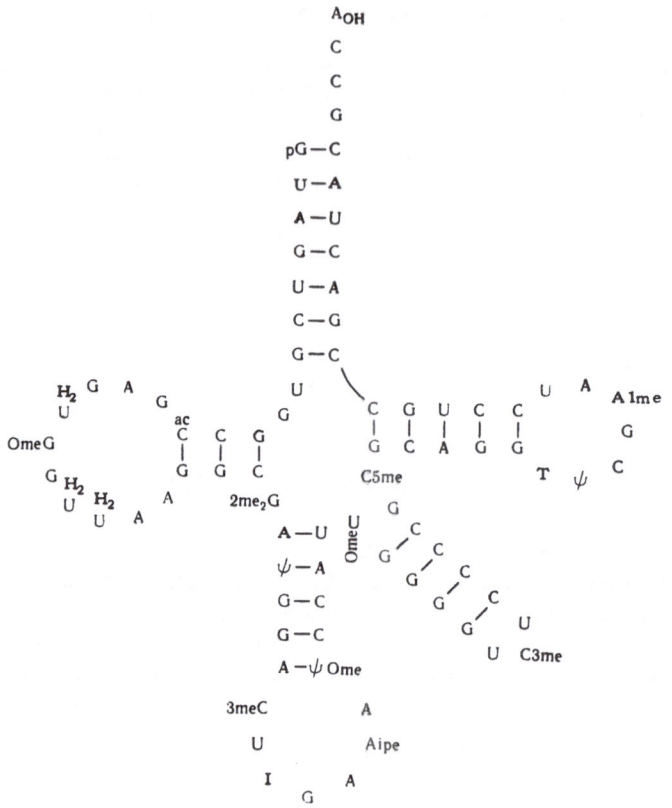

Fig. 42. tRNA^Ser from rat liver in the clover-leaf configuration.

reverse-phase chromatography (Weiss and Kelmers, 1967), gave a specimen of approximately 90% purity.

The primary structure of tRNA^Phe was determined by the usual methods of total (Katz and Dudock, 1969) and partial (Dudock and Katz, 1969) hydrolysis by two RNases with subsequent fractionation and analysis of all the resulting fragments. The nucleotide composition of tRNA^Phe from wheat germ is given below:

Nucleotides, principal	Moles/mole of tRNA	Nucleotides, minor	Moles/mole of tRNA	Nucleotides, minor	Moles/mole of tRNA
A	16	ψ	3	7meG	1
G	20	H_2U	3	2me₂G	1
U	7	T	1	OmeG	1
C	19	1meA	1	OmeC	1
		2meG	1	Y*	1

The structure of tRNA^Phe from wheat germ and the products of partial hydrolysis used for its reconstruction are illustrated in Fig. 43 (facing page 138).

Like tRNA^Phe from yeast, the molecule of wheat germ tRNA^Phe consists of 76 nucleotides, 14 of which are minor components. Y* (Y_w in Fig. 43) is very similar to the Y nucleoside found in tRNA^Phe from yeast (RajBhandary et al., 1967), and identified by Nakanishi et al. (1970). However, they differ in certain properties. It is postulated that the differences between the Y residues in yeast and wheat-germ tRNA^Phe are concerned with the side groups, because their fluorescence spectra are identical (Yoshikama et al., 1968). Phenylalanine tRNA from wheat germ is shown in Fig. 44 in the clover-leaf configuration. The model has the standard characteristics; an unusual feature is the presence of a pair formed by two purines (A and G) in the duplex segment joining the ends of the molecule.

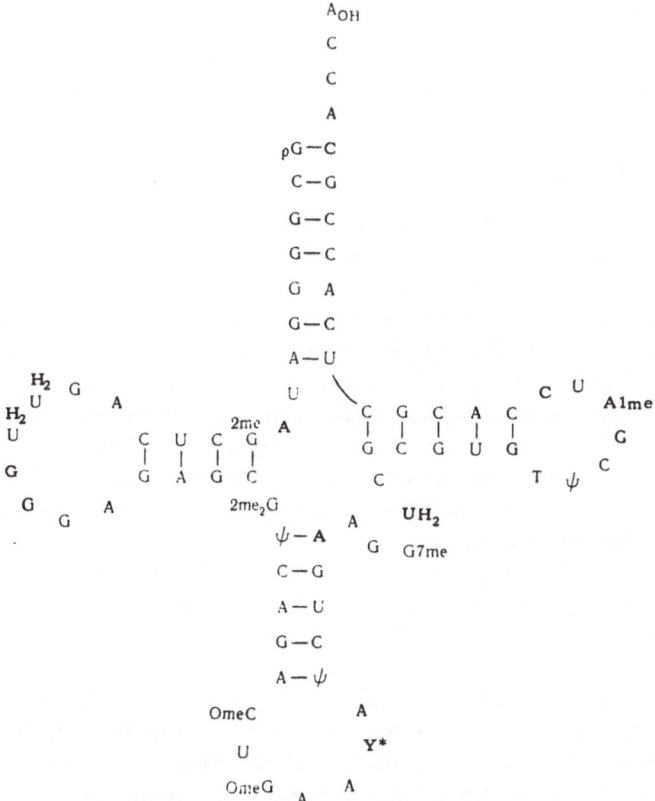

Fig. 44. tRNA^Phe from wheat germ in the clover-leaf configuration.

The phenylalanine tRNAs from bakers' yeast, wheat germ, and *E. coli* are compared on page 202.

Aspartate tRNA from Brewers' Yeast (Keith et al., 1970; Gangloff et al., 1971)

Fractions rich in aspartate-specific tRNA have been isolated from brewers' yeast by countercurrent fractionation in the system of Holley et al. (1963), as was described originally in 1967 (Dirheimer and Ebel, 1967). Final purification of the tRNA^{Asp} was achieved either by chromatography on DEAE-cellulose and subsequent fractionation on hydroxylapatite (Dirheimer et al., 1967), or by chromatography on BD-cellulose at pH 4 (Gangloff et al., 1970).

The tRNA^{Asp} molecule consists of 75 nucleotides, including eight minor components:

Nucleotides, principal	Moles/mole of tRNA	Nucleotides, minor	Moles/mole of tRNA
A	13	ψ	3
G	23	H_2U	2
U	13	T	1
C	18	5meC	1
		1meG	1

Together with tRNA^{Val} from *T. utilis,* tRNA^{Asp} has the shortest chain of all tRNAs which have been studied, and it is the first tRNA to be sequenced with a terminal uridine at the 5′-end.

The products of exhaustive guanyl-RNase hydrolysis were fractionated initially on DEAE-cellulose at pH 7.5 in 7 M urea under the conditions described by Madison et al. (1967a), and after removal of the urea, by high-voltage (800–1200 V) electrophoresis for 8–16 h on DEAE-paper impregnated with 7% formic acid by the method of Sanger et al. (1965) as adapted for analysis of nonradioactive material by Gangloff et al. (1970). The oligonucleotides obtained by the action of pyrimidyl RNase were fractionated only by high-voltage electrophoresis.

The nucleotide composition of the oligonucleotides was determined by hydrolysis in 10% piperidine followed by fractionation in a thin layer of cellulose in the two-dimensional system suggested by Baguley and Staehelin (1969).

To determine the structure of the oligonucleotides, hydrolysis by the complementary RNases and snake-venom phosphodiesterase (after alkaline phosphatase) was used. Determination of the structure of some of the oligonucleotides required partial hydrolysis by snake-venom phosphodiesterase and also the use of micrococcal nuclease and U_2-RNase (Uchida et al., 1970).

Analysis of the oligonucleotides did not yield sufficient overlaps, and accordingly partial hydrolysis by RNases was also carried out. Under the mildest conditions (pH 6.5, 0 °C, Mg^{2+}), guanyl RNase ruptured one bond, leading to the formation of two halves, one of which contained the 5'-terminal pUp and the other the tetranucleotide G-G-A-Gp, formed from the 3'-terminal G-G-A-G-C-C-A_{OH} as the result of removal of the sequence C-C-A_{OH}. The oligonucleotide composition and distribution of oligonucleotides between the two fragments are shown in Table 15.

To determine the complete structure of tRNA^{Asp}, additional partial hydrolyses were carried out with RNases at 0 °C, resulting in different sequences ranging from 17 to 33 nucleotides in length. Their isolation and subsequent analysis gave sufficient overlaps to allow reconstruction of the entire molecule.

The clover-leaf formula for tRNA^{Asp} is illustrated in Fig. 45; 5meC, just as in yeast tRNA^{Phe}, occupies position 28 counting from the 3'-end of the

TABLE 15. Oligonucleotides of Total Ribonuclease Hydrolysates of tRNA^{Asp} from Brewers' Yeast and Their Distribution between Two Large Fragments Formed by Limited Action of RNase T_1

Pyrimidyl-ribonuclease digest			Guanyl-ribonuclease digest		
	moles/mole of tRNA			moles/mole of tRNA	
Oligo(mono)nucleotides	tRNA^{Asp}	fragment	Oligo(mono)nucleotides	tRNA^{Asp}	fragment
		1	2			1	2
A_{OH}	1	—	—	C-C-A_{OH}	1	—	—
Cp	12	3	8	Gp	7	3	4
Up	3	2	1	U-C-1meG > p	1	—	1
ψp	3	2	1	C-Gp	3	1	2
5meCp	1	—	1	U-Gp	2	1	1
1meG-Cp	1	—	1	A-Gp	1	—	1
G-Cp	3	1	2	U-C-Gp	1	—	1
G-Up	4	1	2	H_2U-C-A-Gp	1	1	—
A-A-H_2Up	1	1	—	C-ψ-U-Gp	1	1	—
A-A-Up	1	—	1	C-C-A-Gp	1	—	1
pUp	1	1	—	A-U-5meC-Gp	1	—	1
G-A-Up	1	1	—	A-A-U-Gp	1	1	—
A-G-Up	1	1	—	A-U-A-Gp	1	1	—
G-G-H_2Up	1	1	—	pU-C-C-Gp	1	1	—
A-G-A-Up	1	—	1	U-U-ψA-A-H_2U-Gp	1	1	—
G-G-G-Cp	1	1	—	T-ψ-C-A-A-U-U-C-			
G-G-A-Gp	0	—	1	C-C-C-Gp	1	—	1
A-G-A-A-Up	1	1	—				
G-G-A-G-Cp	1	—	—				
G-G-G-G-Tp	1	—	1				

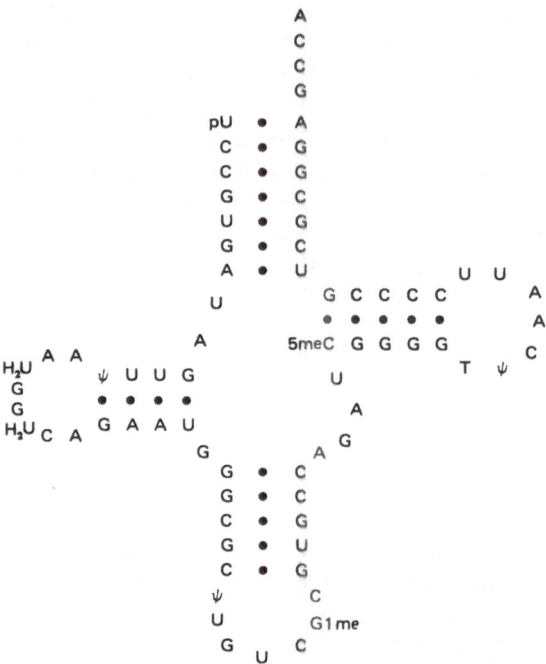

Fig. 45. Nucleotide sequence of yeast tRNA^Asp in the clover-leaf configuration.

molecule and it is included in the duplex segment of the Tψ-branch, in which it forms one of the five pairs. Like the leucine and tryptophan tRNAs of *E. coli*, tRNA^Asp has the sequence A-A-H₂U, and not the sequence A-G-H₂U as in most other tRNAs. This difference makes it impossible for the nucleotides in positions 14 and 15 to pair with ψ54 and C55, as the model of Cramer et al. (1968) envisages. However, the replacement of C in position 47 by U permits the tertiary structure postulated by Levitt (1969), in which a hydrogen bond is formed between A15 and U47. The anticodon of tRNA^Asp is evidently the G-U-C triplet, the location of which is analogous to that of the anticodon in other tRNAs. According to the wobble hypothesis, the G-U-C triplet can pair with both aspartic acid triplets, GAC and GAU. A minor component lies on the 3′-side of the anticodon in tRNA^Asp, but as an exception to an almost universal rule, it is a derivative of guanine, and not of adenine, namely, 1meG.

Tryptophan tRNA from *E. coli* (Strain CAJ64) (Hirsh, 1970)

The structure of the tryptophan tRNA of *E. coli* was determined at the Laboratory of Molecular Biology in Cambridge (Hirsh, 1970). Only a brief notice has so far been published, in which it is stated that the molecule consists of 76 monomers and has the following nucleotide composition:

Nucleotides, principal	Moles/mole of tRNA	Nucleotides, minor	Moles/mole of tRNA
A	14	ψ	1
G	22	H_2U	3
U	11	T	1
C	20	7meG	1
		4thioU	1
		OmeC	1
		mtpA	1

A clover-leaf configuration with standard characteristics can be ascribed to this molecule (Fig. 46). The interesting feature distinguishing tRNA[Trp] is that it has A instead of G in position 15 and U instead of C in position 48; similar exceptions are tRNA[Leu] from *E. coli* (Dube et al., 1970) and tRNA[Asp] from *S. cerevisiae* (Keith et al., 1970). This coordinated replacement (G15→ A15 and C48→U48) is evidence in support of Levitt's suggested model for the tertiary structure of tRNA (Levitt, 1969), in which it is postulated that the nucleotides of tRNA lying in positions 15 and 48 are linked by hydrogen bonds.

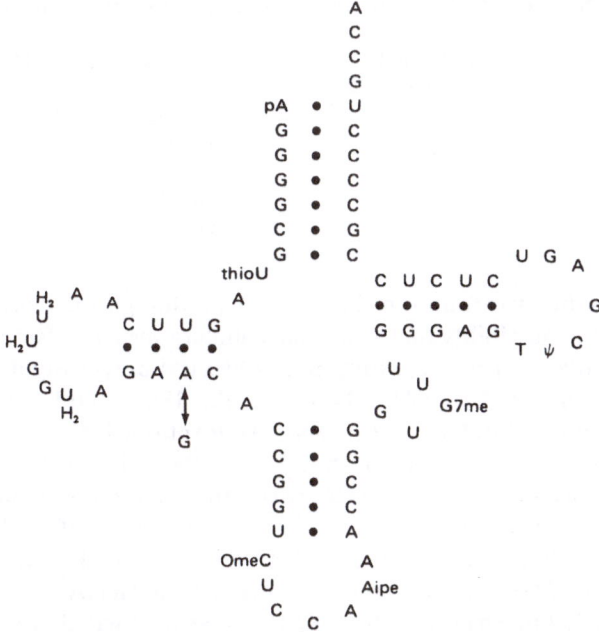

Fig. 46. Nucleotide sequence of tRNA[Trp] from *E. coli* in the clover-leaf configuration.

The structure of the tRNA^{Trp} of both the Su$^+$ and the Su$^-$ strains of *E. coli* has been determined. The only difference between them is that Su$^-$ has G instead of A in position 24. They both have a CCA anticodon, which, in accordance with the wobble hypothesis, can only recognize the codon UGG; strange as it may seem, both tRNAs can read the UGA triplet, though that from Su$^+$ is more effective.

Leucine tRNA from *E. coli* B (Dube et al., 1970)

^{32}P-labeled tRNA^{Leu} was isolated from *E. coli* B, grown in the presence of ^{32}P, by electrophoresis in polyacrylamide gel. Final purification and separation from the only tRNA accompanying tRNA^{Leu} on the polyacrylamide gel were achieved by chromatography on BD-cellulose; the tRNA^{Leu} was eluted by NaCl in a concentration of 0.7 *M*. Enzymic hydrolysis, separation of the hydrolysis products, and their subsequent analysis were undertaken by methods developed previously (Dube et al., 1969a; Dube and Marcker, 1969; Cory and Marcker, 1970). The products of partial ribonuclease hydrolysis were separated by homochromatography (Brownlee et al., 1968). Analysis of a large number of partial hydrolysis products unequivocally established the nucleotide sequence in tRNA^{Leu} (Fig. 47).

The molecule consists of 87 nucleotides, the largest number yet found in a transfer RNA. The tRNA^{Leu} molecule includes the following components:

Nucleotides, principal	Moles/mole tRNA	Nucleotides, minor	Moles/mole tRNA
A	15	ψ	3
G	27	H_2U	3
U	12	T	1
C	24	OmeG	1
		G*	1 (1meG or 2meG)

No 4-thiouridine was found in the molecule, although the workers cited do not rule out the possibility that it may have undergone conversion into uridine during the isolation process, as has been observed previously during the analysis of tRNA$_f^{Met}$ and tRNA^{Met} (Dube et al., 1969a; Dube and Marcker, 1969; Cory and Marcker, 1970). As was mentioned above, unlike most tRNAs which have been studied, in tRNA^{Leu} the G in position 15 from the 5'-end is replaced by A. By contrast with other tRNAs, in which there is a C residue between the Tψ-loop and the extra branch, in tRNA^{Leu} there is a U residue. A similar coordinated substitution is found in the tryptophan tRNA of *E. coli* (Hirsh, 1970) and in the aspartate tRNA of brewers' yeast (Keith et al., 1970). The possibility of pairing, as postulated in Levitt's model (Levitt, 1969), is thus also confirmed in the case of this tRNA.

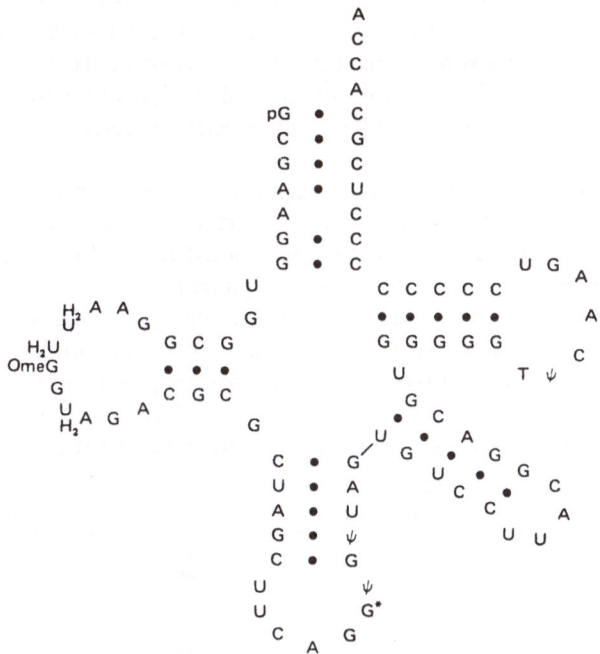

Fig. 47. Nucleotide sequence of tRNA^{Leu} from *E. coli* in the clover-leaf configuration.

"Denaturable" Leucine tRNA from Bakers' Yeast (Kowalski et al., 1971)

The primary structure of tRNA$_3^{Leu}$ of *S. cerevisiae*, i.e., the most important of the three fractions, accounting for 40 % of the leucine-acceptor activity of bakers' yeast (Lindahl et al., 1966), has been determined. This tRNA^{Leu} can be detected in a denatured metastable state, in which its biological and physical properties differ considerably from those of the native form (Fresco et al., 1966). This fact served as the basis for the method used for its isolation (Kowalski and Fresco, 1971). Since it was intended to use Sanger's methods to determine its primary structure, it was necessary to obtain small quantities of the tRNA with high radioactivity. Highly active specimens of tRNA had previously been isolated from *E. coli,* phages, and mammalian cells. Kowalski and Fresco (1971) described an effective method of incorporating ^{32}P into the nucleic acids of bakers' yeast and a method of isolating "denaturable" leucine tRNA in a highly purified state. To enable incorporation of the label to take place throughout the log-phase into all nucleic acids present in the cell, a small quantity of inoculum was grown in the presence of label on the smallest possible quantity of phosphate. The low-molecular-weight RNAs were isolated by the phenolic method and subsequently purified on DEAE-cellulose. The tRNA$_3^{Leu}$ was then converted into the denatured state (Lindahl et al., 1967a) and separated from the main mass of the tRNA on Sephadex G-100. The final purification and, in particular, separation from 5S and residues of other tRNAs were achieved by electrophoresis in polyacrylamide gel (Fig. 48).

The primary structure was determined by Sanger's methods, using homochromatography to fractionate the partial hydrolysis products.

The nucleotide sequence was reconstructed by analogy with previously known structures, following the rules derived from the clover-leaf model and the coding properties of tRNA$_3^{Leu}$. To begin with, only the data from total RNase hydrolysates were used, and the reconstructed formula was then confirmed by results obtained from partial hydrolysis products (Fig. 49).

The molecule of tRNA$_3^{Leu}$ consists of 85 nucleotides, including 12 minor components, and it has the following nucleotide composition:

Nucleotides, principal	Moles/mole of tRNA	Nucleotides,[a] minor	Moles/mole of tRNA
A	21	ψ	3
G	19	H_2U	2
U	15	T	1
C	18	OmeG	1
		5meC	1
		C*, probably acetylated derivative of C	1
		G*, probably 2meG	1
		G$^+$, probably 1meG	1
		G**, probably 2me$_2$G	1

[a]Minor components were identified mainly from their electrophoretic mobility and by analogy with other tRNAs.

It is assumed that the low mobility of tRNA$_3^{Leu}$ from *S. cerevisiae* and of tRNA^{Leu} from *E. coli* (Dube et al., 1970), which have no significant features of structural similarity, is the result of the comparatively large size of their molecules (85 and 87 nucleotides respectively).

In the clover-leaf configuration of tRNA^{Leu} (Fig. 50) there is an unusually large number of A-U base pairs. The structure of the Tψ-loop is completely identical with that of tRNA^{Tyr} and tRNA^{Met} of *E. coli*, suggesting that this loop participates in functions common to all transfer RNAs.

The anticodon CAA corresponds to the coding properties previously established for tRNA^{Leu} (Lindahl et al., 1967b). The structure of the anticodon branch of tRNA^{Leu} is an exception in that a derivative of G, and not of A, lies on the 3'-side of the anticodon. Otherwise, tRNA^{Leu} does not infringe any of the previously established rules but, on the contrary, conforms to them. The mere fact that the structure of its molecule could be reconstruct-

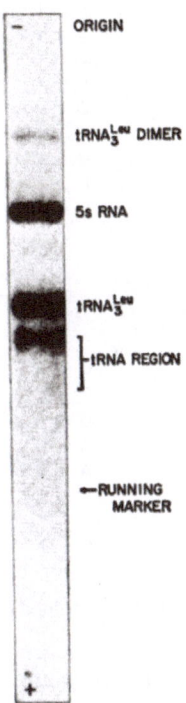

Fig. 48. Electrophoresis of tRNA$_3^{Leu}$ -enriched material on 10% polyacrylamide gel.

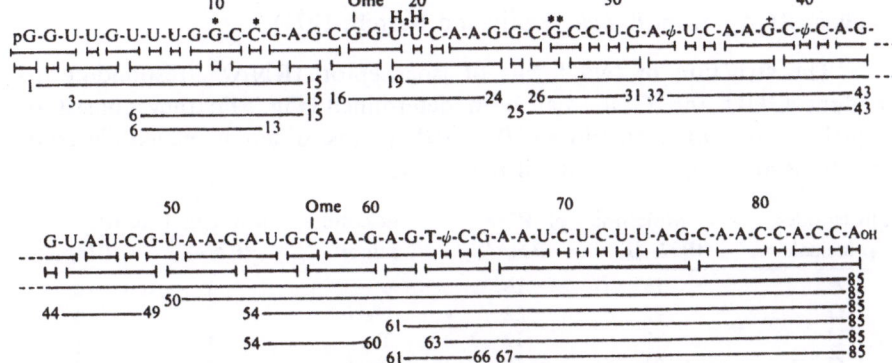

Fig. 49. Nucleotide sequence of tRNA$_3^{Leu}$ from bakers' yeast and partial hydrolysis products used for its reconstruction.

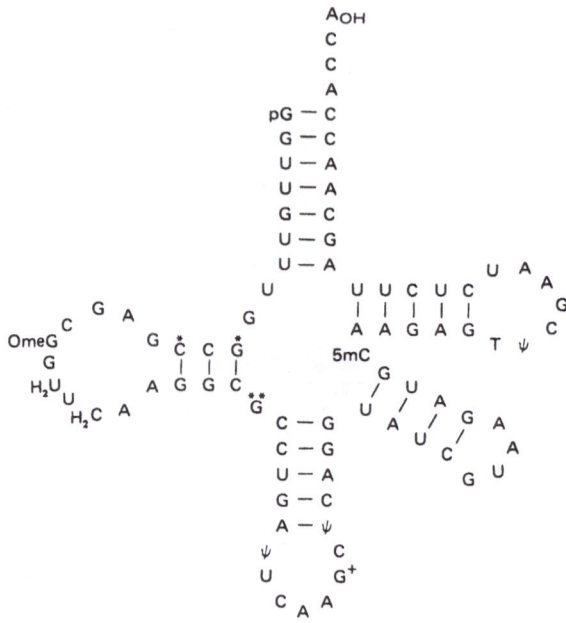

Fig. 50. Nucleotide sequence of tRNA$_3^{Leu}$ folded into the clover-leaf configuration.

ed on the basis of total RNase hydrolysates, using the rules established for other tRNAs, is sufficient confirmation of this fact.

I give below the nucleotide composition and clover-leaf configurations of tRNAs whose structure has been established, but by methods whose details still await publication.

Valine tRNA$_2$ from *E. coli* (Yaniv and Barrell, 1971)

The structure of two forms of isoacceptor tRNA$_2^{Val}$, responding to codons GUU and GUC, has been determined (Fig. 51); they consist of equal numbers of nucleotides (77), but they have different nucleotide compositions affecting three pairs of nucleotides:

Nucleotides, principal	Moles/mole of tRNA		Nucleotides, minor	Moles/mole of tRNA	
	tRNA$_{2A}^{Val}$	tRNA$_{2B}^{Val}$		tRNA$_{2A}^{Val}$	tRNA$_{2B}^{Val}$
A	13	16	ψ	1	1
G	25	22	H$_2$U	4	4
U	9	12	T	1	1
C	21	18	thioU	1	1
			7mcG	1	1
			N	1	1

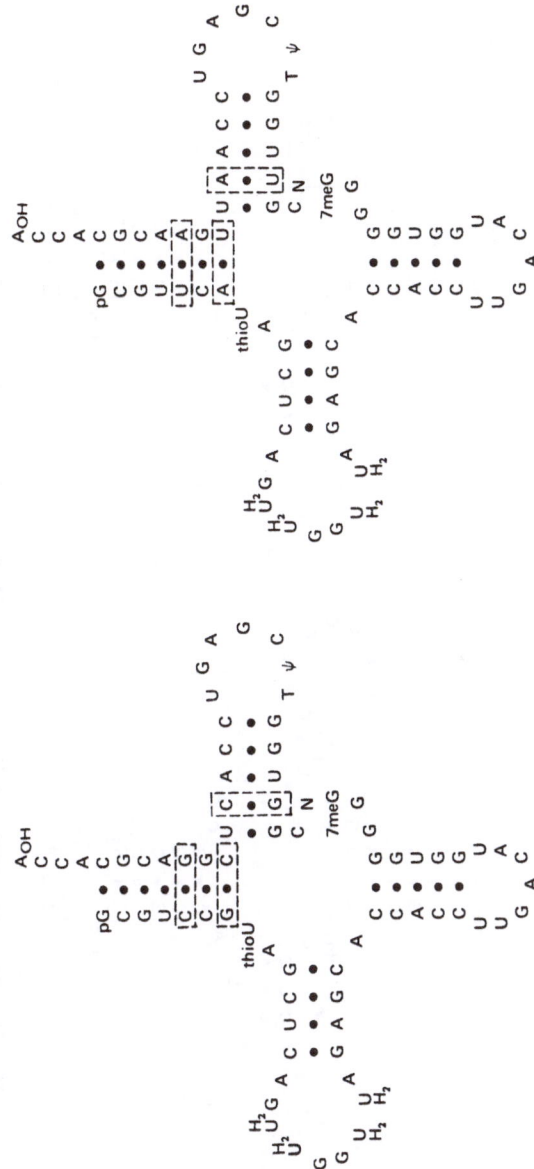

Fig. 51. Nucleotide sequence of tRNA$_{2A}^{Val}$ and tRNA$_{2B}^{Val}$ of *E. coli* in the clover-leaf configuration.

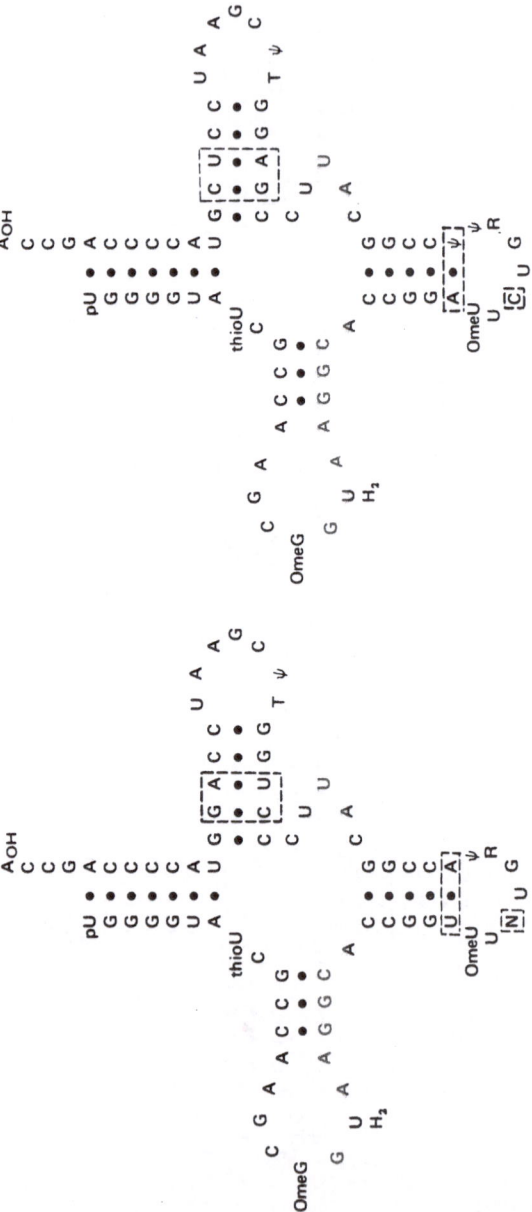

Fig. 52. Nucleotide sequence of tRNA$_I^{Gln}$ and tRNA$_{II}^{Gln}$ from *E. coli* in the clover-leaf configuration.

Glutamine tRNA$_I$ and tRNA$_{II}$ from *E. coli* (Folk and Yaniv, 1972)

The structural formulas of the two isoacceptor glutamine tRNAs (tRNA$_I^{Gln}$ and tRNA$_{II}^{Gln}$) are given in Fig. 52. Their molecules, like the molecules of tRNA$_{2A}^{Val}$ and tRNA$_{2B}^{Val}$, differ in three pairs of nucleotides, and also in the first nucleotide of the anticodon. The nucleotide composition of tRNA^{Gln} (altogether 78 nucleotides) is given below:

Nucleotides, principal	Moles/mole of tRNA tRNA$_I^{Gln}$	tRNA$_{II}^{Gln}$	Nucleotides, minor	Moles/mole of tRNA tRNA$_I^{Gln}$	tRNA$_{II}^{Gln}$
A	14	14	ψ	2	3
G	22	22	H$_2$U	1	1
U	10	9	T	1	1
C	23	24	thioU	1	1
			OmeG	1	1
			OmeU	1	1
			R	1	1
			N	1	—

Glycine tRNA from *E. coli* (Squires and Carbon, 1971)

The clover-leaf configuration of the tRNA^{Gly} from *E. coli* is shown in Fig. 53. The molecule consists of 76 nucleotides, including seven minor components (although the evidence for the presence of 4thioU is not conclusive):

Nucleotides, principal	moles/mole of tRNA	Nucleotides, minor	moles/mole of tRNA
A	14	T	1
G	23	H$_2$U	3
U	11	ψ	1
C	21	4thioU	(1)
		7meG	1

This exhausts for the moment the list of transfer RNAs whose structure has been established. Our next task will be to compare these structures. The detection of the features they have in common and the differences among them is an essential step toward the understanding of the functional topography of this class of nucleic acids.

COMMON FEATURES AND DIFFERENCES IN STRUCTURE OF INDIVIDUAL tRNAs

The existence of common features in the architecture of molecules of different tRNAs, postulated initially as a logical hypothesis, subsequently progressing to a working hypothesis in structural research, is now an experi-

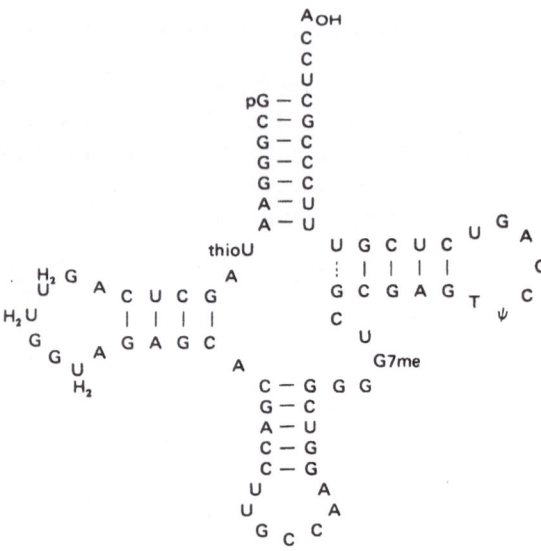

Fig. 53. Nucleotide sequence of tRNA^{Gly} from *E. coli* in the clover-leaf configuration.

mentally verified fact. Let us compare the described structures of those tRNAs which have been elucidated and examine the extent of this similarity.

The primary structures of individual tRNAs have already been compared many times and from many different points of view (Jukes, 1966; Zachau et al., 1966d; Zachau, 1968a; Baev et al., 1967a; Madison, 1968; Ebel, 1968), but until recently most attention has been paid to yeast tRNAs. At the present time, with the accumulation of much new material and, in particular, with the elucidation of the structure of transfer RNAs from other sources (*E. coli* and mammalian and plant tissues), and also of a number of isoacceptor tRNAs, such a comparison not only has not lost its previous importance but, on the contrary, has acquired fresh interest. Recently two detailed surveys have been published, one based on 15 (Zachau, 1969), the other of 18 (Stae-helin, 1971) known structures.

The structural unity is apparent, to begin with, in the existence of segments of identical or, at least, similar structure in the molecules of individual tRNAs. These segments are shown in Table 16 as sequences containing H_2U, the anticodon, and the "universal" oligonucleotide.

An even more striking manifestation of the universality of tRNA architecture is the equality of the distances between certain points of the tRNA molecule when unfolded into a straight line. This is clear from the results in Table 17, where the length of the molecules and their 5'-halves and the distance between certain points of the molecule, unfolded into a straight line, are expressed as numbers of nucleotides.

TABLE 16. Homologous Units in the Molecules of Transfer RNAs

Specificity	Source	Unit containing			
		5'-end	Dihydrouridine	Anticodon[a]	"Universal" oligonucleotide
Ala	S. cerevisiae	pGp	A-G-H_2U-C-G-G-H_2U	U-I-G-meI	G-T-ψ-C-G-A
Ser$_1$	"	pGp	A-G-H_2U-OmeG-G-(H_2U)$_2$	U-I-G-A-ipeA	G-T-ψ-C-A-A-
Ser$_2$	"	pGp	A-G-H_2U-OmeG-G-(H_2U)$_2$	U-I-G-A-ipeA	G-T-ψ-C-G-A
Tyr	"	pCp	A-G-(H_2U)$_2$-OmeG-G-(H_2U)$_3$	U-G-ψ-A-ipeA	G-T-ψ-C-G-1meA
Val$_1$	"	pGp	A-G-H_2U-C-G-G-(H_2U)$_2$	U-I-A-C-A	G-T-ψ-C-G-1meA
Phe	"	pGp	A-G-(H_2U)$_2$-G-G-G	U-OmeG-A-A-Y	G-T-ψ-C-G-1meA
Asp	"	pUp	A-A-H_2U-G-G-H_2U	U-G-U-C-1meG	G-T-ψ-C-A-A
Leu$_3$	"	pGp	G-C-OmeG-G-(H_2U)$_2$	U-C-A-A-G*	G-T-ψ-C-G-A
Val$_1$	T. utilis	pGp	A-G-(H_2U)$_2$-G-G-H_2U	U-I-A-C-A	G-T-ψ-C-G-1meA
Ile	"	pGp	A-G-(H_2U)$_2$-G-G-(H_2U)$_2$	U-I-A-U-thrA	G-T-ψ-C-G-1meA
Tyr	"	pCp	A-G-(H_2U)$_2$-OmeG-G-(H_2U)$_3$	U-G-ψ-A-ipeA	G-T-ψ-C-G-1meA
Tyr Su$_{+1,2}^{-}$	E. coli	pGp	—	U-G*-U-A-mtpA	G-T-ψ-C-G-A
Tyr Su$_{III}^{+}$	"	pGp	—	U-C-U-A-mtpA	G-T-ψ-C-G-A
fMet	"	pCp	A-G-C-C-U-G-G-H_2U	U-C-A-U-A	G-T-ψ-C-A-A
Met	"	pGp	A-G-(H_2U)$_2$-OmeG-G-(H_2U)$_2$	U-C*-A-U-A*	G-T-ψ-C-G-A
Val$_1$	"	pGp	A-G-C-H_2U-G-G-G	U-oacU*-A-C-meA	G-T-ψ-C-G-A
Phe	"	pGp	A-G-H_2U-C-G-G-H_2U	U-G-A-A-mtpA	G-T-ψ-C-G-A
Gln$_1$	"	pUp	G-C-OmeG-G-H_2U-A-A-G	U-N*-U-G-R*	G-T-ψ-C-G-A
Gln$_2$	"	pUp	G-C-OmeG-G-H_2U-A-A-G	U-C-U-G-R*	G-T-ψ-C-G-A
Gly	"	pGp	A-G-(H_2U)$_2$-G-G-H_2U	U-G-C-C-A	G-T-ψ-C-G-A
Leu	"	pGp	A-A-(H_2U)$_2$-OmeG-G-H_2U	U-C-A-G-G-*	G-T-ψ-C-A-A
Trp	"	pGp	A-A-(H_2U)$_2$-G-G-H_2U	U-C-C-A-mtpA	G-T-ψ-C-G-A
Val$_2$	"	pAp	A-G-(H_2U)$_2$-G-G-(H_2U)$_2$	U-G-A-C-A	G-T-ψ-C-G-A
Ser	Rat liver	pGp	A-G-H_2U-OmeG-G-(H_2U)$_2$	U-I-G-A-ipeA	G-T-ψ-C-G-1meA
Phe	Wheat germ	pGp	A-G-(H_2U)$_2$-G-G-G	U-OmeG-A-A-Y*	G-T-ψ-C-G-1meA

[a] Unidentified nucleosides are marked by asterisks.

TABLE 17. Characteristic Parameters of Molecules of Transfer RNAs

Parameter	Saccharomyces cerevisiae							Torulopsis utilis			Escherichia coli										Rat liver	Wheat germ
	Ala	Asp	Leu$_3$	Phe	Ser	Tyr	Val	Ile	Tyr	Val	fMet	Gln	Gly	Leu	Met	Phe	Trp	Tyr	Val	Val$_2$	Ser	Phe
Length																						
of whole molecule	77	75	85	76	85	78	77	77	78	75	77	78	76	87	77	76	76	85	76	77	85	76
of 5'-half (to first nucleotide of anticodon)	36	34	35	34	35	36	35	35	36	35	34	36	34	35	35	34	34	35	34	35	34	34
Length of segment																						
from 3'-end to T	23	23	23	23	23	23	23	23	23	23	23	23	23	23	23	23	23	23	23	23	23	23
from 5meC* to T	—	7	7	—	7	7	7	—	7	7	—	—	—	—	—	—	—	—	—	—	7	—
from 5'-end to 1meG	9	—	—	—	—	—	9	9	9	—	—	—	—	—	—	—	—	—	—	—	—	—
from 5'-end to H$_2$U**	18	16	19	16	16	16	16	16	16	16	—	19	16	16	16	16	16	—	17	16	16	16
from 5'-end to OmeG	—	—	17	—	17	18	—	—	18	—	—	17	—	18	18	—	17	—	—	—	17	—
from first nucleotide of anticodon to																						
2meG	9	—	9	9	9	9	—	9	9	—	—	—	31	31	—	31	—	—	—	31	9	9
from 3'-end to 7meG	—	—	—	31	—	—	—	—	—	—	31	—	—	—	31	31	—	—	31	31	—	31
from 5'-end to thioU	—	—	—	—	—	—	—	—	—	—	8	8	8	8	8	8	8	8	8	8	—	—
from 5'-end to 2meG	—	—	10	10	10	10	—	10	10	10	—	—	—	—	—	—	—	—	—	—	—	10

TABLE 18. Comparison of the Primary Structures of Transfer RNAs

Saccharomyces cerevisiae							*Torulopsis utilis*			*Escherichia coli*										Rat liver	Wheat germ
Ala	Asp	Leu$_3$	Ser	Tyr	Val$_1$	Phe	Val$_1$	Ile	Tyr	Tyr Su$_{\text{III}}$	Gln$_1$[a]	Gly	Leu	Met	fMet	Phe	Trp	Val$_1$	Val$_{2A}$[b]	Ser	Phe
pG	pU	pG	pG	pC	pG	pG	pG	pG	pC	pG	pU	pG	pG	pG	pC	pG	pA	pG	pG	pG	pG
G	C	G	G	U	G	C	G	G	U	G	G	C	C	G	G	C	G	G	C	U	C
G	C	U	C	C	U	U	U	U	C	U	G	G	G	C	C	G	G	G	G	A	G
C	C	U	A	U	U	G	U	C	U	G	G	G	A	U	G	C	G	U	U	G	G
G	U	G	A	C	U	A	U	C	C	G	G	G	A	A	G	G	G	G	C(U)	U	G
U	G	U	C	G	C	U	C	C	G	G	U	A	G	C	G	G	C	A	C	C	G
G	A	U	G	G	G	U	G	G	G	G	A	A	G	G	A	A	G	A	A	G	A
U	U	U	U	U	U	U	U	U	U	U*	thioU	thioU	U	U*	U*	4thioU	thioU	4thioU	thioU	U	U
—	A	G	—	A	—	A	—	—	—	U*	C	A	G	A	—	A	A	A	A	—	A
1meG	G	G*	G	—	1meG	G	1meG	G	1meG	C	G	G	C	G	G	G	G	G	G	G	—
G	U	C	G	2meG	G	2meG	G	G	2meG	C	—	C	C	C	—	G	—	U	—	C	2meG
C	U	C*	C	C	U	C	U	C	C	C	C	U	G	C	A	C	U	C	U	C	C
G	φ	G	—	—	U	—	—	C	—	A	A	A	U	G	C	U	C	U	C	—	U
C	A	A	acC	C	C	C	C	C	C	C	A	A	A	C	C	C	A	C	A	acC	C
G	A	G	—	—	—	—	—	—	—	G	G	A	—	—	—	A	—	G	—	—	—
U	H₂U	C	G	A	φ	—	φ	—	A	G	C	H₂U	H₂U	—	—	H₂U	—	H₂U	G	G	—
A	G	OmeG	A	A	A	A	A	A	A	A	OmeG	H₂U	H₂U	A	A	H₂U	A	H₂U	A	A	A
G	G	G	G	G	G	G	G	G	G	G	G	OmeG	G	G	G	G	G	G	G	G	G
H₂U	—	H₂U	H₂U	H₂U	H₂U	H₂U	H₂U	H₂U	—	H₂U	G	G	U(H₂U)	C	H₂U	G	C	G	H₂U	H₂U
C	H₂U	H₂U	—	H₂U	C	H₂U	H₂U	H₂U	C	A	H₂U	H₂U	H₂U	C	C	H₂U	H₂U	H₂U	—	—	H₂U
G	—	OmeG	OmeG	G	G	G	G	G	OmeG	OmeG	A	—	—	OmeG	U	G	—	G	—	OmeG	G
G	G	C	G	G	G	G	G	G	G	G	G	A	A	G	G	G	—	G	—	G	G
—	A	A	—	H₂U	—	G	—	—	H₂U	C	G	G	—	G	—	A	G	A	—	G	G
—	G	H₂U	H₂U	H₂U	—	H₂U	H₂U	H₂U	—	C	C	A	A	H₂U	—	H₂U	G	A	G	H₂U	—
H₂U	A	G	H₂U	H₂U	H₂U	—	C	H₂U	H₂U	A	G	C	H₂U	H₂U	H₂U	—	A	G	A	H₂U	—
A	A	G	A	A	A	A	A	A	A	A	G	C	C	A	A	A	A(G)	A	G	A	A
G	U	C	A	A	U	G	U	A	A	A	C	—	C	G	G	G	C	—	—	C	A
C	G	G*	G	G	A	G	G	G	G	G	A	A	G	A	C	A	A	—	A	C	A
G	G	C	G	C	G	G	G	G	G	G	C	C	C	G	U	G	C	G	C	G	G
C	G	C	C	C	C	C	C	C	C	C	C	G	U	C	C	C	C	C	C	C	C
2me₂G	C	U	2me₂G	2me₂G	—	2me₂G	—	2me₂G	2me₂G	G	G	A	A	A	G	A	G	A	A	2me₂G	2me₂G
C	G	G	A	C	A	C	A	—	—	A	G	C	C	U	G	G	C	C	C	A	—
U	C	A	A	A	φ	C	φ	φ	φ	G	U(A)	C	C	A	C	U	C	C	C	φ	φ
C	φ	φ	A	A	C	A	C	G	C	C	OmeU	U	U	U	G	G	OmeC	U	U	G	C
—	U	U	—	—	U	—	U	G	A	U	U	U	—	G	G	U	C	U	G	A	A
C	G	C	G	G	G	G	G	U	G	N*(C)	G	C	C	—	C	—	C	G	A	G	—
C	U	A	A	A	A	C	A	A	A	C	U	C	A	A	A	C	A	C	A	A	A
U	C	A	φ	C	φ	OmeC	φ	C	C	C	G	C	C	G	C	OmeC	φ	A	—	3meC	OmeC
U	1meG	G*	U	U	U	U	U	U	U	U	R*	A	G*	U	U	U	mtpA	U	A	U	U
I	C	C	I	G	I	OmeG	I	I	G	G	φ	A	φ	C*	C	G	A	oncU	U	I	OmeG
G	G	φ	G	φ	A	A	A	A	φ	U	A(φ)	G	A	A	A	A	A	G	G	G	A
C	U	C	A	A	C	A	C	U	A	C	G	φ	U	U	A	C	C	C	G	A	A
meI	G	A	ipeA	ipeA	A	Y	A	thrA	ipeA	mtpA	C	U	U	A*	A	mtpA	C	6meA	U	ipeA	Y
φ	C	A	A	A	C	A	C	A	A	A	G	C	A	A	A	A	G	A	G	A	A
G	C	G	φ	φ	G	φ	G	C	φ	φ	G	G	G	φ	C	φ	G	G	G	Omeφ	φ
G	A	U	C	C	C	5meC	C	G	C	C	—	G	U	G	C	C	—	G	C	C	C
G	—	A	U	A	U	A	A	C	U	U	—	—	G	A	C	C	—	A	G	C	U
A	—	U	U	G	G	G	C	G	C	G	C	—	U	U	G	A	C	—	A	G	A
G	—	C	U	G	A	G	A	A	A	C	A	—	C	G	A	C	—	G	—	U	A
A	G	G	OmeU	A	A	A	A	A	A	C	U	G	C	A	G	G	—	U	G	OmeU	A
G	—	U	G	G	C	G	C	G	C	U	—	U	G	G	U	U	—	G	G	G	G
G	—	A	G	A	7meG	7meG	—	A	A	U	—	—	—	U	7meG	7meG(A)	7meG	—	7meG	G	7meG
—	A	A	G	—	—	—	—	—	—	C	—	—	A	—	—	—	—	—	—	G	—
—	G	C	—	—	—	—	—	A	—	A	C	7meG	C	—	—	—	7meG	—	7meG	G	—
—	A	U	—	—	—	—	—	—	—	G	—	G	—	U	—	—	—	—	—	U	—
—	U	U	U(C)	—	—	—	—	U(C)	—	C(A)	—	U	G	—	C	—	U	—	N*	3meC
—	G	U	—	—	—	—	—	—	—	G	—	A	—	G	—	—	—	—	—	U	—
—	—	5meC	G	—	—	—	—	—	—	A	—	C	C	—	—	U	—	C	—	C	—
—	—	—	C	—	—	—	—	—	—	C	—	G	—	—	—	—	—	—	—	C	—
—	5meC	A	C	—	—	—	—	—	—	U	C	G	U	—	—	G	—	G	—	C	—
—	G	A	C	—	—	—	—	—	—	G(C)	C	C	—	—	—	G	—	G(U)	—	C	—
U*	G	G	G	H₂U	H₂U	U	—	H₂U	H₂U	U	A(U)	G	G	X	U	X	G	U	U	G	H₂U
C	G	A	5meC	5meC	5meC	C	5meC	5meC	5meC	C	G	A	G	C	C	C	A	C	G	5meC	C
U	G	G	G	G	C	C	C	A	A	G	G	G	A	—	C	G	G	G	G	G	C
C	T	T	C	G	C	U	C	G	G	A	T	T	G	C	—	U	T	G	T	C	C
C	φ	φ	A	A	C	C	C	C	G	A	φ	C	φ	T	A	—	U	ψ	C	A	G
G	C	C	G	C	A	U	A	A	C	C	C	φ	G	G	G	C	G	C	C	C	U
G	A	G	G	G	G	G	G	G	G	G	G	C	G	G	G	G	G	G	G	G	G
T	A	A	T	T	T	T	T	T	T	T	A	A	A	T	T	T	T	A	T	T	T
φ	U	A	φ	φ	φ	φ	φ	φ	φ	φ	A	G	A	φ	φ	φ	G	φ	G	φ	φ
C	U	U	C	C	C	C	C	C	C	C	U	U	C	C	C	U	C	U	U	C	C
G	C	C	G(A)	G	G	G	G	G	G	C	C	U	G	A	G	C	G	C	C	G	G
A	C	C	U	A	1meA	1meA	1meA	1meA	1meA	A	C	U	C	A	A	A	U	A	1meA	1meA
U	C	C	G(A)	U	C	U	U	U	A	A	U(A)	G	C	A	A	U	C	A	A	U	C
U	C	U	G	U	C	U	C	C	U	U	C(G)	G	C	U	U	U	C	C(A)	U	U	C
C	G	U	C	C	C	C	C	C	C	C	G	U	C	C	C	C	C	C	C	C	C
C	U	A	C	G	U	A	U	U	G	C	U	U	C	C	C	C	C	C(U)	C	C	A
G	C	G	U	C	G	C	G	G	G	U	A	U	C	C	G	G	G	G	U	G	C
G	G	C	G	C	G	G	G	G	C	C	C	C	C	G	G	A	C	U	G(A)	C	C
A	C	A	C	C	G	G	U	G	C	C	C	C	C	C	G	C	G	C	A	C	C
C	G	A	A	C	C	C	A	C	C	C	C	C	U	C	C	U	C	A	C	C	U
U	G	C	G	G	A	A	U	G	G	C	C	C	U	G	C	C	U	U	G	G	G
C	A	C	U	G	A	A	U	A	G	G	C	A	C	G	C	C	C	C	A	A	A
G	A	U	U	G	G	U	G	G	A	C	U	G	U	C	C	C	U	G	A	C	A
U	—	—	—	A	C	A	A	G	C	A	—	A	G	G	G	G	—	C	—	U	C
C	—	—	—	U	A	U	G	U	C	C	—	C	C	C	—	—	—	C	—	A	C
C	—	—	—	C	U	G	C	C	G	C	—	C	A	C	—	—	—	—	—	C	C
A	—	—	—	G	A	A	A	A	A	A	—	—	A	A	A	—	—	A	—	G	A

C
C
A$_{\text{OH}}$

[a]Parenthetical entries refer to Gln$_2$ where different from Gln$_1$.

[b]Parenthetical entries refer to Val$_{2B}$ where different from Val$_{2A}$.

If the length of the molecule is used as the basis, the structures at present known can be divided into two groups. The chain length of the great majority of tRNAs ranges from 75 to 78 monomers; only 5 tRNAs (tRNA^{Ser} from yeast and rat liver, tRNA^{Leu} from yeast and *E. coli,* and tRNA^{Tyr} from *E. coli*) have a longer molecule, consisting of 85 to 87 nucleotides. Visual inspection of the loops will show that this difference is due mainly to the size of the extra branch (see below), which consists in the first case of only four or five nucleotides, but in the second of 13 to 15 nucleotides, some of which form pairs. The length of the 5′-halves of the molecules of all transfer RNAs is about the same, varying only from 34 to 36 nucleotides. The differences in the lengths of individual tRNAs are thus confined to the 3′-half and, as already mentioned, are due to differences in the extra branch.

The fixed distances between certain nucleotides of different tRNAs, which can be identified by their structure (T, 5meC, H_2U, etc.), function (nucleotides of the anticodon), or position (terminal nucleotides) indicate the existence of three-dimensional organization of the tRNA molecules incorporated in their primary structure. In Table 17 the reference points are nucleotides occurring in the homologous units shown in Table 16; it thus follows that the position of the homologous units themselves in tRNA molecules must also be rigidly fixed.

The unity of structure of the tRNAs is clearly visible from examination of Table 18, where the formulas of the tRNAs so far studied are given. The identical units are placed in identical positions; this can be done by inserting a dash in the shorter tRNAs to replace the absent nucleotides.

Examination of Table 18 shows that the sequence G-T-ψ-C is not the only universal component; the U preceding the anticodon and certain other nucleotides also are universal.

There are isolated exceptions to some of the existing rules; for example, until recently the dinucleotide AG in the H_2U-loop was regarded as "universal," but now three exceptions are known, in which the dinucleotide AA occurs at the corresponding place in the chain (yeast tRNA^{Asp} and tRNA^{Leu}, the tRNA^{Trp} from *E. coli*), so that the only remaining universal component is the A which occupies position 14 from the 5′-end of the molecule.

It is interesting to note that segments which are similar or identical in all cases or in several tRNAs are located mainly in unpaired regions of the clover-leaf model. The similarity becomes even more obvious if modifications are omitted and the minor components are replaced by the corresponding principal nucleotides. This was the procedure followed earlier by Jukes (1966), who compared the primary structures of four yeast tRNAs in an attempt to show how they could have arisen from a single common prototype as the result of mutations. Jukes suggested that the original tRNA consisted of a larger number of nucleotides, and that the shorter tRNAs were formed

TABLE 19. Minor Components of Transfer RNAs (moles/mole tRNA)

	Saccharomyces cerevisiae							Torulopsis utilis			Escherichia coli															Rat liver	Wheat germ
											Tyr		fMet		Met				Gln								
Nucleoside	Ala	Ser	Tyr	Val_1	Phe	Asp	Leu_3	Val_1	Ile	Tyr	Su^+_{iii}	$\overline{Su}_{iii}$	1	2	1	2	Phe	Val_1	1	2	Gly	Leu	Trp	Val_2	Ser	Ser	Phe
ψ	2	3	3	4	2	3	3	4	2	4	2	2	1	1	2	2	3	2	2	3	1	3	1	1	2	2	3
H_2U	2	3	6	4	2	2	2	3	5	6	—	—	1	1	3	4	2	1	1	1	1	3	1	4	3	3	3
T	1	1	1	1	1	1	1	1	1	1	1	1	1	1	1	1	1	1	1	1	1	1	1	1	1	1	1
I	1	—	1	1	1	—	1	1	1	1	—	—	—	—	—	—	1	—	—	—	1	—	—	1	1	—	1
meI	1	—	1	—	1	—	—	—	1	—	—	—	—	—	—	—	—	—	—	—	—	—	—	—	—	—	—
5meC	—	1	1	—	2	—	1	—	1	1	—	—	—	—	—	—	—	—	—	—	—	—	—	—	1	1	—
3meC	—	—	—	—	—	—	—	—	—	1	—	—	—	—	—	—	—	—	—	—	—	—	—	—	2	—	—
acC	1	—	—	1	—	—	1(?)	—	—	—	—	—	—	—	—	—	—	—	—	—	—	—	—	—	1	—	—
1meG	—	1	—	1	1	1(?)	1(?)	—	—	1	—	—	—	—	—	—	—	—	—	—	1(?)	—	—	—	—	—	—
2meG	—	—	1	1	—	1	1(?)	1	—	1	—	—	—	—	—	—	—	—	—	—	—	—	—	—	—	—	1
$2me_2G$	1	1	1	1	—	1	1(?)	—	1	1	—	—	—	—	—	—	—	1	—	—	—	1	—	—	—	—	1
7meG	—	—	—	1	—	1	—	—	1	—	—	—	—	—	—	1	—	—	—	—	1	—	1	1	—	1	1
1meA	—	1	—	1	—	—	—	—	—	—	—	—	1	—	1	—	1	1	—	—	—	1	—	—	1	1	1
6meA	—	—	—	—	—	—	—	—	—	—	—	—	—	—	—	—	1	1	—	1	—	—	—	—	—	—	—
ipeA	—	1	1	—	—	—	—	—	1	1	1	1	—	—	—	—	—	—	—	—	—	—	—	—	1	—	—
mtpA	—	—	—	—	—	—	—	—	—	—	1	1	—	—	—	—	1	—	—	—	—	1	—	—	—	—	—

	4thioU	OmeG	OmeU	OmeC	Omeψ	thrA	C*	G*	A*	Y	X	oacU	R	Total
		1		1						1				14
		1	1	1										17
	1	1				1								9
	1			1										9
		1												9
	1													7
	1	1	1									1		9
	1	1	1		1							1		9
	1										1			7
	1								1		1			10
	1	1				1		1		1				13
	1	1				1		1		1				12
	1			1										5
	1			1										6
	2	1					1							8
	2	1												7
		1												18
					1									13
														12
		1												12
														8
		1		1					1					14
														14
		1												16
		1	1											14
														9
Total	9	14												

from it as a result of deletions. As Madison (1968) rightly points out, the opposite hypothesis could be true, or in other words the original tRNA could have contained fewer nucleotides (76–78), and the corresponding gene could subsequently have increased in size through the addition of extra nucleotides to the DNA.

While stressing the features of structural unity of the tRNAs, the existence and role of their individuality must also be mentioned. The results given in Table 16 and, in particular, in Table 18 demonstrate the numerous differences existing between individual tRNAs; the principle of "unity in variety," which is often observed in the phenomena of life, is also manifested in the structure of biopolymers.

Among the factors which must play an important role in the structure and function of tRNA the minor components must be mentioned. The minor components of the transfer RNAs so far studied are compared in Table 19; only two minor components, ψ and T, are found in every tRNA, and many of these compounds are characteristic of only a few tRNAs. The view that the minor components give the tRNA its features of individuality is evidently correct: each tRNA is distinguished by its own specific assortment of minor nucleotides. The list of these minor components is probably still far from complete, and the study of almost every new tRNA brings the discovery of increasingly complex and hitherto unknown derivatives of the nucleic-acid bases. To create variety against the background of a single structural plan, the device of modification of the structural components of the polynucleotide chain is thus extensively used in nature.

Analysis of the oligonucleotides of RNase hydrolysates of total tRNA and calculation of the frequency of nucleotide pairs led to the conclusion that the minor components show a tendency toward grouping (Venkstern et al., 1963b). Examination of the structures of invididual tRNAs shows that this conclusion corresponds to the facts. Of the nine minor components in tRNA[Ala] from *S. cerevisiae,* four are arranged in pairs; in tRNA[Ser] and tRNA[Phe] from yeast, six of the 14 minor components are in pairs; etc. The H_2U component frequently forms sequences of two or even three residues. In the extra loop (see below) H_2U frequently adjoins 5meC or 7meG; one ψ residue in all tRNAs is next to T.

A striking feature is the analogous arrangement of the minor components in different tRNAs. Some minor components, mainly those which cannot take part in hydrogen bonding, such as 1meG, 2me$_2$G, meI, and so on, are located mainly in the loops and in the sequences linking the paired segments. Many minor components occupy an absolutely identical position (Table 18), and in those tRNAs which do not contain them, the unmodified base is frequently found there. (In some cases another nucleotide is found there; C often replaces H_2U, and A often replaces 2me$_2$G.) The disposition of the minor components in the clover leaf, as Madison et al. (1966b) point out, suggests that the specificity of enzymes whose function it is to modify nucleotides is determined not only by the sequence of the bases in which the particular nucleotide occurs, but also by the three-dimensional structure of the corresponding part of the molecule. This hypothesis can now be taken as proved, in view of work confirming that the spatial organization of the molecule is the basic factor determining the action of the tRNA methylases (Shershneva et al., 1971). This is probably explained, for example, by the fact that H_2U residues are localized in all tRNAs in two positions in the molecule only, in the H_2U- and extra loops; this indicates the existence of a highly specific enzyme (or enzymes) capable of hydrogenating U in those parts of the molecule only. The localization of ψ in nearly every case is confined to the loop of the universal oligonucleotide and to the anticodon branch; the ψ component here may be found either in the single-stranded part or in the double-stranded part, forming an A-ψ pair. The only exceptions to this rule are the yeast $tRNA_Y^{Val}$ and $tRNA^{Asp}$, which have one ψ residue in the H_2U-branch. The absence of minor components close to the acceptor and nonacceptor ends of the molecule, and also in the double-stranded part of the Tψ-branch, is a characteristic feature of all transfer RNAs; the nucleotides here are evidently inaccessible to the modifying action of the enzymes. The arrangement of the minor components in the clover leaf is shown schematically in Fig. 54.

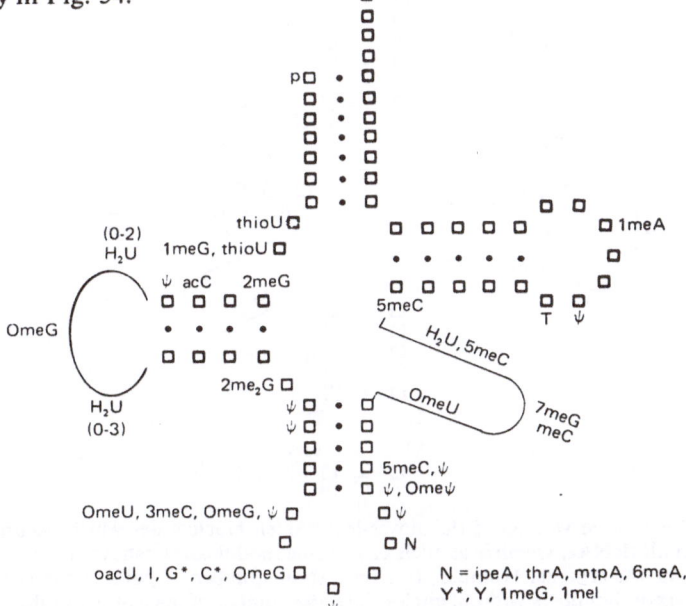

Fig. 54. Arrangement of minor components within the clover leaf.

Although the primary structure of a number of tRNAs has been established experimentally and without a shadow of doubt, the problem of their macromolecular structure is still being discussed in terms of models. The model most widely used at the present time, as has already been mentioned several times above, is the clover leaf suggested by Penswick and Keller in Holley's laboratory (Holley, 1966) as one possible version of a two-dimensional representation of the secondary structure, permitting the formation of the maximum number of Watson–Crick base pairs.

To test the validity of this model, enzymic, chemical, physical, and structural–functional methods have been used; on the whole the results indicate that this model does correctly reflect the arrangement of the hydrogen bonds in the tRNA molecule. The unity of structure is most demonstratively seen in the fact that a clover-leaf configuration can be ascribed to any tRNA, and as yet no exception to this rule has been found. Let us examine the most important features of the structure of this model (Fig. 55).

The acceptor end of the tRNA molecule is formed by the sequence C-C-A$_{OH}$. The hydrogen bonds linking the seven base pairs in this part of the molecule begin from the fifth nucleotide from the 3'-end, which is paired with

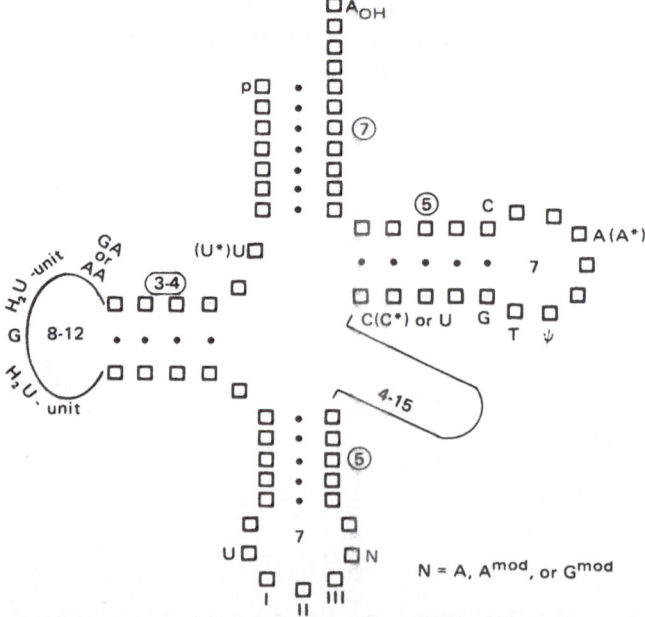

Fig. 55. Generalized version of the clover-leaf model. Nucleotides which occupy the same position in all tRNAs, whether as such or as some modification (shown by an asterisk) are indicated on the scheme. Numbers in circles show number of pairs in the corresponding helical segment. Numbers without circles show the number of nucleotides in the corresponding loop; I, II, and III mark the nucleotides of the anticodon.

the first nucleotide of the nonacceptor end, formed in the great majority of tRNAs by G, in three tRNAs by C, in two by U, and in only one case by A. These are mainly the Watson–Crick pairs, and also in all tRNAs from *S. cerevisiae* and the tRNA$_1^{Val}$ from *T. utilis,* one G-U pair which, however, is possible according to Crick's hypothesis (Crick, 1966). In tRNA$_f^{Met}$ there are only six pairs here, and they start with the second base from the 5'-end and the sixth base from the 3'-end of the molecule; the suggestion has been made that this is associated with the special function of this tRNA. The distinguishing feature of yeast tRNA^{Ala} is that it contains two U residues in this part, while the tRNA^{Phe} of wheat germ contains A and G residues, and the tRNA^{Leu} from *E. coli* A and C residues lying opposite each other.

Besides its paired segment, linking the 3'- and 5'-ends of the molecule, the clover leaf for all tRNAs has three principal branches and one extra branch.

The dihydrouridine branch in all tRNAs is characterized by a small number (3–4) of hydrogen bonds in the helical part and a variable number of nucleotides in the loop itself (8–12). The number of H$_2$U residues in this loop varies from one to five, and the other nucleotides are predominantly purines. The only tRNA in which H$_2$U is not present is tRNA^{Tyr} from *E. coli.* In all tRNAs the H$_2$U-loop is closed by a G-C pair, except in yeast tRNA^{Asp}, in which this position is occupied by the unusual pair G-ψ; in yeast tRNA^{Ser} and tRNA$_3^{Leu}$, the C is modified, while in the tRNA^{Leu} from *E. coli* the pair is reversed.

The structure of the anticodon branch is strikingly uniform; its helical part consists of five pairs of nucleotides linked by hydrogen bonds; as has already been mentioned, the A-ψ pairs occurring in most tRNAs are located entirely in this part of the molecule. G-U base pairs are very rare: U-G in the tRNA^{Ile} from *T. utilis* and yeast tRNA^{Asp}, and ψ-G in the tRNA^{Leu} from *E. coli.* The loop is formed by seven nucleotides, including the three nucleotides of the anticodon. In every tRNA without exception the position on the 5'-side of the anticodon is occupied by U, and with four exceptions, the position on the 3'-side by a minor component which, in the great majority of cases, is an A derivative, and only in four cases (phenylalanine, leucine, and aspartate yeast tRNAs and the leucine tRNA from *E. coli*) a G derivative. In the tRNA^{Val} from *S. cerevisiae,* and *T. utilis,* and tRNA$_2^{Val}$ from *E. coli,* and formylmethionine tRNA from *E. coli,* an unmodified A residue occurs on the 3'-side of the anticodon.

Next follows the extra branch, which differs strongly in composition and size in the different transfer RNAs. The characteristic feature of this branch is the frequent presence of 5meC in it (in six tRNAs), adjacent in some cases to H$_2$U, while 7meG has been identified in nine tRNAs.

It is interesting to note that this same branch in rat-liver tRNA^{Ser}

contains a 3meC residue which, like the 7meG, carries a positive charge in a neutral medium. A sufficient number of nucleotides to allow hydrogen bonding in this segment of the clover leaf is available in five tRNAs which, because of this feature, have the longest molecules (85–87 nucleotides). These are tRNA^{Ser} from yeast and liver, tRNA^{Leu} from yeast and *E. coli,* and tRNA^{Tyr} from *E. coli.* In these cases the extra branch contains 13–15 nucleotides.

The size of the extra branch and the number of pairs in the duplex segment of the H$_2$U-branch were used by Levitt (1969) as the basis for a classification of the transfer RNAs. In terms of these features, all tRNAs can be divided into three classes: class I, with three pairs in the H$_2$U-branch and a short extra branch; class II, with three pairs in the H$_2$U-branch and a long extra branch; class III, with four pairs in the H$_2$U-branch and a short extra branch. These three clasess of tRNA are illustrated schematically in Fig. 56, taken from the survey by Staehelin (1971). Within the framework of these three classes Staehelin compares primary structures by considering each branch of the clover leaf separately, and the results are very clear. Tables from Staehelin's paper (Staehelin, 1971), supplemented by a number of recently determined structures, are reproduced in the Appendix to this book.

The helical part of the Tψ-branch consists in every case of five pairs of nucleic-acid bases which, with four exceptions, consist of Watson–Crick pairs. The exceptions are the tRNA$_f^{Met}$, tRNA^{Val}, tRNA^{Phe}, and tRNA^{Trp} from *E. coli,* each of which has one G-U pair in this segment, and the tRNA^{Met} from *E. coli,* in which A is opposite C. The loop itself contains seven nucleotides, including the T-ψ-C-G sequence of the universal oligonucleotide, after which this branch was named. The exceptions are yeast tRNA$_f^{Ser}$ and tRNA^{Asp}, as well as the tRNA$_f^{Met}$ and tRNA^{Leu} from *E. coli,* which contain the sequence T-ψ-C-A. Thus the sequence T-ψ-C is truly universal, as also is the preceding G, which forms the pair with the C residue at the entrance to the Tψ-loop. As yet no exception has been found to this rule. Another standard feature is that in every case the universal oligonucleotide is followed by A, although, admittedly, in many tRNAs it is methylated. This

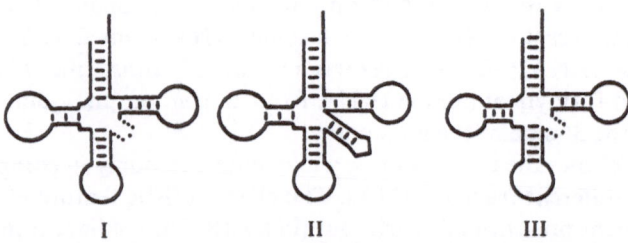

I   II   III

Fig. 56. Schematic representation of the three classes of tRNA. After Levitt (cited by Staehelin, 1971).

uniformity of the structure of the Tψ-loop is weighty evidence in support of the view that this loop plays a role in one of the functions common to all transfer RNAs.

The structure of the duplex segments and loops is thus very similar in the different transfer RNAs, irrespective of their amino-acid specificity or of the source from which they are obtained. This is shown particularly in the tables in the Appendix (Staehelin, 1971).

Let us now briefly examine the data which, taken as a whole, provide the basis for the clover-leaf model as an acceptable two-dimensional version of the structure of tRNA. It has already been mentioned that this model satisfies the requirement of forming the maximum number of hydrogen bonds, and this is true for all tRNAs so far studied. Evidence for this model was obtained in enzyme experiments (Zachau et al., 1966d; Bell and Russell, 1967; Millar and Byrne, 1967). Zachau showed that the sites predominantly and most readily attacked by pyrimidyl and guanyl RNases are located in those parts of the tRNA^{Ser} molecule which are considered to be unpaired in the clover-leaf model (Fig. 25). Confirmation of the clover-leaf model has also been obtained by chemical methods. For instance, bromosuccinimide and HNO_2 (Nelson et al., 1967), water-soluble carbodiimide (Brostoff and Ingram, 1967), dimethyl sulfate (Bollack et al., 1969), kethoxal (Litt, 1969), perphthalic acid (Erdmann et al., 1969), and other reagents react predominantly with nucleotides located, according to this model, in nonhydrogen-bonded segments. Results obtained by chemical methods at the same time indicate that transfer RNAs possess a definite three-dimensional structure. For example, most chemical modifications indicate that the loop containing the universal oligonucleotide is screened (Nelson et al., 1967; Cramer et al., 1968; Bollack et al., 1969); the nucleotides of this loop evidently interact, during formation of the tertiary structure, with nucleotides located in other parts of the molecule (Cramer et al., 1968; Doctor et al., 1969; Igo-Kemenes and Zachau, 1971). Results were obtained which indicate that the extra loop is also screened, so that the reactivity of the 7meG located in this loop is sharply reduced (Igo-Kemenes and Zachau, 1971).

A strong argument in support of the clover-leaf model is the fact that the principles established by models constructed for the first four tRNAs of yeast to be sequenced, when applied to the reconstruction of the primary structure of tRNA^{Phe}, gave a result which corrresponds almost completely to the truth (RajBhandary et al., 1966). In precisely the same way, the correct nucleotide sequence for tRNA^{Ile} of T. utilis was suggested simply on the basis of analysis of total RNase hydrolysates and the rules of the clover leaf (Takemura et al., 1969a).

Measurement of the optical rotatory dispersion of tRNA^{Ala}, according to the interpretation of Vournakis and Scheraga (1966), confirms the clover leaf model; other workers, however, give a different explanation of these results (Cantor et al., 1967; Armstrong et al., 1966). Data for the thermal denaturation of tRNA (Fresco et al., 1963; Felsenfeld and Cantoni, 1964) and for low-angle x-ray scatter (Lake and Beeman, 1967) also agree with the model under discussion.

A double-helical oligonucleotide complex consisting of the 3′- and 5′-terminal sequences has been isolated from a partial guanyl-RNase hydrolysate of tRNA^{Tyr} from *E. coli* (RajBhandary et al., 1969):

$$pG\text{-}G\text{-}U\text{-}G\text{-}G\text{-}G\text{-}G \qquad\qquad (5'\rightarrow 3')$$
$$_{OH}A\text{-}C\text{-}C\text{-}A\text{-}C\text{-}C\text{-}A\text{-}C\text{-}C\text{-}C\text{-}C\text{-}U\text{-}U\text{-}C\text{-}C\text{-}U\text{-}A\text{-}A \qquad (3'\rightarrow 5')$$

The isolation of such a complex is direct proof of the validity of the clover-leaf model, according to which the 3′- and 5′-ends of all tRNAs are linked by hydrogen bonds.

Weighty evidence that Watson–Crick base pairs are distributed in tRNA in accordance with the clover-leaf model is the specificity of aggregation of the "quarters" of the model obtained by cleavage of tRNA$_I^{Val}$ in the dihydrouridyl, anticodon, and T-ψ loops (Mirzabekov et al., 1970). Interaction was observed between only those quarters which, according to the clover-leaf model, are adjoined and linked by an adequate number of Watson–Crick pairs.

Important evidence that the clover-leaf reflects the true arrangement of the Watson–Crick pairs in tRNA molecules has recently been obtained by comparison of the structural and functional changes induced by mutation (Smith et al., 1971). These workers found that single nucleotide substitutions in the tRNA^{Tyr} of *E. coli* Su$_{III}^{+}$ mutants, by interfering with the pairing predicted by the clover leaf, give rise to profound changes in the properties of the corresponding mutants. Reversions, in which substitutions of nucleotides are observed in other segments of the polynucleotide chain, lead to restoration of the original properties only when pairing is restored, despite the fact that the old pair is replaced by a new one (for example, an A-U pair is formed instead of the G-C pair.)

Important evidence in support of the clover-leaf model is given by the presence, and the characteristic location in all tRNAs, of a triplet possessing a number of properties on the basis of which it must be identified as the anticodon, to correspond to the function it performs in protein synthesis. In tRNAs so far studied the triplet is situated in analogous places on the polynucleotide chain, approximately in its center, and it is the most reactive part of the molecule both in enzymic and in chemical reactions (Nelson et al., 1967; Brostoff and Ingram, 1967; Ukita, 1967; Yoshida et al., 1967a,b, 1968).

These triplets are evidently located in the most exposed part of the molecule, which corresponds exactly to the concept of anticodons. Moreover, in every case the presumptive anticodons occupy the position which, according to Fuller and Hodgson (1967), creates a favorable conformation for codon–anticodon interaction: the anticodon triplet is always preceded by two pyrimidines, and as a rule it is followed by a modified nucleoside. The existence of complementary and antiparallel triplets with respect to the corresponding codons in analogous segments of the clover leaf of all transfer RNAs is itself confirmation of the localization of the anticodon. However, this can be determined with absolute certainty only when about a score of structures in which this principle is repeated unchanged are known. At the beginning of the structural investigations various methods were used to identify the anticodon, of which the most widespread was the binding of the aminoacyl-tRNA with trinucleoside-diphosphate–ribosome complexes. These experiments not only helped to locate the anticodon, but also finally proved the validity of the adaptor hypothesis and also of Crick's wobble hypothesis.

A few examples will suffice. Experiments to detect the anticodon of yeast and *E. coli* tRNA^{Tyr} were carried out by Doctor et al. (1966). Specific binding of tyrosyl-tRNA^{Tyr} took place in the presence of U-A-U and U-A-C; in accordance with the wobble hypothesis, the G residue in position 1 of the anticodon can pair with the U and C in position 3 of the codon, so that the anticodon of tRNA^{Tyr} must be the triplet G-U-A or, if U is replaced by ψ, the triplet G-ψ-A. These sequences have actually been found at the apex of the anticodon branch of tRNA.

Two triplets which could play the role of anticodon have been identified in the yeast tRNA^{Phe} molecule; these are the trinucleotides OmeG-A-A and the G-A-A which is located near the acceptor end. The experiments of Söll et al. (1966a) showed that phenylalanyl-tRNA^{Phe} binds with ribosomes in the presence of both U-U-A and U-U-C, so that in accordance with the wobble hypothesis, both these trinucleotides are possible anticodons, for G in position 1 of the anticodon pairs with both U and C in position 3 of the codon. Evidence in support of OmeG-A-A is given by the position of this triplet in the molecule and the ease with which the neighboring bond is split by pancreatic RNase. The presence of OmeG in the anticodon of yeast tRNA^{Phe} is fully reconcilable with the fact that this tRNA is not split into halves by T_1-RNase, not only under mild, but also under much more vigorous conditions than is the case with other tRNAs.

In the case of tRNA$_1^{Val}$, the problem of the anticodon has been unequivocally solved in favor of the sequence I-A-C, since valyl-tRNA$_1^{Val}$ was bound with ribosomes in the presence of three triplets, namely, G-U-U, G-U-C, and G-U-A (Mirzabekov et al., 1967); according to the wobble hypothesis,

I in position 1 of the anticodon can pair with U, C, and A in position 3 of the codon. Important evidence that the anticodon of tRNA$_1^{Val}$ is the triplet I-A-C is the fact that the associated molecule of tRNA$_1^{Val}$ in which the I-A bond of this triplet is broken is not bound with ribosomes in the presence of trinucleotides (Table 20), whereas it still remains capable of aminoacylation.

The most direct evidence for the localization of anticodons in tRNAs was obtained by the study of amber-suppressor mutations of *E. coli* (Goodman et al., 1968). *E. coli* Su$_{III}^+$ mutants are known in which the U-A-G-codon, which is chain-terminating for strains Su$_{III}^-$, is considered to be the codon for tyrosine. The primary structure of the tRNA^{Tyr} of both strains of *E. coli* has been elucidated. It is the same for the tRNA^{Tyr} of both Su$_{III}^+$ and Su$_{III}^-$, except for the first nucleotide of the anticodon which, in the Su$_{III}^-$ mutants is represented by modified guanine (Fig. 32A), and in Su$_{III}^+$ by cytosine, which is the reason that the terminating triplet U-A-G can code for tyrosine (Fig. 32B).

Of no less importance for the identification of the anticodon was the work of Clark et al. (1968b), who isolated a nonadecanucleotide from a partial guanyl-RNase digest of tRNA$_f^{Met}$ from *E. coli,* which was subsequently found to contain the entire anticodon branch:

```
                              G
                             /
                      U    A
                       \   |
                        C  A
                        |  |
                        G –G
                        |  |
                        G –C
                        |  |
                        G  C
                 OmeC /      \ A
                    |         A
                    U          \
                     \C – A – U/
      12
```

TABLE 20. Binding of Yeast tRNA^{Val} and Its Fragments with Ribosomes

Aminoacylated fraction	Used in experiment	Without addition	Bound with ribosomes in the presence of			
			GpUpU	GpUpC	GpUpA	GpUpG
	pmoles		Δ pmoles			
Total	33	1.5	4.2	2.2	2.3	0.5
tRNA$_1^{Val}$	78	11.9	6.7	3.0	4.4	1.2
tRNA$_2^{Val}$	28	1.3	1.6	0.2	2.8	3.1
tRNA$_3^{Val}$	34	4.8	6.2	2.8	3.2	0.2
Mixture of 3'- and 5'-halves of tRNA$_1^{Val}$	49	11.5	−0.4	−0.2	−0.4	−0.2

Experiments using the method of Nirenberg and Leder (1964) showed that this fragment is bound to ribosomes in the presence of initiator triplets A-U-G and G-U-G and, in this respect, it does not differ from the whole tRNA$_f^{Met}$ molecule. This observation, according to which an oligonucleotide occupying a definite position in tRNA, and containing the presumptive anti-codon C-A-U, possesses the same coding properties as the whole tRNA molecule, is direct experimental evidence for the localization of the anticodon in the tRNA$_f^{Met}$ molecule.

Triplets identified as the anticodons in the tRNAs so far decoded, and also the codons with which the corresponding tRNAs must be linked are given in Table 21. In most cases there is complete agreement between the results of studies of the anticodons of the tRNA and the genetic code. The second and third letters of the anticodon are in fact complementary with and antiparallel to the second and third base of the codon. Pairing of the first letter of the anticodon obeys the principles predicted by Crick's wobble hypothesis: I in position 1 of the anticodon determines the ability of the cor-responding tRNAs to react with three codons differing in their 3'-terminal nucleotide (U, C, A); G in position 5' of the anticodon recognizes two codons of the corresponding tRNA, namely those terminating in C and U; finally, anticodons beginning with C recognize only one triplet, namely that possess-ing a 3'-terminal G. The only exceptions to these rules are tRNA$_f^{Met}$, which recognizes not only the codon A-U-G, but also the codon G-U-G, and tRNA^{Trp}, which recognizes the codons U-G-G and U-G-A. Degeneracy of the genetic code is expressed by the fact that the anticodons of the tRNA can recognize codons differing in the last nucleotide from the 3'-end; by contrast, tRNA$_f^{Met}$ recognizes codons differing by the first nucleotide (at the 5'-end) of the triplet.

Otherwise, the principles of codon–anticodon interaction are so rigid that if the triplets with which a given tRNA is linked in the Nirenberg–Leder system are known, its anticodon can be predicted. For example, in the case of the tRNA$_f^{Val}$ from E. coli, it could be reasonably confidently asserted that the minor component in position 1 of its anticodon must be a derivative of U with unchanged coding properties. In fact, it was found to be uridine-5-oxyacetic acid.

The list of tRNAs so far studied and their corresponding anticodons is now quite large, and it enables certain generalizations to be made. The fre-quency with which minor components are found in the anticodons is striking; they are found mainly in position 1, less commonly in position 2, and never in position 3. The importance of this fact is not yet quite clear, as is true of almost everything else about these interesting compounds. The reason for the presence of I in position 1 of the anticodons of the yeast tRNAs could be to provide maximal degeneracy of the code, for in this position it can recognize

TABLE 21. Anticodons of Transfer RNAs

Specificity of tRNA	Source	Index of tRNA	Anticodon	Codons with which tRNA must pair in accordance with wobble hypothesis[a]	Minor component in 3'-position of anticodon
Ala	*S. cerevisiae*		I-G-C	$\substack{U\\GCC\\A}$	1meI
Asp	*S. cerevisiae*		G-U-C	$GA\substack{U\\C}$	1meG
Gly	*E. coli*		G-C-C	$GG\substack{U\\C}$	A
Gln	*E. coli*	1	N*-U-G	CA?	R*
		2	C-U-G	CAG	R*
Ilu	*T. utilis*		I-A-U	$\substack{U\\AUC\\A}$	thrA
Leu	*S. cerevisiae*	3	C-A-A	UUG	G[b]
	E. coli		C-A-G	CUG	G[c]
fMet	*E. coli*		C-A-U	$\substack{A\\G}UG$	A
Met	*E. coli*		C[d]-A-U	AUG	A[e]
Phe	*S. cerevisiae*		OmeG-A-A	$UU\substack{U\\C}$	Y
	Wheat germ		OmeG-A-A	$UU\substack{U\\C}$	Y
	E. coli	1	G-A-A	$UU\substack{U\\C}$	mtpa
Ser	*S. cerevisiae*	2	I-G-A	$\substack{U\\UCC\\A}$	ipeA
	Rat liver		I-G-A	$\substack{U\\UCC\\A}$	ipeA
Trp	*E. coli*	$\substack{Su^+\\Su^-}$	C-C-A	$\substack{UGA\\UGG}$	mtpA
Tyr	*S. cerevisiae*		G-ψ-A	$UA\substack{U\\C}$	ipeA
	T. utilis		G-ψ-A	$UA\substack{U\\C}$	ipeA
	E. coli Su$_\mathrm{III}^-$		G*-U-A	$UA\substack{U\\C}$	mtpA
	E. coli Su$_\mathrm{III}^+$	$\frac{1}{2}$	C-U-A	UAG	mtpA

Table 21. (*Continued*)

Specificity of tRNA	Source	Index of tRNA	Anticodon	Codons with which tRNA must pair in accordance with wobble hypothesis[a]	Minor component in 3′-position of anticodon
Val	*S. cerevisiae*	1	I-A-C	GU$\begin{smallmatrix}U\\C\\A\end{smallmatrix}$	A
	T. utilis	1	I-A-C	GU$\begin{smallmatrix}U\\C\\A\end{smallmatrix}$	A
	E. coli	1	oacU-A-C	GU$\begin{smallmatrix}A\\G\end{smallmatrix}$	6meA
		2A 2B	G-A-C	GU$\begin{smallmatrix}U\\G\end{smallmatrix}$	A

*Unidentified nucleosides.
[a]Codons for most tRNAs, including tRNAs with unidentified nucleotides in the anticodon, determined experimentally.
[b]Probably 1meG.
[c]Probably 1meG or 2meG.
[d]Probably acC.
[e]Probably A acylated in position 6.

3 triplets. The role of the ψ residue discovered in position 2 of tRNA^{Tyr} from *S. cerevisiae* and *T. utilis* could be to provide a stronger codon–anticodon interaction because of its ability to form two hydrogen bonds. In some cases the presence of a minor component in the anticodon is known to modify the coding properties of the corresponding tRNA. For example, it has been known for a long time that the glutamate-specific tRNA of yeast has unusual coding properties; it recognizes only the codon G-A-A and not the codon G-A-G. It is assumed that this is because of the presence of the methyl ester of 2-thiouridine-5-acetic acid in position 1 of the anticodon (SUC); the presence of sulfur in the 2 position of uridine prevents it from pairing with G, and this minor component thus modifies the coding properties of glutamate-tRNA (Yoshida et al., 1970, 1971).

The presence of 5-methylaminomethyl-2-thiouridine in position 1 of the anticodon of the glutamate-specific tRNA from *E. coli* has similar results, because it cannot pair with G on account of the presence of sulfur in the U in position 2 (Ohashi et al., 1970). The following structure is suggested for the anticodon region:

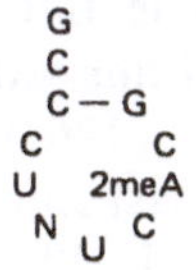
G
C
C — G
C C
U 2meA
N U C

N is 5-methylaminomethyl-2-thiouridine

Examination of Table 21 shows that the anticodons of most tRNAs at present identified by no means account for all the codons of the individual amino acids known. For example, four codons are known for glycine, but the tRNA studied can interact with only two of them; the same applies also to tRNA^{Ser}, which because of the presence of I in the anticodon can react with three triplets, but it does not match the three remaining codons. In these cases it is evident that isoacceptor tRNAs whose anticodons correspond to the remaining codons have still to be discovered. However, in practice the situation is much more complex. In some organisms individual tRNAs which should exist according to tables of the genetic code are found in very small quantities or not at all. In other organisms, on the other hand, the whole necessary assortment of tRNAs has been found. For example, in E. coli (Sundharadas et al., 1968) and now also in rat liver (Staehelin, 1971) tRNAs have been found for all six codons corresponding to serine; on the other hand, in yeast it has not yet been possible to find with certainty a tRNA which would correspond to the codons U-C-G, A-G-U, and A-G-C (Zachau, 1969). These results confirm the long establsihed view that although the genetic code is universal, it is used differently by different organisms (Söll et al., 1966b).

Evidently, depending on the content of tRNA, some codons are used more often than others while in the absence of the corresponding tRNA they are not translated at all. If the relative proportions of the different tRNAs can also vary when the organism is in different physiological states, and this is most probably true, tremendous opportunities are present for the regulation of protein synthesis at the translation level.

Let us now turn again to the clover leaf and examine the structure of the anticodon loop.

Examination of the primary structure of tRNA demonstrates the remarkable uniformity of structure of the anticodon segment: in every case, without exception, the anticodon triplet is preceded by the sequence Py-U; the importance of this phenomenon is not yet known, and no experimental attempts to shed light on it have yet been undertaken.

The nucleosides next to the anticodons on the 3'-side have attracted the closest attention for a long time. This is because they are mainly minor components, and they exhibit the most complex modifications, as the result of which they have been called "hypermodified" (Schweizer et al., 1969). It is also interesting to note that most of these derivatives of the nitrogenous bases are found only in this position in the polynucleotide chain of tRNA, and nowhere else. The most complex compounds of this series are N^6-($\triangle^2$-isopentenyl) adenosine (Biemann et al., 1966; Hall et al., 1966), identified in tRNA^{Ser} from yeast and rat liver and in tRNA^{Tyr} from yeast and T. utilis; its thiomethyl derivative (Burrows et al., 1968; Harada et al., 1968) present in tRNA^{Phe}, tRNA^{Tyr}, and tRNA^{Trp} from *E. coli;* N-(purin-6-ylcarbamoyl) threonine riboside (Schweizer et al., 1969) found in tRNA^{Ile} from *T. utilis;* and finally, the Y residue discovered earlier in tRNA^{Phe} from yeast and wheat germ, but identified only recently because of the complexity of its structure (Nakanishi, 1970). This position is filled only in isolated cases by simple minor components, such as 1meG in yeast tRNA^{Asp} and 6meA in tRNA$_1^{Val}$ from *E. coli.* A few tRNAs (the tRNA$_1^{Val}$ from *S. cerevisiae* and *T. utilis,* and *E. coli* tRNA^{Gly} and tRNA$_f^{Met}$) are exceptions in that an unmodified adenosine is found on the 3'-side of their anticodon. It was formerly held that the neighbor of the 3'-terminal nucleotide of the anticodon could only be A or one of its derivatives. It is now clear, however, that this is not so because this position is occupied by a G derivative in four tRNAs (tRNA^{Phe}, tRNA^{Asp}, and tRNA^{Leu} from *S. cerevisiae* and tRNA^{Leu} from *E. coli*).

It is natural to suppose that the hypermodified minor components, possessing highly reactive functional groups and occupying a key position in the tRNA molecule must play an important role in its function. Since tRNA$_f^{Met}$ has an unmodified A on the 3'-side of its anticodon and recognizes two codons differing in their first nucleotide (from the 5'-end), it has been postulated that modification of this nucleotide prevents the third base of the anticodon from wobbling (Dube et al., 1968). This hypothesis is still viable since in the other tRNAs with an unmodified A as the 3'-neighbor of the anticodon the third nucleotide of the anticodon is C, which according to Crick is unable to wobble and can pair only with G. Schweizer et al. (1969) suggest that the functional groups of the hypermodified nucleotides can play a role in binding the tRNA with ribosomal proteins.

Recent results (Gefter and Russell, 1969) in fact show that differences in the level of modification of the minor component next to the anticodon on the 3'-side in tRNA^{Tyr} of *E. coli* Su$_{III}^+$ influence the binding of the aminoacyl-tRNA with ribosomes. Zachau (1969) postulates that nucleotides next to the anticodon on the 3'-side can play a role in forming the three-dimensional structure of the anticodon branch. The few experimental data relevant to this issue show that the minor component next to the anticodon has evidently no effect on acceptor function, but plays an important role in the adaptor function of the tRNA. Selective chemical modification of the ipeA residue occupying this position in tRNA^{Ser} (Fittler and Hall, 1966; Gefter and Russell, 1969) or removal of the base from the minor component Y in yeast tRNA^{Phe} (Thiebe and Zachau, 1968a) render the corresponding aminoacylated tRNA unable to bind with the mRNA–ribosomal complex.

It thus follows that much is known about the anticodons of individual tRNAs and their localization can be confidently identified. The functional importance of other segments of the clover leaf (with the exception, of course, of the acceptor end of the molecule), however, are unknown and the suggestions that have been made are confirmed by some experimental, facts, and contradicted by others.

Let us now consider the universal oligonucleotide for a moment. The results of investigation of acceptor activity of tRNA$_1^{Val}$ by the "sliced-molecules" method show directly that the universal oligonucleotide does not in this case, at least, play a role in the recognition of the aminoacylating enzyme or in the formation of the necessary conformation for this process; a reassociated molecule not containing the tetranucleotide T-ψ-C-G possesses valine-acceptor activity (Mirzabekov et al., 1969a). The view that this sequence is responsible for the binding of the tRNA with the 50S subunit of the ribosomes has received some experimental confirmation. It has been shown, for example, that not only the whole tRNA molecule, but also its 3'-half in which the universal sequence is located, can undergo nonspecific binding with ribosomes (Mirzabekov et al., 1967). It has further been shown that the tetranucleotide T-ψ-C-G, unlike certain other oligonucleotides, competitively inhibits the binding of tRNA with ribosomes (Ofengand and Henes 1969). Finally, recent experimental findings have confirmed that the nonspecific binding of tRNA with ribosomes probably takes place through a 5S rRNA (Siddiqui and Hosokawa, 1969), in whose composition sequences complementary (and antiparallel) to the universal oligonucleotide have been identified (Forget and Weissman, 1967; Brownlee et al., 1968). Although these results are only indirect, nevertheless, when taken as a whole, they indicate that participation of the universal sequence in the nonspecific binding of tRNA with ribosomes is the most probable hypothesis, and can serve as the basis for further research.

Examination of problems connected with the functional centers is outside the scope of this monograph; I shall therefore give only a brief outline of the approaches used to study them, which are based on their known primary structure. Investigators have so far paid most attention to the recognition site, i.e., the center responsible for specific interaction of the tRNA with aminoacyl-tRNA-synthetases, and most research to study nucleic-acid–protein interaction has been carried out on this model. The methods used in this case, however, can also evidently be used to study interaction between tRNA and other proteins.

The simplest approach which began to be used as soon as the first structural formulas of tRNAs became known is the comparison of primary structures. The basis of this method is that the performance of highly specific functions must require highly specific structural determinants; this applies above all to the aminoacyl-tRNA-synthetases, which aminoacylate only tRNA with an absolutely definite amino-acid specificity and which consequently, must discriminate those structural elements by which tRNAs specific with respect to the various amino acids differ from each other. On the other hand, the possibility of heterologous reactions, i.e., the aminoacylation of tRNAs of the same specificity, but from different sources and differing to a greater or lesser degree in their nucleotide sequence, by the same enzyme prompted the search for common structural features which could provide the basis for this recognition. When an enzyme interacts with all tRNAs of a given type, it is logical to seek common features in their structure which are independent of amino-acid specificity. This applies, for example, to C-C-A-pyrophosphorylase, which is considered to be a protein that lengthens the ends of all tRNAs regardless of their specificity. A segment with a function common to all tRNAs must be less specific still; for this reason it is assumed that the universal oligonucleotide present in all tRNAs so far studied is responsible for the nonspecific binding of the tRNA with ribosomes.

Closely allied to simple comparison of the primary structures are chemical modifications, the object of which in this particular case is to make a strictly determined change in particular nucleotides and then to study the effect of this change on the function of the tRNA. Many investigations of this type have been carried out with minor components; selective modification of those which do not repeat in the molecule can give useful information on their role in a particular function.

A method which has recently achieved wide popularity is that of "sliced molecules," by means of which the properties of tRNA molecules with ruptures in different segments of the chain, molecules from which particular nucleotides or nucleotide sequences have been removed, and also hybrid molecules obtained by joining fragments from different tRNAs, can be investigated. Comparison of the functional properties of native tRNAs and tRNAs modified in this manner provides considerable information about the functional role of a given segment of the molecule.

Another new approach, which I mentioned above in connection with the clover-leaf model, must also be discussed. I refer to the study of the correlation between structure and function undertaken on tRNA mutants at the Laboratory of Molecular Biology in Cambridge. By producing mutants with single substitutions of particular nucleotides in a tRNA with known primary structure, the functional role of that nucleotide can be determined. Some particularly interesting results have been obtained with revertants in which the pair disturbed by the first mutation is restored by a second mutation. The conclusion drawn from these experiments is that, at least in some cases, it is not so much the nature of the base which is important as the presence of nucleotides capable of forming hydrogen bonds in the particular segment of the clover leaf. This is a conclusion of great fundamental importance. A serious limitation of most of the approaches mentioned above is that it is difficult to state whether functional changes associated with a modification are linked with a change (or the removal) of the nucleotide itself or with disturbances which the modification introduces into the three-dimensional structure of the tRNA.

The level of our ignorance regarding the recognition center of aminoacyl-tRNA-synthetase, a problem to which the greatest amount of investigation has been devoted, reflects the number of sequences which have been discussed as possible candidates for this role: the anticodon, the H_2U-loop, pairs in the duplex segment connecting the 3'- and 5'-ends of the molecule, the extra loop, the polypurine sequences, etc. Examination of the results obtained during the testing of these hypotheses (for the actual details, see Zachau, 1969) shows an extremely varied picture; a conclusion which is true for one tRNA is not confirmed for another, and vice versa. It is now considered, therefore, that the process of recognition of the corresponding tRNAs by synthetases is not as universal as was hitherto believed, and that these centers may be in different locations in different tRNAs.

The next important conclusion is that the recognition site of a tRNA can be formed as a result of the participation of several nucleotides or nucleotide sequences, located in different, perhaps distant, parts of the molecule. In other words, the recognition site of the aminoacyl-tRNA-synthetase, like the active center of enzymes, is formed in three dimensions, by folding of the molecule and interaction between its parts. If this is true, and the tertiary structure of the molecule plays a principal role in the formation of the recognition site, as well as sites responsible for interaction with other proteins, completely new approaches are necessary and, in particular, the tertiary structure of the transfer RNAs themselves and of their complexes with enzymes must be determined. The successful progress of research into tRNA crystallization indicates that important results will be obtained in this field very soon.

Thus while knowledge of the primary structure is certainly essential, it is by no means an adequate condition for identification of the various functional centers. It evidently can no longer be disputed that these centers are formed through the formation of a definite three-dimensional structure of the molecule and that they are a part of it. When Holley (1966) originally published the structural formula of tRNA^{Ala}, in addition to the two-dimensional representation of the distribution of Watson–Crick base pairs, he also suggested a three-dimensional model in which the molecule was folded along its longitudinal axis. Some such three-dimensional structure, for example, is the only way of explaining why the anticodon region, and no other region in the single-stranded parts of the sequence, should be the most reactive segment of the molecule. Experiments demonstrating the screening of the lateral loops and, in particular, of the loop containing the universal oligonucleotide, have already been mentioned above.

The existence of a tertiary structure of tRNA was postulated some time ago (Fresco et al., 1960), but it is only recently that this has been verified (Henley et al., 1966; Lindahl et al., 1966; Fresco et al., 1966; Gartland and Sueoka, 1966). According to Fresco's findings, tRNA, like proteins, possesses a tertiary structure; in his opinion this exists in many different versions, but only one of them possesses biological activity. The formation of a stabilized tertiary structure is facilitated by interaction between unpaired segments of the partially helical tRNA molecule in such a way that the duplex segments are fixed in space. This interaction gives the tRNA molecules an absolutely stable conformation, ensuring the much greater specificity of its reaction with proteins during aminoacylation and during interaction with ribosomes. The three-dimensional structure of tRNA is largely determined by ions, especially Mg^{2+}, for as a polyelectrolyte, it reacts precisely to their concentration and nature.

The native form of tRNA possesses asymmetry and a high degree of compactness; it is the same for all tRNA, although each has its own distinctive primary structure. Fresco postulates a definite analogy between tRNAs and globular proteins as regards both their general conformation and the demands which must be satisfied for preservation of their functional activity (Fresco et al., 1968a).

The biologically active conformation of tRNA thus depends above all on its primary structure but, in addition, it is also determined by certain external environmental factors. It thus follows that the primary structure contains most, but by no means all, of the information required to explain the highly specific interactions in which the tRNA participates. Many of these phenomena, which have been associated with primary structure (for example, the existence of several isoacceptor tRNAs) are possibly in fact determined by its tertiary structure. The study of tRNA from this point of view is only in its infancy, and it is not yet possible to decide which form is biologically active and which spatially organized centers are responsible for a particular function. However, there is no doubt that among these centers there must be some which are highly specific and responsible for those functions whereby individual tRNAs differ from one another, such as the centers interacting with the aminoacyl-tRNA synthetases. On the other hand, nonspecific interaction with ribosomes or with the pyrophosphorylase attacking the CCA-end, which is equally characteristic of all tRNAs, could be carried out by centers with a similar three-dimensional configuration and formed with the participation of, for example, the universal oligonucleotide.

The presence of a tertiary structure of tRNA molecules is now firmly established. All the evidence at present available goes to show that a tRNA cannot be a loose combination of double-helical segments; it must be a compact, asymmetrical particle, elongated in shape, with a clearly defined three-dimensional structure (for reference, see Cramer, 1971; Kiselev, 1971).

The problem now is to study the concrete characteristics of the three-dimensional structure, to determine the forces participating in its stabilization, and to understand how the numerous functions with which the tRNA is endowed can be carried out within the limits of this three-dimensional structure.

A number of models of the spatial organization of tRNA have been suggested on the basis of the known nucleotide sequence (Cramer et al., 1968; Connors et al., 1969; Doctor et al., 1969; Fuller et al., 1969; Ninio et al., 1969; Melcher, 1969; Levitt, 1969). All these models are based on the clover leaf and are in fact developments of it. All the results obtained by investigation of tRNAs by physical, chemical, enzymic, and other methods have been used in the construction of these models. Let us examine the models with the most direct bearing on primary structure briefly in order to demonstrate how the results obtained have been used to suggest three-dimensional structures.

Enzymic action, widely used from the very beginning of structural research, has demonstrated differences between the sensitivity of segments of the molecule to nucleases and has thus led to the conclusion that helical segments exist in the molecule together with other unpaired segments. An interesting approach to the discovery of unshielded sequences in polynucleotides with a known primary structure was suggested by Uhlenbeck et al. (1970); it is based on the complementary binding of tri- and tetranucleotides by unpaired segments of the molecule.

It has been concluded from the numerous chemical modifications that only a few nucleotides which, according to the clover-leaf model, are free are capable of reacting; this leads to the distinction between exposed and hidden nucleotides, an important contribution to the study of three-dimensional structure (for reference, see Cramer, 1971).

Important information leading to the elucidation of the spatial organization of tRNA molecules is provided by results showing that interaction can take place between nucleotides which are not neighbors in the polynucleotide chain. The formation of a covalent bond between the eighth and thirteenth nucleotides in most of the tRNAs from *E. coli*, as the result of photochemical oxidation, can only be explained by assuming that these nucleotides are close together in space (Yaniv et al., 1969).

I have already mentioned the new approach based on the study of the consequences of changes in the primary structure of tRNAs induced by genetic mutations (Smith et al., 1971; Cashmore, 1971). Comparison of the reactivity of the nucleotides in the original tRNA and in tRNA modified by mutations has shown that disturbances in the extra loop influence reactivity of the nucleotides in the H_2U-loop and vice versa. This is almost incontrovertible proof that the nucleotides of these two segments of the clover leaf interact in the formation of the three-dimensional structure.

Finally, I must mention the evidence given by nature itself. Coordinated substitution of two bases, which are able to form a base pair, located in distant segments of the molecule, for two other bases, also capable of pairing, is strong evidence in support of the actual existence of the corresponding link. Recent work has revealed the structure of three tRNAs (tRNA^{Leu} and tRNA^{Trp} from E. coli and tRNA^{Asp} from brewers' yeast) which possess one common feature that distinguishes them from all other tRNAs whose structure has previously been determined: in the H$_2$U-loop, in what was previously regarded as the universal G15 position, they have the nucleotide A15; at the same time N48 in these tRNAs is U instead of C. The possibility of interaction between N15 and N48, postulated by certain models (Levitt, 1969) thus still remains. This is a good argument in support of the view that interaction between these nucleotides is a true feature of the three-dimensional organization of tRNA molecules.

Models of the three-dimensional structure of tRNA must correspond to all the experimental results reflecting physical, chemical, and functional properties of the tRNA. A critical analysis of these models lies outside the scope of this book, and I shall merely state that not all models satisfy this criterion equally, and they are not therefore equally probable. Furthermore, the methods so far used to study tRNA conformation could not, in principle, solve all the problems concerned with the three-dimensional structure of tRNA molecules. At the present time only one method which can yield a detailed picture of the three-dimensional arrangement of the individual elements of the tRNA molecule is known, namely x-ray structural analysis. The year 1968, when the first crystalline preparations of tRNA were obtained, must therefore be regarded as the beginning of a new and, probably, the final stage in the study of the three-dimensional structure of tRNA. The efforts of many research groups are now being concentrated on the production of crystals suitable for x-ray structural analysis, and considerable progress has been achieved in this field (for reference, see Cramer, 1971).* The tRNA is crystallized from various solvents in the presence and in the absence of bivalent cations; crystals differing in shape and size and also in their relative nucleic-acid content are obtained; in some cases the results of x-ray structural analysis have enabled the Patterson projections to be calculated; the first successful attempts at crystallizing isomorphic derivatives of tRNA have been made by the introduction of ions of the heavy metals and also by certain chemical modifications of the molecule. An important conclusion essential to the study of tRNA crystals is that, as had been postulated, the spatial

*See the paper in Nature 219:1209 (1968).

organization of this class of nucleic acids is uniform; unity of morphology of tRNA molecules is clear from the fact that crystals can be obtained from a bulk tRNA preparation (Fresco et al., 1968b), and also that mixed crystals can be obtained from tRNAs of different amino-acid specificity (Blake et al., 1970).

The rapid development of this line of research raises the hope that very soon we shall know the details of the molecular organization of tRNA, and these, in turn, will shed light on its functional centers.

It follows from the facts described above that tRNA is characterized by homology of structure, which is evidently manifested at all levels of organization of the molecule: the unity of the common structural plan of the transfer RNAs is the result of their similar function, on the one hand, and the result of their common origin on the other, and differences among them are due to sebsequent divergence in the course of evolutionary development. The study of tRNA structure can therefore shed light not only on the relationships between the structure and function, but also on the evolution of these biopolymers.

STRUCTURE OF THE ISOACCEPTOR tRNAs

The heterogeneity of transfer RNAs specific with respect to one amino acid is an extremely wide subject with many aspects; we shall examine it principally from the point of view of primary structure, i.e., we shall discuss only those investigations whose results provide information on the chemical structure of the molecules of isoacceptor tRNAs.

The existence of several physically separable tRNAs specific with respect to one amino acid is a widespread phenomenon. The multiplicity of the tRNAs is revealed by their countercurrent distribution (Goldstein et al., 1964; Apgar and Holley, 1964; Mirzabekov et al., 1965b), by fractionation on MAK columns (Glebov et al., 1965; Zaitseva et al., 1966; Sueoka and Yamane, 1962; Melchers and Zachau, 1965), and also by the use of enzymic (Berg et al., 1961; Bennett et al., 1963) and chemical (Berg et al., 1962) methods for this purpose. As a result of these investigations, most tRNAs specific with respect to different amino acids were successfully separated into several fractions, and there is now hardly a single amino acid with only one corresponding tRNA. In *E. coli*, for example, five leucine tRNAs (Kelmers et al., 1965; Apgar and Holley, 1964), three (Goldstein et al., 1964) or four (Ishikura and Nishimura, 1967) serine tRNAs, two tyrosine, two phenylalanine, and three valine tRNAs (Goldstein et al., 1964), have been found. In yeast, two alanine, two or three valine, and two, three, or even five serine tRNAs have been found (Bergquist, 1966a), and so on. An example is the

$tRNA_2^{Val}$ from bakers' yeast, which was apparently homogeneous when the countercurrent method was used (Mirzabekov et al., 1965b), but which was found to consist of at least three fractions when the methods of chromatography on BD-cellulose and reverse-phase chromatography were used (Kryukov et al., 1971). The same subdivision into subfractions was observed by Staehelin and his collaborators (Müller et al., 1971; Staehelin, 1971) for rat-liver $tRNA^{Ser}$, and in this case reverse-phase chromatography demonstrated the nonhomogeneity of the two peaks obtained by distributive chromatography, so that the number of isoacceptor serine-specific tRNAs in rat liver has risen to five.

Various methods have been used to investigate the reasons for and chemical basis of the multiplicity of the tRNAs. These have included aminoacylation of the tRNA by homologous and heterologous enzymes; determination of the ability of isoacceptor tRNAs to transfer an amino acid to identical or different places on synthetic and natural templates; the study of binding of aminoacyl-tRNAs with ribosomes by the use of various polynucleotides and trinucleotides. However, it can be confidently stated that the most direct road to the understanding of the multiplicity of the tRNAs is the study of their primary structure.

Heterogeneity of the tRNAs was originally ascribed to degeneracy of the genetic code, and in some cases it has in fact been shown that different tRNAs transferring the same amino acid respond to different coding triplets. However, there is no doubt now that degeneracy of the genetic code is by no means the only cause of tRNA multiplicity. This follows, *inter alia,* from the fact that for many amino acids the number of coding triplets established is smaller than the number of corresponding tRNAs and, consequently, not only the anticodons of the tRNAs, but also certain other structural elements in their molecule must be different.

On the other hand, the increasing wealth of experimental evidence shows that this problem of the isoacceptor tRNAs must be approached with caution, and that the detection of acceptor activity with respect to a given amino acid in several places on the elution profile is not a firm enough basis for the conclusion to be drawn that several isoacceptor tRNAs exist. In some cases the reason for the presence of two tRNAs was found to be that one of them lacked the terminal adenosine residue (Lebowitz et al., 1966; Makman and Cantoni, 1966; Lebowitz et al., 1967; RajBhandary et al., 1966), which was replaced during aminoacylation by the pyrophosphorylase present in unpurified samples of aminoacyl-tRNA synthetases. Another possible cause of erroneous multiplicity of tRNAs may be the aggregation of their molecules (Schleich and Goldstein, 1964; Röschenthaler and Fromageot, 1965; Zachau, 1968b; Adams and Zachau, 1968). Transfer RNAs which are different conformers also can separate into several peaks. Gartland and Sueoka (1966), for example, showed that under certain conditions the two tyrosine tRNAs from *E. coli* can be converted from one into the other.

A similar situation exists with the $tRNA_3^{Val}$ of *S. cerevisiae*, which can be clearly separated from the other valine tRNAs by countercurrent distribution, but subsequently becomes indistinguishable from $tRNA_1^{Val}$ (Krutilina et al., 1970). A phenomenon of the same order is denaturation of tRNA, which is partially or completely reversible, and under certain conditions may give the impression that isoacceptor tRNAs exist (Lindahl et al., 1966; Fresco et al., 1966). In some cases, therefore, multiplicity of the isoacceptor tRNAs, so far as primary structure is concerned, turned out to be merely apparent.

From the point of view of primary structure, the existence of isoacceptor tRNAs may result from the following causes: differences in the anticodons; differences in other parts of the molecules with identical anticodons; differences both in anticodons and in other parts of the molecules. In the future, when the localization not only of the anticodon, but also of the other functional centers of tRNA has been established, it will be possible to give a more complete analysis of the last two cases, but all that can be said at present is that a change has taken place in one or other segment of the tRNA molecule.

As was mentioned above, in the study of isoacceptor tRNAs attention was at first concentrated on the anticodon. Investigation showed that isoacceptor tRNAs obtained by fractionation of a tRNA react, when tested by the methods of Nirenberg and Matthaei (1961) and Nirenberg and Leder (1964) with different polynucleotides and trinucleotides (Weisblum et al., 1962; von Ehrenstein and Dais, 1963; Bennett et al., 1965; Kellog et al., 1966; Mirzabekov et al., 1967; Galizzi, 1967). For example, the results of experiments with $tRNA^{Asp}$ from *E. coli*, in which trinucleoside diphosphates differing from one another in the third component were used as templates, can be considered. The results in Table 22 show that $tRNA_1^{Asp}$ binds with the triplets G-A-U and G-A-C, while $tRNA_3^{Asp}$ binds with G-A-U only; $tRNA_1^{Asp}$ and $tRNA_3^{Asp}$ thus recognize codons which differ in their third nucleotide; in accordance with Crick's wobble hypothesis, it can be predicted that the anticodon of $tRNA_1^{Asp}$ must be the trinucleotide G-U-C, while the anticodon of $tRNA_3^{Asp}$ must be A-U-C, i.e., these isoacceptor tRNAs differ from each other in the first nucleotide of their anticodon. In this case no infomation is available regarding the rest of the polynucleotide chain. The structure of one yeast $tRNA^{Asp}$ has recently been discovered; its anticodon is actually the trinucleotide G-U-C. A less definite result has been obtained with $tRNA_2^{Asp}$.

Rather more is known about the valine-specific tRNAs. I have already mentioned that the binding of valyl-$tRNA^{Val}$ by trinucleoside-diphosphate–

TABLE 22. Binding of ^{14}C-Asp-tRNA^{Asp} from *E. coli* by Ribosomes in the
Presence of Trinucleoside Diphosphates (Galizzi, 1967)

Trinucleoside diphosphates	^{14}C-Asp-tRNA$_1^{Asp}$	^{14}C-Asp-tRNA$_2^{Asp}$	^{14}C-Asp-tRNA$_3^{Asp}$
		Δ pmoles	
G-A-U	2.38	0.63	1.29
G-A-C	1.57	0.12	0.22
G-A-A	0.02	—	0
G-A-G	0.05	−0.01	0.01
Without trinucleoside diphosphate, pmoles	0.15	0.27	0.30

ribosomal complexes (Table 22) has demonstrated different anticodons in tRNA$_1^{Val}$ and tRNA$_2^{Val}$: I-A-C and ψ-A-C (or U-A-C), respectively. The I-A-C anticodon of tRNA$_1^{Val}$ has been identified not only indirectly, but also by direct structural determination. tRNA$_2^{Val}$ has not yet been analyzed, but there is evidence that the differences between tRNA$_1^{Val}$ and tRNA$_2^{Val}$ are not confined to the anticodons. Grachev et al. (1965) have shown that oligonucleotides removed from the 3'-end of these two isoacceptor tRNAs by guanyl ribonuclease from actinomycetes are different. So far as the rest of the polynucleotide chain is concerned, the problem still awaits experimental solution. As I mentioned above, tRNA$_2^{Val}$ has been separated into several fractions, and these are at present undergoing structural analysis (Kryukov et al., 1971). Structural investigations have shown that the primary structure, including the anticodon, of tRNA$_3^{Val}$ is the same as for tRNA$_1^{Val}$; this is confirmed by the fact that valyl-tRNA$_3^{Val}$ is bound with ribosomes in the presence of the same trinucleotides as tRNA$_1^{Val}$ (Table 20). Its distinct separation from tRNA$_1^{Val}$ on fractionation by the countercurrent method can probably be explained by differences in their tertiary structure, which gradually disappear in the course of subsequent analysis (Krutilina et al., 1970).

The presence of two valine-specific tRNAs with differences at the acceptor end has also been shown for tRNA from rat liver (Ishida and Miura, 1965) and *Torulopsis utilis* (Miyazaki et al., 1967).

Credit for the first complete unraveling of the structure of two isoacceptor tRNAs, as has already been stated, goes to Zachau (Zachau et al., 1966b). Contrary to expectation, he found that the anticodons of these two principal forms of tRNA^{Ser} from brewers' yeast are the same, and that the differences between them occur in the three nucleotides located in accordance with the clover-leaf model in unpaired segments of the molecule, namely in the loop of the universal oligonucleotide and in the extra loop (Fig. 23). Compared with the 85 nucleotides composing the tRNA^{Ser} molecule, this is an extremely small change; however, because of the shielding of the universal oligonucleotide loop and of the extra branch (Igo-Kemenes and Zachau, 1971), i.e., their probable interaction with other segments of the molecule, it could well be that replacement of nucleotides in this segment may result in considerable changes in the tertiary structure of the molecule. Be that as it may, the differences in conformation of the molecules are the most likely explanation why physical separation of these tRNAs, with such a similar primary structure, is possible.

Similar results were obtained by Goodman et al. (1968) for tRNA^{Tyr} from *E. coli*. They found that tRNA$_1^{Tyr}$ and tRNA$_2^{Tyr}$ have the same anticodons, and the differences between them occur in nucleotides situated in the extra loop (Fig. 32A). Investigation of tRNA$_f^{Met}$ also revealed two components differing in only one nucleotide in the extra loop (7meG→A) (Fig. 34). It does not seem likely that the replacement of H$_2$U by U in the molecules of yeast tRNA^{Ala} and of tRNA^{Met} from *E. coli* can be explained by the existence of two isoacceptor tRNAs differing in these components, but rather that it reflects different stages in the modification of the corresponding tRNAs. As recent work has shown, replacement of H$_2$U in yeast tRNA^{Ala} by U has no effect on its aminoacylation, and alanyl-tRNA-synthetase recognizes both forms of tRNA^{Ala} equally well (Kuo and Keller, 1968).

Differences in the degree of tRNA modification depending on the conditions used have been frequently reported (Goodman et al., 1968; Cory et al., 1968; Favorova et al., 1968; Gefter and Russell, 1969), and this is unquestionably one reason for the multiplicity of the tRNAs.

The study of the properties of isoacceptor tRNAs with identical anticodons but differing in one or more nucleotides in unpaired segments of the molecule is of great interest, but the functional significance of the differences is still unknown. There are several examples in which differences between isoacceptor tRNAs occur in double-stranded segments of the molecule. This is the case, in particular, with tRNA$_{2A}^{Val}$ and tRNA$_{2B}^{Val}$ from *E. coli*, the primary structures of which were unraveled only recently (Yaniv and Barrell, 1971). These two tRNAs (Fig. 51) differ in altogether six nucleotides in the helical segments of the molecule (the 3′-5′-stem and the helical segment of the Tψ-branch). It is interesting to note that all six nucleotides are so arranged that the formation of the Watson–Crick pairs required by the clover-leaf pattern is undisturbed; complete pairs are replaced, and in the particular example with which we are concerned the three GC pairs in tRNA$_{2A}^{Val}$ are replaced by three AU pairs in tRNA$_{2B}^{Val}$.

Since the structure of tRNA$_1^{Val}$ from *E. coli* has been determined earlier (Yaniv and Barrell, 1969), it is now possible to compare three isoacceptor tRNAs, one of which, tRNA$_1^{Val}$, differs from the other two also in its coding properties. Since it has the anticodon oacU-A-C, it recognizes the triplets

G-U-A and G-A-G, whereas the anticodon of $tRNA_{2A}^{Val}$ and $tRNA_{2B}^{Val}$ is the triplet G-A-C which, as would be expected, recognizes the triplets G-U-U and G-U-C. The structural similarity between $tRNA_{1A}^{Val}$ and $tRNA_{2B}^{Val}$ is the less than that between $tRNA_{2A}^{Val}$ and $tRNA_{2B}^{Val}$ (**12a**); of the 76 nucleotides composing the molecule of $tRNA_1^{Val}$, only 53 are common with $tRNA_2^{Val}$; in addition, $tRNA_{2A}^{Val}$ and $tRNA_{2B}^{Val}$ have common nucleotides in another 18 positions. In this example isoacceptor tRNAs with identical anticodons differ less in their primary structure than tRNAs with different anticodons.

However, the next example, concerning the glutamine tRNAs from *E. coli* does not confirm this hypothesis (Folk and Yaniv, 1972). Like $tRNA_{2A}^{Val}$ and $tRNA_{2B}^{Val}$, $tRNA_1^{Gln}$ and $tRNA_2^{Gln}$ (Fig. 52) differ in six nucleotides, and just as in the case of $tRNA_2^{Val}$, these are nucleotides lying opposite each other in the clover-leaf model, and their coordinated replacement does not disturb the pairing; it is interesting to note that in this case the three pairs do not even differ in composition (except for the one substitution $U \rightarrow \psi$); the corresponding bases in the duplex segment merely change places.

By contrast with $tRNA^{Val}$ 2A and 2B, the anticodons of $tRNA_1^{Gln}$ and $tRNA_2^{Gln}$ differ in the first nucleotide, although the nature of N has evidently not yet been settled, and the question of whether the two glutamine tRNAs have identical or different coding properties still remains unanswered.

The discovery of isoacceptor tRNAs in which substitutions of nucleotides occur in the duplex segments and are coordinated so that the possibility of hydrogen bonding is undisturbed is of great interest. To begin with, it shows once again that the arrangement of hydrogen bonds in the clover-leaf model is not an artifact but reflects the true features of structure of the tRNA molecule. In addition, a study of the functional properties of these isoacceptor tRNAs may provide useful information about functional centers. The origin and biological significance of these tRNAs are also of considerable interest.

Serine-specific liver isoacceptor tRNAs have been studied by Staehelin and co-workers (Staehelin et al., 1969; Baguley et al., 1970; Müller et al., 1971; Staehelin, 1971). Staehelin points out that the heterogeneity of the

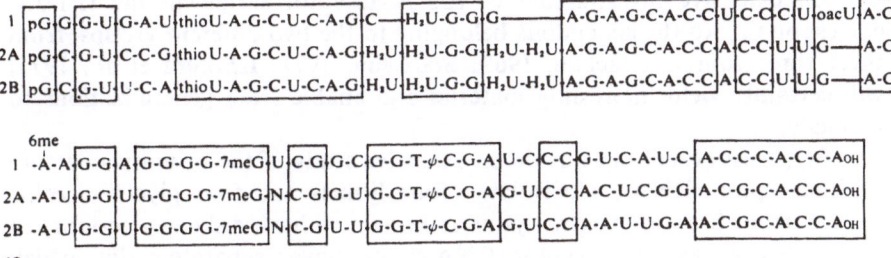

12a

tRNAs is much greater than was hitherto suspected. Fractionation of tRNA^{Ser} initially gave three fractions binding with ribosomes in the presence of different triplets and, consequently, differing in their anticodons; tRNA$_1^{Ser}$ responds to the three codons UCA UCU and UCC, tRNA$_2^{Ser}$ to UCG, and tRNA$_3^{Ser}$ to the codons AGU and AGC; these three tRNAs would thus be sufficient for the six codons known for serine. However, subsequent fractionation showed that these tRNAs are not homogeneous, but contain subfractions. Five fractions have been obtained and their structural analysis has shown that they differ from each other not only in their anticodons, but also in several nucleotides in the duplex segments of the clover leaf. It is interesting to note that in this case also there is coordinated replacement of two opposite nucleotides in the clover leaf, thus not disturbing the possibility of hydrogen bonding. In all these three cases (valine and glutamine tRNAs from *E. coli* and serine tRNAs from rat liver) such substitutions are found in all duplex segments as well as that of the H$_2$U-branch.

Staehelin (1971) concludes from his investigations of rat-liver tRNA^{Ser} that the heterogeneity of this tRNA is very great. He suggests that structural investigations so far undertaken have been done on one of the many isoacceptor tRNAs present in the cell, and he does not rule out the possibility that in some cases a mixture of tRNAs of very similar structure, and therefore very difficult to separate from one another, has been analyzed.

Fractionation of the serine-specific tRNAs from *E. coli* on DEAE-Sephadex and BD-cellulose and by reverse-phase chromatography has led to their separation into three fractions (Ishikura and Nishimura, 1968; Ishikura et al., 1971). The tRNA$_1^{Ser}$ responded to codons U-C-A and U-C-G, and less effectively to U-C-U, while tRNA^{Ser} 3A and 3B responded to A-G-U and A-G-C. Uridine-5-oxyacetic acid and 2-methylthio-N^6-isopentenyladenosine have been identified in tRNA$_1^{Ser}$, and thrA in tRNA$_{3A}^{Ser}$; tRNA$_{3B}^{Ser}$ is very similar to tRNA$_{3A}^{Ser}$ and contains a minor component similar to thrA (Ishikura et al., 1971). It is interesting to note that all three serine-specific tRNAs from *E. coli* are aminoacylated by enzymes from bakers' yeast and rat liver, so that a definite structural similarity may exist between the serine-specific tRNAs from these three sources.

An extensive investigation of the isoacceptor serine-specific tRNAs, corresponding to the six codons belonging to the two different groups, from these three sources (Zachau, 1969; Staehelin, 1971; Ishikura et al., 1971) will certainly yield interesting material and enable fresh generalizations to be made.

Bergquist (1966b) has made an extensive study of the primary structure of isoacceptor tRNAs. He isolated four lysine- and four glycine-specific tRNAs from brewers' yeast. Lysine is known to have only two codons, so that in this case each isoacceptor tRNA cannot have a separate codon, which suggested that other structural differences may exist. Experiments on the binding of glycyl-tRNA by trinucleotide–ribosomal complexes showed that all four tRNAs react about equally to the triplets G-G-U, G-G-C, and G-G-G. The same result was also observed with the four lysine tRNAs, which responded equally to poly(A), poly(A,A,G), and various copolymers containing A-G. The optical rotatory dispersion spectra of the four forms of tRNA^{Gly} showed only slight differences, indicating differences in the primary structures of these tRNAs. This was also demonstrated by direct structural analysis, but as yet only fragmentary results have been obtained and no definite conclusions can be drawn.

Bergquist considers that the isoacceptor tRNAs which he studied may arise from different organelles of the cell. For example, the mitochondria may have their own set of lysine-specific tRNAs, with the same facility for recognition as the cytoplasmic tRNAs, but with a slightly different primary structure. There is now experimental evidence that the mitochondria in fact contain specific tRNAs which differ from the extramitochondrial tRNAs in certain properties (Barnett and Brown, 1967; Suyama and Eyer, 1967; Buck and Mass, 1968; Epler, 1969).

In Bergquist's opinion, a change in primary structure sufficient to enable physical separation of isoacceptor tRNAs is bound to be important for the transmission of information. He proposes for these tRNAs the term "superfluous" rather than "degenerate." However, it is difficult to accept the presence of "superfluous" compounds in the cell; there is no doubt that with the passage of time the presence of this large "surplus" of isoacceptor tRNAs in the cell, as well as the role of each individual change in their primary structure, will be explained. It is possible that a change in the multiplicity of the tRNAs by a change in their synthesis and modification may play a regulatory role in the cell.

SPECIES DIFFERENCES BETWEEN INDIVIDUAL tRNAs

The species specificity of the tRNAs has been studied for a long time, mainly by the methods of homologous and heterologous aminoacylation. Extensive investigations in this direction were undertaken by Belozerskii and co-workers (Zaitseva et al., 1964; Glebov et al., 1965; Zaitseva et al., 1966). Their work demonstrated differences in the degree of species specificity of tRNAs specific with respect to different amino acids, ranging from its complete absence to absolute species specificity of both tRNA and aminoacylating enzyme. It is interesting to note that as the complexity of organization of the vertebrates increases, the specificity of the tRNAs and activating enzymes as a rule also increases. The work of Belozerskii and his collaborators led to the very important conclusion that besides the common features characterizing the structure of tRNAs and enzymes, there are also specific features which are found only in a given species of organism. Structural investigations of tRNAs have completely confirmed this conclusion.

It is already possible at this stage to draw certain comparisons between tRNAs on the taxonomic plane, because the nucleotide sequence of a number of tRNAs of identical amino-acid specificity, but obtained from different sources, has been determined. For example, we know the primary structure of serine tRNA from brewers' and bakers' yeast and also from rat liver; the nucleotide sequence has been established for valine- and tyrosine-specific tRNAs from *S. cerevisiae, T. utilis,* and *E. coli* and the phenylalanine tRNA from *S. cerevisiae, E. coli,* and wheat germ, and leucine-specific tRNA from *E. coli* and *S. cerevisiae.* This material, of course, is still inadequate to allow extensive generalizations to be made, but some preliminary conclusions are perhaps possible. For example, it can be accepted that the primary structure of tRNAs specific with respect to one amino acid, and obtained from different but taxonomically closer sources, differs less than that of tRNAs obtained from more distant sources. This can be seen by comparing the structural formula of tRNA^{Ser} from brewers' yeast (Zachau et al., 1966a) with results obtained by Neelon et al. (1967) for tRNA^{Ser} from bakers' yeast, i.e., tRNA from another strain of *S. cerevisiae.* All the oligonucleotides of total pyrimidyl- and guanyl-RNase hydrolysates of bakers' yeast tRNA^{Ser} were found to be identical with the oligonucleotides of brewers' yeast tRNA$_2^{Ser}$ (Zachau et al., 1966b). Determination of the oligonucleotide sequence in the polynucleotide chain is still incomplete in the case of bakers' yeast, but where it has been carried out, the structure proved to be the same. The identical assortments of minor components in tRNA^{Ser} from brewers' and bakers' yeast reduce the number of possible alternative structures and make the conclusion that they are completely identical even more likely.

There is only one point of disagreement between the results obtained by Zachau and Neelon, namely that only one tRNA^{Ser} is found in bakers'

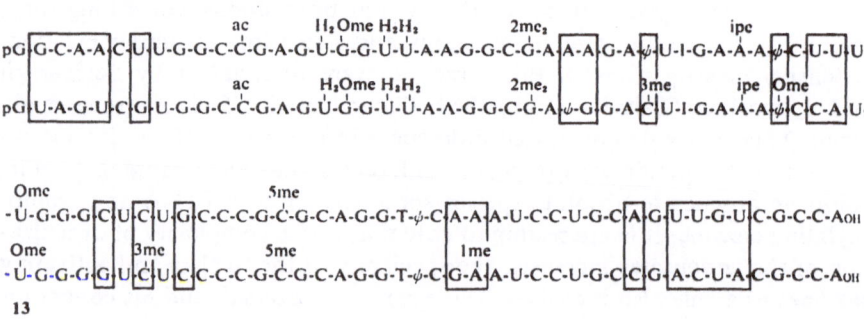

13

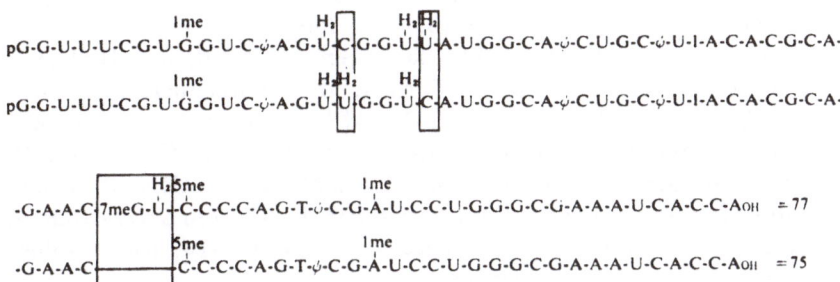

yeast, unlike in brewers' yeast. Hence, in this case we have the complete identity of the primary structures of tRNAs aminoacylated by one amino acid (serine) and isolated from two different, but closely related sources (bakers' and brewers' yeasts).* On the other hand, the primary structure of rat-liver tRNA^{Ser} (Staehelin et al., 1968) differs from that of yeast tRNA^{Ser}. The structural formulas (13) of tRNA^{Ser} from yeast (top line) and from rat liver (bottom line) are shown on page 198. Segments which differ are boxed.

Comparison shows that these sequences differ in 21 nucleotides, and in five cases the differences affect minor components:

Yeast tRNA^{Ser}	A	ψ	ψ	C	A
Rat-liver tRNA^{Ser}	ψ	3meC	Omeψ	3meC	1meA

In three cases the minor components of rat-liver tRNA^{Ser} consist of methylated derivatives of the principal nucleic-acid bases present in yeast tRNA.

The majority of the differences between yeast and rat-liver tRNA^{Ser} occur in the duplex segments of the clover leaf, but in every case, despite these modifications, hydrogen bonding is possible. The structure of the H$_2$U-branches of the two serine-specific tRNAs is identical. Their anticodons also are the same (I-G-A,) but the sequences to the left and right of the anticodons are alternatives of the same structure (A-3meC-U and ipeA-A-Omeψ in rat-liver tRNA^{Ser} and A-ψ-U and ipeA-A-ψ in yeast tRNA^{Ser}).

Work on the isolation and determination of the nucleotide sequence of tRNAs from *T. utilis,* belonging to the *Fungi imperfecti,* is being undertaken in Japan; from the investigations so far completed (Takemura et al., 1968, 1969a,b; Hashimoto et al., 1969) it is clear that the tRNAs of *T. utilis* are comparable with tRNAs from other sources.

Examination of the formulas (14) for tRNA^{Val} from *S. cerevisiae* (top line) and *T. utilis* (bottom line) shows that they differ in only four positions.

*The structure of the tRNA^{Val} from the same two sources is also evidently identical or very similar (Bonnet et al., 1971; unpublished data from the laboratories of Ebel and Baev).

Instead of the sequence A-G-H_2U-C-G-G-H_2U-H_2U-A-U-G in the H_2U-loop of *S. cerevisiae*, in the corresponding loop of *T. utilis* the sequence A-G-H_2U-H_2U-G-G-H_2U-C-A-U-G has been identified, i.e., H_2U and C have apparently changed places. Further, the undecanucleotide A-A-C-7meG-H_2U-5meC-C-C-C-A-Gp from tRNA$_1^{Val}$ of *S. cerevisiae* is represented in *T. utilis* by the nonanucleotide A-A-C-5meC-C-C-C-A-Gp, for the H_2U and 7meG residues are not present. The molecule of tRNA$_1^{Val}$ from *T. utilis* thus consists of 75 nucleotides, as compared with 77 in *S. cerevisiae*, and it differs from the latter in the absence of 7meG and H_2U (one residue of each) and in the sequences given above.

The similarity between *S. cerevisiae* and *T. utilis* also extends to the second tRNA^{Val}: analysis has shown that, like the tRNA$_2^{Val}$ from *S. cerevisiae*, the tRNA$_2^{Val}$ from *T. utilis* differs from tRNA$_1^{Val}$ in its anticodon and its acceptor end.

Examination of the structural formula established by Yaniv and Barrell (1969) for tRNA$_1^{Val}$ from *E. coli* shows that it resembles the structures of the tRNA^{Val} from *S. cerevisiae* and *T. utilis* no more than it does a tRNA of different amino-acid specificity. Not only the duplex segments but also the nucleotides of the loops and also the whole assortment of minor components are different. The anticodon also is different, because the anticodon is oacU-A-C, whereas in yeast tRNAs the anticodon contains inosine (I-A-C). Judging from the fact that the tRNA$_1^{Val}$ which was investigated by Yaniv and Barrell, and accounts for 80% of all valine-acceptor activity in *E. coli*, responds to the trinucleotides G-U-A and G-U-G, it corresponds more with the tRNA$_2^{Val}$ of *S. cerevisiae*, which is present in smaller quantity and which binds with ribosomes in the Nirenberg–Leder system in the presence of the same triplets (Mirzabekov et al., 1967). The primary structure of this tRNA has not yet been determined.

Comparison of yeast tRNA$_2^{Val}$ with the tRNA^{Val} 2A and 2B of *E. coli*, which have the G-A-C anticodon and very similar structural formulas (see the section on isoacceptor tRNAs) likewise reveals no marked features of similarity; however, it is interesting to note that in all forms of tRNA^{Val} the fourth pair in the stem connecting the ends of the molecle is an A-U pair, while the second and fifth pairs in the double-helical segment of the anticodon branch are G-C pairs; in all forms of tRNA^{Val} regardless of their source, the sequence G-C-A at the bend between the H_2U- and anticodon branches is also the same. This similarity may have some definite relationship to the amino-acid specificity of the corresponding tRNAs.

The same picture as in the case of the valine tRNAs is also observed when tyrosine tRNAs from the same three sources (15) are compared: tRNA^{Tyr} from *T. utilis* (the top line) and *S. cerevisiae* (the middle line) differ

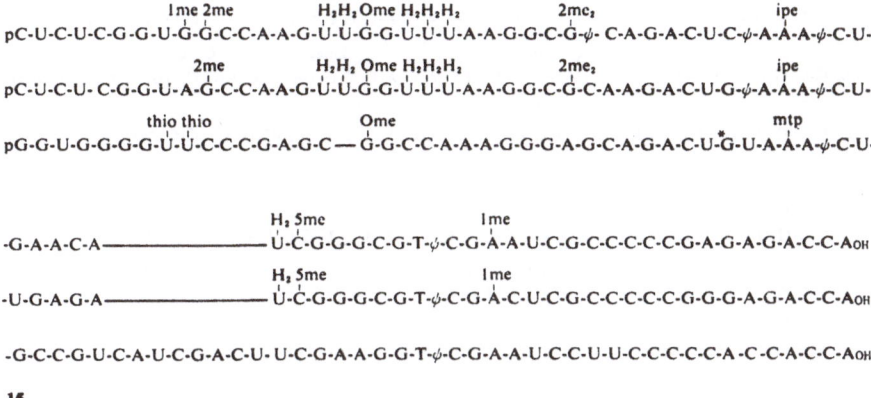

15

in only seven positions, whereas the tRNA^Tyr from *E. coli* (bottom line) has a completely different structure.

Comparison of tRNA^Tyr from yeast and *E. coli* showed that they possess absolute species specificity, i.e., they are aminoacylated only by their own, and not by the heterologous enzyme (Doctor and Mudd, 1963; Doctor et al., 1966, 1967; Doctor, 1967). However, binding with ribosomes is stimulated in both cases by the same trinucleotides, namely U-A-U and U-A-C. These observations suggest that there must be differences between the structures of tRNA^Tyr from yeast and *E. coli*, but they do not influence their coding properties and affect mainly the recognition site of the aminoacyl-tRNA-synthetase. Comparison of the primary structures of the tRNAs from these two sources does not conflict with the conclusions drawn from the indirect findings. The primary structure of the tyrosine tRNAs from yeast and *E. coli* is different, and their features of similarity with each other are only slightly greater than with tRNAs specific with respect to different amino acids. The greatest similarity is observed in the anticodon branch; two nucleotides in the anticodon itself are different but in such a way that its coding properties remain unchanged (substitutions $G \rightarrow G^*$ and $\psi \rightarrow U$). The tyrosine tRNAs from yeast and *E. coli* differ substantially in their content of minor components. tRNA^Tyr from *S. cerevisiae* and, in particular, from *T. utilis* are rich in minor components (14 and 18) respectively, whereas the tyrosine tRNAs from *E. coli* are distinguished by their extremely low content of minor components (6–7).

The available material now enables a comparison to be made of the phenylalanine tRNAs from three sources (**16**): *S. cerevisiae* (RajBhandary et al., 1967), wheat germ (Dudock et al., 1969), and *E. coli* (Barrell and Sanger, 1969a).

    2me   H₂H₂   2me₂   Ome Ome   5me

pG-C-G-G-A-U-U-U-A-G-C-U-C-A-G-U-U-G-G-G-A-G-A-G-C-G-C-C-A-G-A-C-U-G-A-A-Y-A-ψ-C-U-G-

    2me   H₂H₂   2me₂   Ome Ome

pG-C-G-G-G-G-A-U-A-G-C-U-C-A-G-U-U-G-G-G-A-G-A-G-C-G-ψ-C-A-G-A-C-U-G-A-A-Ẏ-A-ψ-C-U-G-

    4thio   H₂   H₂   mtp

pG-C-C-C-G-G-A-U-A-G-C-U-C-A-G-U-C-G-G-U-A-G-A-G-C-A-G-G-G-G-A-ψ-U-G-A-A-A-A-ψ-C-C-C-

    7me   5me   1me

-G-A-G-G-U-C-C-U-G-U-G-T-ψ-C-G-A-U-C-C-A-C-A-G-A-A-U-U-C-G-C-A-C-C- A_OH

    7me H₂   1me

-A-A-G-G-U-C-G-C-G-U-G-T-ψ-C-G-A-U-C-C-A-C-A-C-G-C-U-C- A-C-C-G-C-A-C-C-A_OH

    7me

-C-G-U-G-X-C-C-U-U-G-G-T-ψ-C-G-A-U-U-C-C-G-A-G-U-C-C-G-G-G-C-A-C-C-A_OH

16

The structures of the tRNA^{Phe} from *S. cerevisiae* (top line) and wheat germ (middle line) are very similar, and there are only 16 differences, mainly in the duplex segments. The H₂U-branch, the loop of the universal oligonucleotide, and the anticodon loop are absolutely identical except for modification of the minor component Y. Much greater differences are found in tRNA^{Phe} from *E. coli* (bottom line). In this case differences exist not only in the duplex, but also in the single-stranded segments; the assortments of minor components and also the anticodons are different; however, the changes in anticodon do not affect codon–anticodon interaction (OmeG-A-A and G-A-A).

The last comparison which can be made on the basis of the available material is that between the leucine tRNAs from *E. coli* (Dube et al., 1970) and yeast (Kowalski et al., 1971). Quite apart from the source of isolation, in this case a high degree of similarity might not be expected, because the leucine-specific tRNAs whose structures have been determined have anticodons with different 3′-terminal nucleotides, and as a result of this they recognize codons possessing different 5′-terminal nucleotides and evidently belonging to a group of widely differing isoacceptor tRNAs. In fact the similarity is small, but nevertheless some features of it are worth mentioning. The leucine-specific tRNAs from *E. coli* and yeast each have two tetranucleotides which, disregarding modifications, are identical:

E. coli	U-G-G-C	G-C-G-C
Yeast	U*-G-G-C	G-C-G*-C

These sequences lie on the 3′- and 5′-sides of the H₂U-branch, and they probably produce a marked similarity in this segment of the clover leaf in these two tRNAs.

The second essential analogy lies in the large size of the accessory branch, permitting the formation of Watson–Crick pairs. In view of results indicating that the nucleotides of the H_2U- and accessory branches interact spatially (Smith et al., 1971; Cashmore, 1971) it seems that during the formation of the tertiary structure of this tRNA a similar three-dimensional conformation may arise.

The last feature of similarity to be mentioned is that in both cases there there is a 1meG residue on the 3'-side of the anticodon, and this is still quite rare.

The examples given above show that the primary structures of the tRNAs are more similar in taxonomically related species, and tRNAs of more distant species differ considerably from each other. This is clear from the identity of the structure of tRNA^{Ser} from bakers' and brewers' yeast, and the very slight differences between the valine and tyrosine tRNAs from *S. cerevisiae* and *T. utilis* and between the phenylalanine tRNAs from *S. cerevisiae* and wheat germ. Meanwhile, the tRNAs of the same specificity from *E. coli,* a phylogenetically isolated group of organisms, as a rule have sharply different primary structures.

Differences are found in the minor components of tRNAs of different species, which is not surprising in view of the long-established species specificity of the modifying enzymes, although admittedly at present this applies only to the methylases. For instance, the tRNA from *E. coli* differs from yeast tRNA in its much lower content of methylated nucleosides and of dihydrouridylic acid, the absence of inosine and N_2-dimethylguanosine, and the presence of 4-thiopyrimidines.

It is interesting to note that most changes in the nucleotides of tRNAs of the same amino-acid specificity but isolated from different sources, disregarding minor components, are located in the duplex segments of the clover leaf; as a rule both nucleotides are replaced in such a way that a new Watson–Crick base pair is formed. As Dudock et al. (1969) rightly point out, this is evidence that the pairing of nucleotides of tRNA as in the clover-leaf model is a reflection of their primary structure, otherwise it would have to be postulated that mutation of the DNA, leading to a change in one of the bases of the tRNA, would be accompanied by mutation in a completely different locus.

As the results given above show, some tRNAs from different sources have different anticodons, although they accept the same amino acid; for instance, in the phenylalanine tRNAs from *S. cerevisiae* and wheat germ the codon is the trinucleotide OmeG-G-A, whereas in that from *E. coli* it is G-G-A; in the tyrosine tRNAs from *S. cerevisiae* and *T. utilis* the anticodon is the triplet G-ψ-A, whereas in that from *E. coli* it is G*-U-A. It will be seen that these changes do not affect codon–anticodon interaction, nor do they conflict with the view of the universality of the genetic code.

There is likewise no contradiction in cases in which the anticodons of tRNAs of the same amino-acid specificity, but obtained from different sources, differ in their 5′-terminal (for example, tRNA^{Val} form *S. cerevisiae* and *E. coli*) or their 3′-terminal (for example, tRNA^{Leu} from *S. cerevisiae* and *E. coli*) nucleotides. In the first case, codons belonging to the same group are recognized, i.e., those differing only in their 3′-terminal nucleotide. It is evident that in the second case, although the tRNAs are of the same amino-acid specificity, they recognize different codons: in the case of yeast tRNA^{Leu} the codon UUG, and in the case of tRNA^{Leu} from *E. coli*, the codon CUG. Thus in this case also there are no contradictions: the code is universal, but its concrete use differs in different species, because the various isoacceptor tRNAs are represented differently. The importance of this in regulation may be that different codons corresponding to isoacceptor tRNAs are read at different intensities.

Chapter 4

MINOR COMPONENTS OF tRNA*

Characteristic structural elements of the tRNAs are the minor, rare, atypical, unusual, or accessory components which are derivatives of the four principal nucleic-acid bases. These compounds have attracted considerable attention for various reasons. To begin with, there was the problem of their origin: it was not understood how a tRNA, containing so many compounds of this type, can be synthesized by the complementary principle on a DNA consisting largely of four principal nucleotides. Investigators studying the primary structure of the nucleic acids showed great interest in these minor components, for they saw these compounds as convenient natural reference points making determination of the nucleotide sequence so much easier. Finally, research workers are still puzzled by the role of the minor components in tRNA.

With the improvement of precise analytical methods, and also with the rapid progress made in the isolation of individual tRNAs and the study of their primary structure, the number of the known minor components continues to grow steadily, and it already exceeds thirty. It is true that there have been cases when what have been described as minor components later turned out to be artifacts (Hemmens, 1964), and the existence of others has still to be confirmed by subsequent research.

The minor components vary widely in their chemical structure, and their classification is a complicated task. Methylated, deaminated, acetylated, sulfur-containing, and many other types of derivatives of the nucleic-acid bases are now known. If the chemical nature of the substitutent or its position in the molecule is used as the basis of classification, the minor components

---

*A full survey of the modified nucleosides in nucleic acids known at the present time can be found in R. H. Hall's book *The Modified Nucleosides in Nucleic Acids* (Columbia University Press, New York, 1971).

can be subdivided as follows. The most numerous and the longest-known group consists of compounds of the four principal nucleic-acid bases methylated in the ring or in the exocyclic amino group. The second group contains the so-called 2'-O-methylribosides, in which the 2'-hydroxyl group of the ribose is methylated. A unique position among the minor components is occupied by ψ, in which the uracil and ribose residues are linked, not by an ordinary N-glycoside, but by a C-glycoside bond. The compound 5,6-dihydrouridine, a reduced derivative of uridine, recently discovered in nucleic acids, also stands apart (Madison and Holley, 1965). Minor components containing sulfur have been found (Lipsett, 1965; Carbon et al., 1965). If the principle of classification of the minor nucleotides on the basis of the chemical nature of the substituent is carried to its end, a whole new series of groups is obtained, but some of them contain only one compound; these minor components include isopentenyladenosine, acetylcytidine, 5-hydroxyuridine, etc. Finally, minor nucleotides with two or even three substituents are known: compounds methylated in the heterocyclic ring and also in the ribose residue (Nichols and Lane, 1966): 2-methylthio-N^6-(Δ^2- isopentenyladenosine) (Harada et al., 1968); U with a sulfur atom in position 2 and a methylcarboxymethyl group in position 5 (or 6) (Baczynskyj et al., 1968); and so on. This method of classification is thus virtually impracticable at the present time.

Szer (1966) proposed classifying minor components by their ability to form complexes. This approach is particularly valuable today because of a tendency to ascribe a role to modified nucleotides in the formation or, at least, the reinforcement of the secondary structure of tRNA, and such a classification could help with the elucidation of this problem. In his first group, Szer includes compounds with increased affinity for complementary bases (ψ, 5meC, 5meU). His second group contains nucleosides forming much weaker complexes with complementary bases (methylamino-derivatives of adenosine, cytidine and, conjecturally, guanosine). The third group is formed by nucleosides incapable of interacting with complementary bases (N-methyl-U, N-methyl-C, H_2U).

The most logical classification of the minor nucleotides, however, would appear to be that based on their role in the tRNA, but this is still unknown. There is also the possibility that their classification can be based on the position occupied by the minor components in the polynucleotide chain of the tRNA, and in the future this will probably be linked up with the origin and role of these rare bases.

I shall adhere to the classification which I have used previously (Venkstern, 1964): the minor components are grouped depending on the nucleic-acid bases by whose modification the derivative is obtained; the 2'-O-methylnucleosides and the thio-derivatives are placed in separate sections.

This arrangement is not only convenient for description, but is also justified by the following considerations. In practical work with the minor nucleotides it is easier to identify the base than the modification which has taken place. For analytical purposes it is more convenient to have data for all derivatives of a particular base grouped together. In addition, the principal nucleic acid bases and their various derivatives are often located in corresponding segments of the polynucleotide chain in different tRNAs, most probably because of their common origin, and this classification thus reflects the genesis of the minor components to a considerable degree.

As a final word on classification it must be pointed out that doubts are now being expressed about the inclusion of T, ψ, and H_2U among the minor components. These compounds are found in all the individual tRNAs which have so far been studied [with the exception of H_2U, which is absent in the tRNA^{Tyr} from *E. coli* and the tRNA^{Gly} from *Staphylococcus epidermidis,* which contains a single minor component, 4thioU (Stewart et al., 1971)]. If, therefore, the minor bases are regarded as giving the tRNAs their features of individuality, this cannot apply to compounds, especially T and ψ, which occur as components of the universal oligonucleotide. Moreover, the content of ψ and H_2U in the molecules of individual tRNAs is fairly considerable, reaching 4–6 moles per mole of tRNA in some of them. The presence of these modified components in the molecules of tRNA may be presumed to have some special significance. It is interesting that all three compounds are derivatives of U.

The chief difficulty when working with minor components is that they are present in small amounts. Even before methods for isolating individual tRNAs existed and all work was carried out on total preparations, calculations showed that the individual minor components are not present in all polynucleotide chains of tRNA. This has been fully confirmed by the study of individual tRNAs. Many minor components have still been found only in a single tRNA (meI in yeast tRNA^{Ala}, Omeψ in rat-liver tRNA^{Ser}), others are found in only two or three, but not in all individual tRNAs. Naturally it was impossible to analyze these compounds when only a total preparation could be investigated, but as the individual tRNAs were studied they began to be found. Nevertheless, the isolation of minor components in quantities sufficient for analysis is still a difficult task.

The second difficulty is that many minor components differ only slightly in their chemical structure and chromatographic and spectrophotometric properties from the corresponding principal nucleosides, and it is these properties which lie at the basis of their isolation and identification (guanine and 1-methylguanine, adenine and 1-methyladenine). Furthermore, many of the minor components are highly labile; for a long time this hindered their detection and identification (H_2U, acC, Y). Finally, the modified bases have a considerable effect on the velocity of enzymic (and sometimes chemical) cleavage of their phosphodiester bonds, and this is a serious complication in structural research.

Minor components have been isolated from nucleic acids in the form of bases, nucleosides, and 2'-, 3'-, and 5'-nucleoside monophosphates; this fact, together with others, indicates that the minor nucleotides are regular components of the polynucleotide chain and participate in the formation of a normal 3',5'-phosphodiester bond.

The question of whether minor bases are specific components of tRNA or whether they are present in all nucleic acids still remained unanswered until quite recently. These compounds have now been found in all types of nucleic acids except messenger RNAs. However, the content of minor components is greatest and their variety widest in the transfer RNAs.

THE ORIGIN OF THE MINOR COMPONENTS

Despite the great variety of the minor components of tRNA, essentially the same group of problems arises during the study of their biosynthesis: (1) what is the source of the chemical group whose attachment causes the modification; (2) does the modification take place at the polynucleotide level or is an already modified monomer incorporated into the polynucleotide; and (3) what are the mechanisms which control biosynthesis of the minor component?

I have the impression that the answer to the second and third questions is the same for all minor components, although experimental proof has been obtained in by no means every case: modification of the principal nitrogenous bases, leading to the formation of minor components, takes place after formation of the polymer (at the polynucleotide level), through the action of enzymes (or enzyme systems) which recognize certain elements of the three-dimensional structure of the tRNA. Naturally the answer to the first question cannot be equally universal, because the modifications which result in the formation of the different minor components are extremely varied. Most commonly they consist of the addition of chemical groups, but they may also take the form of reorganization (ψ), reduction (H_2U), deamination (I), etc. Cases of the addition of two or more chemical groups at several positions of the nucleoside or the successive modification of one position by several enzymes are known. In these cases the question not only of the origin of the chemical groups arises, but also of the order in which they are added by different enzymes.

Let us now look at the experimental material relating to the biosynthesis of minor components which is available at the present time.

The problem of the origin of the methylated nucleosides can be regarded as solved. Fleissner and Borek (1962) discovered enzymes, which were called nucleic acid methylases, with the property of transferring a CH_3-group to bases composing a polynucleotide chain. As a result it became evident that the methylated components are not incorporated into the nucleic acid molecules, but are formed in them as the result of the transmethylation reaction after the polynucleotide chain has been formed. An important role in the elucidation of this problem was played by the discovery of an *E. coli* mutant which unlike the ordinary auxotrophs, continued to synthesize all types of nucleic acids when protein synthesis was blocked (Borek et al., 1955). This mutant (*E. coli* K-12 W6), auxotrophic with respect to methionine, was used in an extensive series of investigations to study methylation of nucleic acids, and the principles governing this process, with which we are now familiar, were discovered.

The universal source of CH_3-groups for the methylated components of the nucleic acids, as in all other processes of biological methylation, is *S*-adenosylmethionine, a product of the enzymic activation of methionine (Cantoni, 1951). The nucleic acids synthesized by *E. coli* K-12 W6 in the absence of methionine differ from those usually formed by having a lower content of methyl groups. These submethylated nucleic acids were used as the substrate for the investigation of enzymic methylation.

Three groups of methylases were found, to correspond with the three types of nucleic acids: tRNA-, rRNA-, and DNA-methylases (Gold et al., 1963a). As was to be expected, none of these groups of methylases is homogeneous, but all include several enzymes of different specificity: they methylate different nucleic acid bases and also introduce CH_3-groups into the same base at different positions.

The tRNA-methylases were purified and fractionated by several investigators. Hurwitz et al. (1964a,b), for instance, separated methylases from *E. coli* into six fractions: one methylates U; a second methylates C; three fractions possess activity against guanine; and the adenine-methylase is concentrated in one fraction but catalyzes the formation of three different derivatives of adenosine (2meA, 6meA, 6me$_2$A). This last fraction is evidently not homogeneous, and purification of the corresponding enzyme has not yet been completed. Fractionation of methylases from extracts of *S. cerevisiae* led to the isolation of eight separate fractions. Four methylated derivatives of guanine, 1-methyladenine, 5-methylcytosine, and thymine were found among the methylation products; three fractions methylated uracil only and possessed no other activity (Björk et al., 1968). The properties of tRNA-methylases from rat liver were studied by Rodeh et al. (1967). These workers concluded that enzymes from mammalian tissues have particularly strict specificity, for they are inactive in heterologous systems (see below). Among the reaction products of homogolous methylation, 5meC, 1meA, T, 7meG, and 2meG (or 2me$_2$G) were found. On the other hand, Simon et al. (1967) demonstrated the possibility of heterologous methylation with enzymes from rat and cat brain.

Transfer RNA methylases also were isolated from rat liver by Baguley and Staehelin (1968a). The enzyme, which was obtained by chromatography on DEAE-cellulose, transferred a CH$_3$ group from S-adenosylmethionine to the amino group of guanine and to position 1 in the adenine residue of submethylated tRNA from E. coli. After further fractionation on Sephadex G-200, the enzyme methylating adenine in position 1 was obtained in an almost pure form (Baguley and Staehelin, 1968b).

Normal, completely methylated tRNA, although not accepting methyl groups in the presence of the homologous enzyme, can incorporate excess methyl groups (supermethylation) under the influence of heterologous enzyme systems (Gold et al., 1963b). It was concluded from these results that the methylases possess both species and strain specificity (Srinivasan and Borek, 1963; Gold et al., 1963b; Svensson et al., 1963; Rodeh et al., 1967; Simon et al., 1967; Phillips and Kjellin-Straby, 1967). These results of structural analysis suggest that the specificity of the methylases is determined by the structural features of tRNAs of different species, on which the number and position of the methyl groups in their molecule depend. Accordingly, of the large number of residues of any given nucleotide in the polynucleotide chain, only those few which occupy a strictly definite position in it will be methylated. The methylases evidently distinguish not only the base, but also its position in the polynucleotide chain, i.e., they are able to recognize nucleotide sequences or, what is more likely, particular spatially organized sites. The problem of recognition of nucleotide sequences by methylases has been investigated by Baguley and Staehelin (1968a,b; 1969). Comparison of nucleotide sequences surrounding 1meA in tRNA from rat liver and yeast, and also from yeast methylated in vitro by liver enzyme, showed that 1meA iş found in all cases in the following six sequences: G-1meA-U; A-1meA-U; G-1meA-C; G-1meA-A-U; A-1meA-A-U, and G-1meA-A-U. Since the sequences surrounding 1meA in yeast tRNA are identical with the sequences methylated by the liver enzyme in yeast in vitro, it is evident that it is not the nucleotide sequence or, at least, not that sequence alone which determines which nucleotides will be modified. These workers conclude that the leading role belongs to the tertiary structure of the molecule, which is also evidently the principal factor concerned in recognition of tRNAs by aminoacyl-tRNA-synthetase (Lindahl et al., 1966; Gartland and Sueoka, 1966). Comparison of oligonucleotides containing 1meA isolated in the experiments of Baguley and Staehelin with sequences identified in individual tRNAs led these workers to conclude that methylation of A in position 1 takes place in the universal oligonucleotide loop. It is in fact localized in this loop in all transfer RNAs in which this minor component is found.

The role of conformation in tRNA methylation has been demonstrated by the work of Shershneva et al. (1971). Methylation of tRNA$_1^{Val}$ from yeast by a partially purified enzyme fraction from liver or hepatoma led to the incorporation of a methyl group into one of the oilgonucleotides in the 3'-half of the molecule. The attempt to methylate the isolated 3'-half was unsuccessful. However, methylation reached virtually the original level when the 3'- and 5'-halves were mixed, in which case, as is known, an associated molecule with similar characteristics to the native molecule is formed (Baev et al., 1967b,c). These experiments indicate that for the enzyme to act, it is not sufficient to have long sequences on the 3'- and the 5'-sides of the base to be methylated, but the nucleotide to be methylated must be an integral element in a particular three-dimensional structure. At the same time, these results provide the answer to the old problem: what happens first, the formation of the tertiary structure or the methylation of the tRNA molecule? If a particular conformation of the molecule is an essential condition for methylase action, clearly it must be formed first, and only then will the methylases begin to act and incorporate methyl groups into the various segments of the molecule in accordance with their specificity.

Interaction between specific methylases and nucleic acids thus determines the high degree of specificity of distribution of the CH_3-groups in the polynucleotide chain. This probably applies to all modifying enzymes and to all minor components formed with their participation.

The reaction of methylation of tRNA is irreversible and can be expressed by the equation

$$tRNA + S\text{-adenosylmethionine} \xrightarrow{\text{tRNA-methylase}} methyl\text{-}tRNA + S\text{-adenosyl-}$$
$$homocysteine$$

S-Adenosylhomocysteine is an inhibitor of the methylation reaction, and its action is competitive in character and can be abolished by increasing the quantity of S-adenosylmethionine. S-Adenosylmethionine cannot be replaced by any other donor of CH_3-groups.

Methylases of tRNA are extremely widespread, and are found in practically all tissues so far investigated, except spermatozoa. They are found, in particular, in the soluble part of the cytoplasm (Burdon et al., 1967; Culp and Brown, 1968; Burdon, 1970). Not only species and strain, but also tissue specificity are evidently exhibited by the tRNA-methylases (Christman and Borek, 1967).

Changes in the methylase activity have been reported in the brain of mammals during development (Glasky and Simon, 1966; Simon et al., 1967); during metamorphosis of insects (Baliga et al., 1965); in the adult liver compared with the embryonic liver (Hancock, 1967a,b); and during the life cycle of the mold *Dictyostelium discoideum* (Pillinger and Borek, 1969). These results suggest that the process of methylation may play a regulatory role in protein synthesis. The tRNA-methylases change the structure of one of the principal instruments of protein synthesis, and this must have some influence on the process.

Results suggesting a role of methylation in carcinogenesis have also been obtained. The RNA of tumor tissues is known to have the highest content of methylated bases (Bergquist and Matthews, 1965; Wainfan et al., 1966; Viale et al., 1967). Under the influence of alkylating carcinogens, moreover, a high content of methylated bases comparatively rarely found in the nucleic acids of normal tissues, is found in tRNA (Magee and Farber, 1962); methylation takes place rapidly (Craddock and Nakai, 1962; Silber et al., 1966) and the utilization of methionine in the cells is increased in leukemia (Weisberger et al., 1954); tRNA-methylase activity is increased during neoplastic growth (Hancock, 1967a,b, 1968; Brown and Attardi, 1965; Tsutsui et al., 1966, 1967; Mittelman et al., 1967; Silber et al., 1967; Baguley and Staehelin, 1968b); the relative proportions of different methylated bases are changed, and methylated nucleotides not characteristically found in normal tissue are formed under the influence of enzymes from transplanted tumors (Tsutsui et al., 1966, 1967); the elution profile of tRNA from Novikoff's hepatoma differs from that of tRNA from normal liver (Baliga et al., 1969); all these observations indicate a causal connection between carcinogenesis and the process of methylation of nucleic acids. The work of Baguley and Staehelin (1968b) has shown that activity of 1-methyladenine-methylase is higher in the leukemic spleen than in the healthy spleen, but the substrate specificity of the enzyme is unchanged.

Since the problem of changes in the content of minor components in tumor cells remains unsolved, it is continually being investigated by new methods. For instance, Randerath et al. (1971) have investigated it by thin-layer chromatography, using tritiated tRNA (Randerath et al., 1969). Although these workers were unable to obtain any definite increase in the percentage content of minor components in the tRNA of brain tumors even by this highly sensitive method, they do not rule out the role of minor components in neoplastic growth. They draw attention to the central position of tRNA in the translation of the genetic code and point out that even a very slight modification of any single tRNA may have a considerable influence on the course of protein synthesis.

If the distribution of CH₃-groups in tRNA is a variety of code, used for example in the recognition of aminoacyl-tRNA-synthetases, their redistribution under the influence of foreign methylases is bound to have some effect on this process. The suggestion has repeatedly been made (Srinivasan and

Borek, 1963; Tsutsui et al., 1966) that foreign methylating agents and, in particular, oncogenic viruses, introducing heterologous methylases into the cell where they methylate tRNAs in new positions, may be natural carcinogens. The possibility thus is not ruled out that factors inducing irregular methylation of tRNA may be initiators of malignant growth.

It has recently been postulated that a change in methylase activity can result from the action of inhibitors of various types (Pillinger and Borek, 1969). In fact, inhibitors of different nature have been found in a wide variety of tissues, and they are being intensively investigated at the present time. Kerr (1970) has described the presence of an inhibitor of tRNA methylases, which is evidently protein in nature, in the organs of higher animals; the inhibitor is much less active in tumor tissues and is absent altogether from embryonic tissues. Later, Kerr (1971) succeeded in separating the inhibitor into a high-molecular-weight protein fraction and a low-molecular-weight fraction and showed that the protein component is not present in embryonic or tumor tissues. It is possible that the increased methylase activity in these tissues can be explained by the absence of this inhibitor. An active inhibitor of tRNA-methylases of polysaccharide nature has been found in spermatozoa (Sheid and Wilson, 1970a,b), hitherto considered to be the only cells without methylase activity. A protein inhibitor has also been found in *Dictyostelium discoideum* (Sharma and Borek, 1970). The suggestion has been made that inhibitors of tRNA-methylases may perform regulatory functions in differentiating tissues, excluding some tRNAs and including others.

An interesting observation with regard to the formation of minor components has been made by Goodman et al. (1968); during an investigation of the primary structure of tRNA^{Tyr} from *E. coli* they observed that its normal minor components were absent if the cells were preliminarily starved; the following residues were left unmodified: G in the anticodon, usually found as an unknown derivative of G; A on the 3'-side of the anticodon, and also the G which, under normal conditions, is converted into OmeG. A similar phenomenon was reported by Cory et al. (1968) in a study of the structure of tRNA^{Met} from *E. coli*. Here, for the first time, changes in the modification of precisely localized minor components were observed in individual tRNAs depending on the conditions of growth of the cells.

Observations of interest from this point of view were made by Sarkar and Comb (1966), when investigating submethylated tRNA from *E. coli* and tRNA from ribosomes of *Blastocladiella emersonii;* in experiments *in vitro* they observed modification of U, leading to the formation of the universal oligonucleotide. The following sequences were identified in the guanyl-RNase hydrolysates of these tRNAs: U-U-C-G, U-ψ-C-G, T-U-C-G, and T-ψ-C-G. These workers regard this discovery as proof of the formation of the tetranucleotide T-ψ-C-G at the polynucleotide level from the sequence

U-U-C-G by methylation of one uridyl residue (U→T) and isomerization of the other (U→ψ), in the process of formation of tRNA from its precursor. The existence of precursors of tRNA, differing in the greater length of their molecule and in the low degree of methylation and low content of ψ, can now be taken as established. The properties of the precursor and the process of its conversion into tRNA are now being studied (Burdon and Clason, 1969; Bernhardt and Darnell, 1969; Burdon, 1970; Smillie and Burdon, 1970).

The problem of the origin of the CH₃-group in the ribose residue of RNA has not received such detailed study. Nevertheless, experiments *in vivo* (Brown and Attardi, 1965; Dubin and Günalp, 1967; Phillips and Kjellin-Straby, 1967; Nichols and Lane, 1967, 1968, 1969) have shown that methionine is also the source of the CH₃-group in ribose. Experiments *in vitro* with an enzyme isolated from *S. cerevisiae* and tRNA from *E. coli,* showed (Björk et al., 1968) that methylation of the 2'-OH-group of ribose takes place at the polynucleotide level, and that *S*-adenosylmethionine is the donor of CH₃-groups.

Gefter (1969) observed transfer of the CH₃-group of *S*-adenosylmethionine to the 2'-OH-group of the guanosine residue of the GG sequence in tRNA^{Tyr} under the influence of a cell-free extract from *E.coli.* He also showed that *S*-adenosylmethionine is a donor of CH₃-groups for the synthesis of 2-methylthio-N^6-isopentenyladenosine in tRNA from *E. coli.*

Enzymes transferring a CH₃-group to ribose have not been isolated, but it is logical to suppose that this group of methylases is not homogeneous but consists of several enzymes. A simple method of determining their activity has been published (Baskin and Dekker, 1967).

Results have been obtained which show that the methyl group for esterification of 5-carboxymethyluridine and 2-thio-5-carboxymethyluridine is also donated by *S*-adenosylmethionine (Tumaitis and Lane, 1970; Kwong and Lane, 1970). Just as in the case of mtpA, these workers consider it more likely that the first act in the biosynthesis of 2-thio-5-carboxymethyluridine is carboxymethylation in position 5, and the second act is thiolation in position 2 (Kwong and Lane, 1970).

The origin of the H₂U in the polynucleotide chain of tRNA has not yet been finally settled. According to the findings of Roy-Burman et al. (1967), dihydrouridine triphosphate can be incorporated into RNA by DNA-dependent RNA-polymerase. The reaction velocity is low, but according to calculations it is sufficient to account for the comparatively small quantity of H₂U residues present in tRNA. Nevertheless, Roy-Burman et al. consider that the highly specific distribution of H₂U in tRNA could more easily be produced by hydrogenation of particular U residues after the formation of the polynucleotide chain, and they consider that this is the more likely mechanism. The same opinion is held by other investigators (Srinivasan and Borek, 1966).

The probability of formation of H_2U in tRNA by hydrogenation of particular U residues at the polynucleotide level is confirmed by the discovery of two components with only one difference—replacement of one U residue by H_2U—in preparations of individual tRNAs (tRNA^{Ala} from *S. cerevisiae* and tRNA^{Met} from *E. coli*); it is assumed that in these cases the modification is incomplete.

The problem of biosynthesis of ψ (5-ribosyluracil) is particularly interesting because of its unusual structure, and its investigation is still proceeding. It has been shown that both the pyrimidyl and the ribose components of uridine are direct precursors of ψ in yeast tRNA (Robbins and Hammond, 1962). This is evidence that ψ is formed as a result of intramolecular reorganization of U, but it is still not clear whether this takes place before or after the incorporation of uridine into the polynucleotide chain. Hall and Allen (1960) have postulated that ψ is formed from uridine by a mutase mechanism in accordance with the following reversible reaction:

1-ribosyluracil + 1,5-diribosyluracil $\rightleftharpoons$ 1,5-diribosyluracil + 5-ribosyluracil

Evidence in support of this hypothesis was the discovery of free 1,5-diribosyluracil (Lis and Lis, 1962).

Later Lis and Lis (1964) isolated this compound from alkaline and phosphodiesterase hydrolysates of tRNA, a fact which indicated its existence as a component of the polynucleotide chain, and the postulated mechanism was thus transferred to the polynucleotide level. However, according to the observations of Dlugajczyk and Eiler (1966), the spectrum of the compound isolated from tRNA was not identical with the spectrum of the synthetic preparation of 1,5-diribosyluracil. Evidence in support of the formation of ψ after assembly of the polynucleotide chain was obtained by Weiss and Legault-Demare (1965). These workers showed that an essential condition for ψ formation is the preliminary synthesis of RNA, and that agents inhibiting polymerization of ribonucleotides, such as actinomycin D, also inhibit ψ formation. The conclusion that ψ is synthesized at the polynucleotide level was also reached by Ginsberg and Davis (1968) on the basis of experiments in which mutants of *E. coli*, auxotrophic with respect to U, were grown in the presence of 2-$^{14}C\psi$.

Further important evidence in support of ψ formation by reorganization of U at the polynucleotide level was obtained by a study of the tRNA^{Tyr} of mutant Su$^-$12. Much of the ψ in position 40 in tRNA^{Tyr} and paired with A32 consisted of uridine (Abelson et al., 1970). These workers postulated that this mutant is less able to carry out the reorganization of U$\rightarrow\psi$, so that sometimes up to 20% of the U remains unmodified.

The suggestion has been made that I and 1meI are formed from A and 1meA respectively, present in the polynucleotide chain, by a deamination process (Crick, 1966; Srinivasan and Borek, 1966). Deaminases of this type, by modifying the nucleotides of the anticodon, could be responsible for degeneracy of the code, to the extent that the information is altered by the conversion A→I.

Kamen (1969) has shown that synthesis of I in tRNA is independent of the content of free inosinic acid in the cell, and on this basis he also has concluded that I in tRNA is formed at the polynucleotide level by deamination of certain adenyl residues in the tRNA precursor, and not by incorporation of ITP into tRNA during its assembly. Kamen and Spengler (1970) later showed that the addition of cold hypoxanthine to ^{14}C-adenine during growth of *E. coli* B-94 does not lead to a decrease in label in the inosine of the tRNA. This is evidence against the incorporation of I-monomers into the polynucleotide chain during construction, and in support of their formation by deamination of particular A residues at the polynucleotide level.

The formation of sulfur-containing nucleotides of tRNA has been studied independently by two groups of investigators (Hayward and Weiss, 1966; Peterkofsky and Lipsett, 1965; Lipsett and Peterkofsky, 1966). At first the tRNA-thiolating system of *E. coli* was studied by determining the conditions necessary for the introduction of sulfur into tRNA in general, i.e., into all thionucleotides. Experiments *in vitro* demonstrated the transfer of sulfur from ^{35}S-cysteine to tRNA in the presence of a water-soluble extract from *E. coli* and from other tissues. Analysis of the alkaline hydrolysate of the ^{35}S-tRNA thus obtained revealed several reaction products containing ^{35}S. The corresponding enzyme was purified and its properties were then studied (Hayward et al., 1966; Lipsett and Peterkofsky, 1966; Lipsett et al., 1967; Norton and Lipsett, 1967).

Work has now begun on the investigation of the biosynthesis of individual thio-derivatives. For instance, Abrell et al. (1971) have described the isolation and purification of a sulfur-transfer system from *E. coli* giving only one reaction product, namely 4thioU. This system was separated into two enzymes, the first of which had an absolute requirement of ATP, bivalent metal ions, tRNA, and a sulfhydryl derivative. The second enzyme used cysteine as its source of sulfur and the product of the first reaction as the receptor; tRNA either from *E. coli* or from yeast, but no other polynucleotide, could act as the sulfur acceptor. This indicates that a definite conformation of the molecule is evidently also essential for the action of these enzymes.

The results described above thus show that thionucleotides in tRNA are most probably formed by an enzymic mechanism at the polynucleotide level at the expense of sulfur from cysteine. However, the question arises: where is the sulfur incorporated if the experiment is carried out with native, i.e., completely thiolated, tRNA and with the homologous enzyme system? In fact, in the discussion of this problem, doubts have been expressed whether a simple exchange reaction takes place. Its possibility has been discussed, but it is considered unlikely, first, because the enzymes readily incorporate sulfur into tRNA with low thionucleotide content, and second, because the scale of the incorporation into normal tRNA is so slight (0.1 % of all existing molecules) that it resembles the finishing of incomplete chains more than the replacement of sulfur already in them. These workers postulate that many specific factors are responsible for the first stage of the reaction, namely activation of nucleotides in particular positions for thiolation; the second factor may be nonspecific, and may transfer sulfur to any preactivated position of any nucleotide.

Considerable progress has also been made by the second group, which is studying the sulfur-transfer system of *B. subtilis* (Wong et al., 1970). This evidently differs from the corresponding system of *E. coli,* because in this case the more active S donor is β-mercaptopyruvate rather than cysteine; also, the reaction product may include other compounds in addition to 4thioU. It is interesting to note that heterologous thiolation has also been observed: incorporation of sulfur by extracts from *E. coli* into liver tRNA, and even into yeast tRNA which, as has previously been shown, does not contain sulfur under ordinary conditions (Hayward and Weiss, 1966).

Many workers have studied the mechanism of ipeA biosynthesis (Hall et al., 1967b; Fittler et al., 1968a,b; Peterkofsky, 1968; Vickers and Logan, 1970; Bartz et al., 1970). All problems concerned with ipeA, including its biosynthesis, are described in detail in the surveys by Hall (1970, 1971). The source of the Δ^2-isopentenyl group has been shown to be mevalonic acid, which provides a five-carbon unit for the biosynthesis of all isoprenoid compounds. During growth of bacteria (Fittler et al., 1968a; Peterkofsky, 1968) or of a cell culture (Chen and Hall, 1969) on medium containing [2-^{14}C] DL-mevalonic acid, the label is found in the ipeA of tRNA and only there. The same result has been obtained for yeast and rat liver (Fittler et al., 1968b).

It was comparatively easy to answer the question whether the isopentenyl group is added to the A residue in the polynucleotide chain, because a tRNA lacking in an isopentenyl group was obtained and was used as the substrate in these investigations. By treating tRNA with permanganate the isopentenyl groups can be removed, without any apparent injury to the molecule as a whole (Kline et al., 1969). The tRNA thus obtained was used as the substrate for the homologous enzyme, and Δ^2-isopentenylpyrophosphate, an intermediate in the reaction utilizing the isopentenyl group of mevalonic acid, was used as donor of the isopentenyl group. The reaction product was identified as N^6-(Δ^2-isopentenyl)adenosine (Fittler et al., 1968b; Kline et al., 1969).

In this way it was shown that the isopentenyl group of mevalonic acid is transferred to an adenosine residue present in the polynucleotide chain by a specific enzyme which is widely distributed.

So far ipeA has been found in four individual tRNAs, and in every case it occupies the analogous position, next to the anticodon on the 3'-side. The factors determining the choice of this particular A residue by the enzyme as the object for its action are not precisely known; however, the structural similarity in the anticodon branch of the tRNAs containing ipeA is considerable, and there is no doubt that a specific receptor site for the isopentenyl group is formed here.

A sequence consisting of three A residues, the middle one of which is modified, is found in tRNA^{Ser} from *S. cerevisiae* and rat liver and also in the tRNA^{Tyr} from *S. cerevisiae* and *T. utilis* on the 3'-side of the anticodon loop. In all four cases, the first two pairs of the helical segment of the anticodon branch are also identical (except that in rat-liver tRNA^{Ser}, the ψ is methylated in the ribose moiety), increasing the similarity still further:

$$
\begin{array}{cc}
\text{N} & - \text{N} \\
\text{G} & - \text{C} \\
\text{A} & - \psi \\
\text{N} \qquad & \text{A} \\
\text{U} \qquad & \text{ipeA} \\
\text{N} \qquad & \text{A} \\
\text{N} &
\end{array}
$$

The fact that enzymes from yeast and rat liver can incorporate a significant quantity of isopentenyl groups into tRNA from *E. coli* untreated with permanganate is regarded by the workers cited as proof of the presence of molecules in which modification has not yet taken place, and they consider this to be an indication that transcription is independent of, and more rapid than, modification (Kline et al., 1969). However, the possibility that species specificity of the enzymes transferring the isopentenyl group plays a role in this case cannot be ruled out.

A system for investigating heterologous synthesis of ipeA in tRNA from *E. coli* has been developed by Vickers and Logan (1970). A yeast extract incorporated one isopentenyl group for every ten molecules of tRNA.

As was mentioned above, *S*-adenosylmethionine is the source of the methyl group for mtpA (Gefter, 1969). There is also experimental evidence that mtpA is formed by the consecutive addition of isopentenyl- and methylthio groups. This is shown by the fact that after incomplete modification the compound N^6-($\varDelta^2$-isopentenyl)adenosine is found in tRNA^{Tyr} from *E. coli*, but not 2-methylthioadenosine (Gefter and Russell, 1969).

It follows from the facts described above that tRNA synthesis is a multistage process in which the first step is polymerization of the nucleotides by RNA polymerase on the DNA template; followed by trimming of the tRNA precursor and modification of some of the nucleotides by various enzymes bringing about the thiolation, methylation, reduction, and conversion of the N–C bond into a C–C bond, and so on. The action of modifying enzymes is determined by the structure of the tRNA, because they can recognize particular nucleotide sequences or three-dimensionally organized sites. The result of the action of these enzymes is a specific assortment of minor components, arranged in a definite manner, and the stabilization of a rigid secondary structure of the molecule. There is reason to suppose that the formation of minor components is a later refinement to the simpler and older process of synthesis of the polynucleotide chain. The general structural plan of tRNA and the analogous arrangement of the modified nucleotides in their molecule are direct indications that the different tRNAs were formed from some common prototype, and that they acquired their distinctive individual features largely through the subsequent modifying action of enzymes. Enzymic modification of preformed macromolecules is not confined to RNA. DNA molecules also contain methylated nucleotides, formed by the action of specific methylases. The suggestion has been made that the action of the DNA methylases may play an important role in the formation and evolution of species (Vanyushin et al., 1968). Protein biosynthesis is also completed by processes of methylation, phosphorylation, acetylation, etc., of the polymer with the resulting formation of structures which cannot be directly coded by DNA.

THE ROLE OF THE MINOR COMPONENTS

It is already quite clear that there is no such thing as a role of the minor components in general. The various minor components, with their different chemical structures and differing in their ability to form hydrogen bonds, cannot perform identical functions in the tRNA molecule. Furthermore, even the same minor component, if occupying different positions in the polynucleotide chain, may perform different functions.

Studies of the primary structure of individual tRNAs have led to certain generalizations regarding the minor components, and as experimental data continue to accumulate, these may be confirmed or, on the contrary, disproved. For example, it can now be said that the minor components exhibit definite species specificity. Yeast tRNAs have a large number (9–16) of minor components and they are rich in I and H_2U. The tRNA from *E. coli* is much poorer in most minor nucleotides, particularly in methylated nucleotides (6–7); but it is rich in thiopyrimidines. Species differences in minor components must be expected, because the methylases, the only enzymes modifying RNA which have been studied to any great extent, possess species specificity. The existence of modified nucleotides related to the species specificity of the tRNA is thus very probable. On the other hand, some minor components, and even their combinations, are located in analogous segments of all tRNAs regardless of their source of origin. These include the T-ψ pair in the universal oligonucleotide, which is evidently related to a function common to all tRNAs, conjecturally to their nonspecific binding with ribosomes.

The characteristic arrangement of the modified nucleotides in the clover leaf is remarkable, and it shows much in common when individual tRNAs are compared (Fig. 54). For instance, 7meG is always located in the extra branch; 2me$_2$G invariably occupies the same position between the anticodon branch and the branch of dihydrouridylic acid; 4thioU has been identified in only one segment of the clover leaf of tRNA from *E. coli,* that joining the dihydrouridylic acid branch and the helical segment formed by the ends of the molecule; and so on. Other modified nucleotides are strictly individual; the first which must be mentioned here is the minor component next to the anticodon on the 3'-side and consisting of a modified adenosine in nearly all tRNAs: meI in yeast tRNA^{Ala}; ipeA in yeast tRNA^{Ser} and tRNA^{Tyr}; mtpA in tRNA^{Tyr} from *E. coli;* and so on. It can be supposed that these "hypermodified" (Schweizer et al., 1969) minor components give the anticodons still more individuality, and thus minimize mistakes during coding. As was mentioned above, participation of the nucleotide next to the anticodon on the 3'-side in codon–anticodon interaction has been demonstrated in tRNA^{Ser}, tRNA^{Phe}, and tRNA^{Tyr}. Selective modification of the ipeA in tRNA^{Ser} by treatment with iodine does not affect acceptor activity but decreases the binding of the seryl-tRNA^{Ser} with ribosomes in the Nirenberg–Leder system (Fittler and Hall, 1966; Hall, 1967). Modification of ipeA by bisulfite likewise did not affect the acceptor activity of yeast tRNA^{Tyr}, but impaired the ability of tyrosyl-tRNA^{Tyr} to bind with the mRNA–ribosome complex (Furuichi et al., 1970). After selective removal of the base from nucleoside Y by treatment with acid, yeast tRNA^{Phe} still remains capable of aminoacylation, but completely loses its ability to recognize the codon, it does not bind with ribosomes in the presence of poly U, and it does not transfer phenylalanine to a growing polynucleotide chain (Thiebe and Zachau, 1968a).

Gefter and Russell (1969) have isolated suppressor tyrosine tRNAs (Su$_{III}^+$) with the same primary structure but differing in the degree of modification of the nucleoside on the 3'-side of the anticodon. All three forms were found to be capable of accepting tyrosine, but in the absence of an isopentenyl group in the minor component on the 3'-side of the anticodon, they do not bind with ribosomes in the Nirenberg–Leder system.

These experiments point to the undoubted role of the minor component next to the anticodon on the 3'-side in codon–anticodon interaction. Removal of the isopentenyl group or of the whole base from the Y residue probably changes the conformation of the anticodon loop so profoundly that the codon–tRNA–ribosome complex can no longer be formed.

Ghosh and Ghosh (1970) have in fact shown that removal of Y changes the configuration of the anticodon loop of tRNA[Phe]. This affects the relative orientation of the wobbling nucleoside, and as a result it affects the coding properties of the anticodon.

In this discussion of the role of minor components, the work of Stewart et al. (1971) must not be overlooked. They showed that the tRNA[Gly] of *Staphylococcus epidermidis*, which participates in the synthesis of peptidoglucan, but which cannot function in protein synthesis, contains only one minor component, namely 4-thiouridine. The discovery of this tRNA is of great interest and its study may shed light on many problems connected with the minor components. Can the molecule be given a clover-leaf configuration? What features of its structure prevent the modifying action of the enzymes? Is its incompetence in protein synthesis explained by the absence of the universal nucleotide? What are the structural features which enable glycyl-tRNA-synthetase to recognize this tRNA? As these workers rightly point out, the minor components, with the possible exception of 4thioU, evidently have no significant role in this process.

Up to the present time, investigators who have studied the role of minor components have concentrated their attention on methylated derivatives of the nucleotides which, as has already been mentioned, constitute the most widely distributed and best studied group of modified nucleotides.

The study of the functional role of methylated nucleotides in tRNA has not been as fruitful as the investigation of their structure and synthesis, and for this reason there are no clear ideas on this subject, although from what is known it is clear that the methylated bases must play some specific role.

This problem has been studied by various methods, and with the development of knowledge and the improvement of techniques it has become clear that much of the earlier work must be disregarded. This applies, in particular, to cases in which preparations of unfractionated submethylated tRNA from *E. coli* K-12 W6, which are mixtures of approximately equal quantities of methylated and nonmethylated molecules (Starr, 1963), have been used. As Yamane and Sueoka (1964) have shown, it is possible for an amino acid to be transferred from one tRNA to another in the systems used to study incorporation of amino acids into the polypeptide chain, so that differences in, e.g., acceptor activity among individual tRNAs cannot be detected in these systems. Accordingly, methylated and nonmethylated tRNAs must be physically separated from each other, and only then can they be investigated. Most current investigations take this factor into account.

The acceptor activity of submethylated tRNA has been studied in several laboratories (Starr, 1963; Littauer, 1964; Peterkofsky et al., 1964); it has been shown to be comparable with the activity of tRNA containing a normal number of methylated nucleotides, but in nearly every case there were slight differences indicating that the methyl groups—directly or indirectly—modify this function. For instance, Peterkofsky and co-workers (Peterkofsky, 1964; Peterkofsky et al., 1966) showed that synthetase from *E. coli* can aminoacylate both normal and submethylated tRNA^{Leu}, whereas synthetase from yeast is much less active against submethylated tRNA. Revel and Littauer (1966) investigated the acceptor function of tRNA^{Phe} and observed that the optimum concentration of Mg^{2+} for submethylated tRNA is much lower than for normal tRNA. Shugart et al. (1968a) isolated four submethylated fractions of tRNA^{Phe} from a methionine auxotroph mutant of *E. coli* 58–161, and concluded from their experiments that there is a correlation between the degree of methylation and acceptor activity. In a later investigation, the same workers (Shugart et al., 1968b) found a marked decrease in the acceptor activity of tRNA as the result of a decrease in the number of methyl groups, and restoration of acceptor activity with respect to phenylalanine and histidine to the original level after methylation with homologous enzyme.

Although few results have been obtained in relation to transport function, they are analogous: submethylated tRNA can transfer amino acids, but differences are always observed between its activity and the activity of a a tRNA with a normal content of methyl groups. In the system of Nirenberg and Leder (1964), methylated leucyl-tRNA binds with ribosomes in the presence of poly(U,G) and poly(U,C), while the submethylated leucyl-tRNA does so only in the presence of poly(U,C) (Capra and Peterkofsky, 1966, 1968; Peterkofsky et al., 1966). Investigation of tRNA^{Phe} has shown that absence of CH$_3$-groups makes it more ambiguous for the coding of genetic information, for it recognizes not only poly(U), like tRNA^{Phe} with a normal set of methyl groups, but also poly(U,C), to which normal tRNA^{Phe} does not bind (Littauer et al., 1963). More recent experiments (Revel and Littauer, 1966; Littauer et al., 1966) have also shown that the methyl groups of tRNA are concerned with reading the code. On the other hand, Fleissner (1967) observed no increase in the number of coding mistakes in an investigation of submethylated tRNA^{Phe} or of tRNA with an excess of methyl groups introduced with the aid of heterologous enzyme. These workers attribute the slight decrease in activity with respect to the transfer of phenylalanine to the polypeptide chain in two of the submethylated fractions into which tRNA^{Phe} could be separated to their partially denatured state. Methyl groups in fact modify the physical parameters of tRNA. This is shown, first, by the possibility of separation of tRNAs differing in their content of methyl groups: tRNA^{Leu} (Lazzarini and Peterkofsky, 1965; Capra and Peterkofsky, 1968), tRNA^{Phe} (Revel and Littauer, 1966), etc. Second, a decrease in absorption at 260 nm has been found after introduction of CH$_3$-groups into submethylated tRNA with the aid of homologous enzyme (Borek and Christman, 1965). A hypochromic effect of the same order is observed during chemical methylation of polyadenylic acid (Ludlum et al., 1964). Methylation, whether enzymic or chemical, evidently endows the structure of nucleic acids with a higher degree of orderliness. The same conclusion was reached some years ago by Shugar and Szer (1962), who compared the properties of polyuridylic and polythymidylic acids. The melting temperature of polyuridylic acid is 8.5° C, which indicates that this polymer exists in solution as a random coil. Polythymidylic acid, which can be regarded as the completely methylated derivative of polyuridylic acid, has a melting temperature of 36° C, indicating a high degree of orderliness. Comparison of the melting curves of double-stranded complexes also showed that methylation of one partner increases the melting temperature by 19–20° C (Shugar and Szer, 1962; Szer, 1965).

Methylation of the nitrogenous bases at atoms participating in hydrogen bonding abolishes the template function of synthetic ribonucleotides. Examples of such ribonucleotides are poly-N^1-methyluridylic and poly-N^1-methylcytidylic acids (Szer, 1966) and poly-N^1-methyladenylic acid (Ludlum, 1966). This has led some workers to regard methylation as a possible mechanism of blocking the template activity of polynucleotides (Nakada 1965).

It can be accepted on the basis of the foregoing facts that methyl groups help to determine the biological properties of tRNA. The influence of the methyl groups is most probably indirect, and is exerted chiefly on the secondary structure of the molecule. The degree and character of the change in functions of tRNAs of different specificity varies depending on the number, type, and location of the methylated bases. The remarks which have been made about minor components in general apply equally to methylated bases: methylated nucleotides located in different segments of the polynucleotide chain do not necessarily perform identical functions. Now that the primary structure of several tRNAs is known, the time has arrived for a differential approach to the individual methylated nucleotides, but in view of what has been said already, the scarcity of the results obtained in the past is not surprising.

A promising approach to the study of the functions of minor components is their selective chemical modification. If the primary structure is known, modification of any minor component, especially of one present in the molecule in the proportion of 1 mole per mole of tRNA, can provide useful information about its functional role. An essential condition in such research is that the verification of the site and degree of the resulting modification must be carried out by the same methods and with the same degree of meticulous care as determination of the primary structure. This is particularly important in the case of minor components which repeat in the molecule, because the reactivity of nucleotides depends on their location in the polynucleotide chain, a fact which has been used to study the conformation of tRNA. In addition, chemical reagents as a rule do not possess absolute specificity, and their action may spread to some extent to other nucleotides, so that this also requires experimental verification.

Several agents which selectively modify one or more minor components are known; however, very few investigations in which modification has been accompanied by fractionation of the reaction products and the necessary structural studies have yet been carried out. Some of these can be mentioned.

Japanese workers have shown that acrylonitrile cyanoethylates I, ψ, and to a lesser degree U in yeast tRNA (Yoshida and Ukita, 1965a, 1968). In high salt concentrations about two-thirds of the ψ residues are resistant to acrylonitrile, whereas the modification of I continues (Yoshida and Ukita, 1965b). This reaction has acquired special interest in connection with the identification of I in the anticodons of some tRNAs. A study of the effect of cyanoethylation on the biological properties of various tRNAs has shown that the ability of valyl- and alanyl-tRNAs, which contain I in their anticodon, to bind with mRNA–ribosome complexes is lost; the modification had virtually no effect on binding of lysyl- and phenylalanyl-tRNAs (Yoshida et al., 1967b, 1968a). Modification with acrylonitrile thus showed that the I residue in tRNA^{Val} and tRNA^{Ala} plays an important role in codon recognition, thus indirectly confirming identification of the anticodon.

Yoshida et al. (1968b) later investigated the cyanoethylation reaction on purified tRNA^{Ala} from bakers' yeast. They showed that as the result of mild cyanoethylation, under conditions safeguarding the secondary structure, only the I in the anticodon is modified; tRNA modified in this way completely lost its ability to bind with ribosomes in the presence of the trinucleotides G-C-U, G-C-C, and G-C-A, but still preserved its acceptor activity. The cyanoethylation reaction thus confirmed that the anticodon of tRNA^{Ala} is the triplet I-G-C, and that it does not participate in the acceptor function.

The use of the same reagent has shed some light on the secondary structure of the tRNA molecule and the localization of its ψ residues. The fact that modification of ψ in tRNA in the presence of 0.5 M NaCl or 2 mM MgCl$_2$ takes place in two stages suggested that ψ occurs in tRNA in two different conformational segments. Modification of the tRNA of T. utilis by acrylonitrile in a high salt concentration, followed by hydrolysis of the modified tRNA with guanyl RNase and fractionation of its oligonucleotides, showed that the ψ residue in the oligonucleotide T-ψ-C-G is resistant to acrylonitrile and that, consequently, this oligonucleotide lies in a stereochemically protected segment of the molecule (Yoshida and Ukita, 1966; Yoshida et al., 1967b, 1968b). The same conclusion has also been drawn from the results of modification by other agents (Nelson et al., 1967; Brostoff and Ingram, 1967; Litt, 1969; Bollack et al., 1969; Erdmannn et al., 1969).

Rake and Tener (1966) cyanoethylated tRNA with acrylonitrile in aqueous solutions of dimethylsulfoxide. They showed that a relationship exists between the water content, the secondary structure, and the degree of modification of ψ. Total cyanoethylation, accompanied by loss of the secondary structure, prevented aminoacylation by 90%. These workers concluded that ψ maintains the secondary structure of tRNA, and this is essential for aminoacylation to take place. Since all tRNAs behaved identically in these experiments with respect to secondary structure and aminoacylation, there is evidently one common ψ residue, characteristic of all tRNAs, which plays a decisive role both in secondary structure and in acceptor function. Possibly this is the ψ residue in the universal oligonucleotide.

This has been proven by the subsequent experiments of Ofengand and collaborators. Choosing as their test object $tRNA_f^{Met}$ from *E. coli,* which has only one ψ residue, namely that in the universal oligonucleotide, and carrying out the modification at 60° C, Ofengand et al. (Siddiqui et al., 1970) showed that under these conditions the acrylonitrile reacts with these ψ residues and disturbs the tertiary structure of the molecule, leading to total loss of its acceptor activity. The development of these experiments on halves of tRNA molecules has enabled conclusions to be drawn regarding the three-dimensional organization of the $tRNA^{Met}$ molecule (Siddiqui and Ofengand, 1971; Krauskopf and Ofengand, 1971).

Using the method of modification by acrylonitrile, Ofengand (Ofengand, 1965, 1967; Ofengand et al., 1966) observed the total loss of acceptor activity of tRNA from *E. coli;* the suggestion was made that this result is attributable to cyanoethylation of the thioU residues, with which tRNA from *E. coli* is richly supplied. The role of thioU residues for the acceptor and transfer functions of tRNA from *E. coli* has also been investigated by Japanese workers (Saneyoshi and Nishimura, 1967; Nishimura et al., 1967a). After treatment with cyanogen bromide, the acceptor activity of tRNA from *E. coli* relative to lysine and glutamic acid was reduced by 80–90%, but its activity relative to phenylalanine, tyrosine, and valine was reduced by only 30–40%; acceptor activity relative to other amino acids, for example histidine, was unchanged. The change in transfer function occurred completely differently; for example, histidyl-tRNA ceased to bind with ribosomes, while tyrosyl-tRNA bound, not with the poly(A_5, U)–ribosome complex, as in the control, but with the poly(U_4, G)–ribosome complex. These workers consider that the reason for the change in recognition of the codon was modification of the tRNA anticodon, in particular because the action of cyanogen bromide on previously aminoacylated tRNA led to the same result. They postulated that the second letter of the anticodon of $tRNA^{Tyr}$ from *E. coli* must be thioU, and treatment with cyanogen bromide fixes its enolic form; as a result, the thioU can now pair with G but not with A. The primary structure of tRNA^{Tyr} from *E. coli* has now been determined; the thioU derivative was not in the anticodon, but a 2thioA was found to the right of it (Goodman et al., 1968); the facts observed must therefore be explained otherwise. Results obtained by Japanese investigators confirm that the 3'-neighbor of the anticodon has a direct bearing on codon–anticodon interaction.

Cerutti and co-workers (Cerutti et al., 1966; Rottman and Cerutti, 1966; Cerutti and Miller, 1967) developed a method of reductive degradation of H$_2$U with sodium borohydride:

$$\text{(structure)} \xrightarrow{\text{NaBH}_4} O=\overset{\overset{\displaystyle NH_2}{|}}{C}-\underset{\underset{\displaystyle H}{|}}{N}-CH_2-\underset{\underset{\displaystyle H}{|}}{CH}-CH_2OH$$

Molinaro et al. (1968) studied this reaction and concluded that H$_2$U is not essential for the acceptor function of tRNA, or for the binding of aminoacyl-tRNA with ribosomes. As they rightly point out, this is confirmed indirectly by the fact that some tRNAs have no H$_2$U in their composition (e.g., the tRNA^{Tyr} from *E. coli*).

Cerutti (1968) himself drew a different conclusion regarding the functional role of H$_2$U. In his experiments, NaBH$_4$ reduced the valine-acceptor activity of a total preparation of tRNA by 50%, but left serine-acceptor activity unchanged. Cerutti attributes this difference to the H$_2$U residue in the 3'-half of tRNA$_1^{Val}$, but absent in tRNA^{Ser}. However, it has been shown that replacement of the U residue in positon 48 of yeast tRNA^{Ala} by H$_2$U does not affect aminoacylation; yeast alanyl-tRNA synthetase recognizes both forms of tRNA^{Ala} equally (Kuo and Keller, 1968). Igo-Kemenes and Zachau (1969) confirmed and supplemented the results obtained by Cantoni's group (Molinaro et al., 1968). They also showed that treatment of tRNA with NaBH$_4$ does not affect restoration of the CCA-end.

Reagents which act on several components of tRNA, and not on only one, are also known. They include, for example, water-soluble carbodiimide, which reacts with ψ and I, and also with U and G (Knorre et al., 1966; Girshovich et al., 1966; Brostoff and Ingram, 1967).

It is clear from the facts described above that chemical modifications are widely used to study the role of minor components; however, the results of these experiments can be accepted as reliable only if they are based on the known primary structure and are accompanied by analysis of the modified polynucleotide.

Let us now turn to an examination of the individual groups of minor components, paying attention principally to results obtained by studying individual tRNAs, and avoiding, as far as possible, repetition of matter described previously (Venkstern, 1964).

ADENOSINE DERIVATIVES

Four methylated derivatives of adenosine have long been known (17): 1-methyladenosine, 2-methyladenosine, N^6-methyladenosine, and N^6-dimethyladenosine. Their formulas are given below.

1-Methyladenosine   2-Methyladenosine N^6-Methyladenosine

17

N^6-Dimethyladenosine N^6-(Aminoacyl)adenosine N^6-(N-formyl-α-aminoacyl)-adenosine

Isolation of these compounds as bases, nucleotides, and nucleosides, and their chromatographic and spectrophotometric properties have been described by several workers (Adler et al., 1958; Littlefield and Dunn, 1958a,b; Dunn, 1961). Information on the content of methylated adenosines in tRNAs from bacteria and animals has been described by Dunn et al. (1960) and by Zaitseva et al. (1962). Let us now look in greater detail at the A derivatives which have been identified in individual tRNAs in the course of structural research.

Doubts were long expressed about the presence of 1-methyladenosine in tRNA because of the readiness with which it is transformed into N^6-methyladenosine. However, the work of Dunn (Dunn, 1961, 1963; Dunn et al., 1960) confirmed the presence of 1-methyladenosine, at least in the case of total tRNA. Since then, 1meA has been identified in several individual tRNAs from S. cerevisiae and T. utilis, in tRNA^{Ser} from rat liver, and in tRNA^{Phe} from wheat germ. In all these cases, 1meA occupies the same posi-

tion in the polynucleotide chain; it is on the 3′ side of the universal oligonucleotide. The slight difference in chromatographic and spectrophotometric
properties of 1meA and A in acid and alkaline media does not always allow
them to be distinguished with confidence. The principal feature permitting
reliable identification is migration of the methyl group from the N^1-position
on the exocyclic amino-group, with the resulting conversion of 1meA into
6meA (Macon and Wolfenden, 1968), clearly distinguishable from A in all
its chromatographic and spectrophotometric properties. Migration of the
methyl group is known to take place at the base and nucleotide levels; in
the case of the nucleotide it takes place more readily, probably because of the
increased basicity of the exocyclic amino-group relative to N^1 under the
influence of the phosphate radical (Zakharyan et al., 1968). Migration probably takes place more readily still in oligonucleotides and polynucleotides,
because the partial conversion of 1meA into 6meA in oligonucleotides has
frequently been described under comparatively mild conditions (Madison
et al., 1967a; Venkstern et al., 1968a,b; RajBhandary et al., 1968b). Naturally
migration of the methyl group, although facilitating identification in cases
when it is carried out intentionally, is highly undesirable if it takes place
spontaneously: all the fragments containing 1meA thereby become
"duplicated" to give both isomers.

A study of the reduction of 1meA in oligonucleotides and in the native
tRNA molecule with sodium borohydride led to the conclusion that this
minor component falls into the screened category; it is postulated that 1meA,
like other positively charged nucleosides (7meG), stabilizes certain segments
of the tRNA molecule by its interaction with negatively charged phosphate
groups (Igo-Kemenes and Zachau, 1969, 1971).

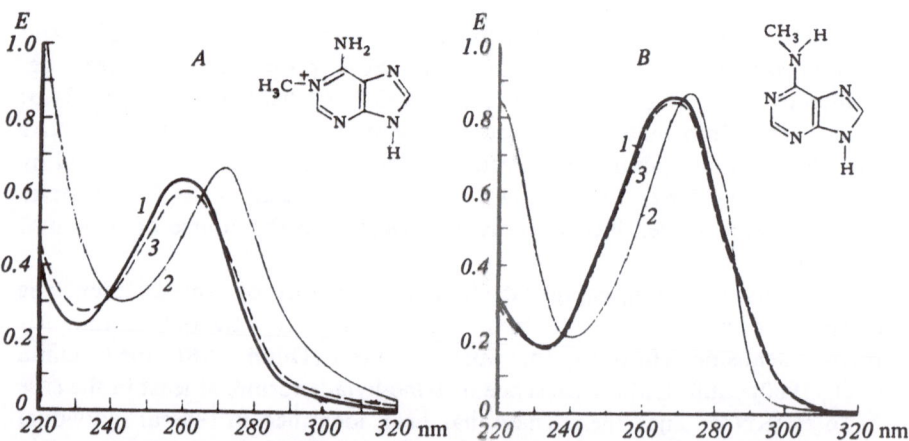

Fig. 57. Absorption spectra of 1-methyl-6-aminopurine (A) and 6-methylaminopurine (B):
(1) in 0.1 N HCl; (2) in 0.1 N KOH; (3) after bromination.

TABLE 23. Spectrophotometric Parameters and Chromatographic Mobility on Paper of Adenine and Its Derivatives

Compound	Acid medium			Alkaline medium			R_f in solvent[a]			
	λ_{max}, nm	λ_{min}, nm	$\varepsilon_{260} \times 10^{-3}$	λ_{max}, nm	λ_{min}, nm	$\varepsilon_{260} \times 10^{-3}$	1	2	3	4
Adenine	263	229	13.1	269	237	12.3	0.37	0.61	0.28	0.35
1-Methyladenine	259	228	11.7	270	239	14.4	0.39	0.53	0.20	0.25
6-Methylaminopurine	267	232	15.1	273	239	—	0.49	0.72	0.42	0.58

[a]Solvents: (1) isopropanol–11.6 N HCl–H_2O (680:176 to 1 liter H_2O); (2) isopropanol–H_2O (70:30), NH_3 in gaseous medium; (3) n-butanol–98% HCOOH–H_2O (77:10:13); (4) n-butanol–H_2O (86:14), NH_3 in gaseous medium.

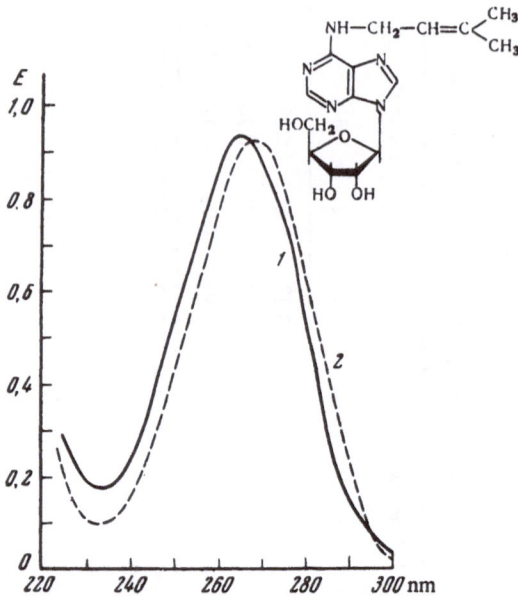

Fig. 58. Absorption spectrum of N^6-isopen-
tenyladenosine: (1) in 0.1 N HCl; (2) in
H_2O.

An A derivative, provisionally identified by its discoverers as 1meA or
6meA, was discovered in the tRNA$_1^{Val}$ from *E. coli* on the 3′-side of the anticodon (Yaniv and Barrell, 1969). It was later shown that this minor component
is in fact 6meA (Harada et al., 1969; Saneyoshi et al., 1969). This was thus
the first case of identification of 6meA in an individual tRNA and it gave
final confirmation to the observations of Dunn et al. (1960), who concluded
that 6meA is present in tRNA from *E. coli* and is not formed in it during
treatment. The spectra of 1meAde and 6meAde are given in Fig. 57 and the
R_f values and some spectrophotometric parameters of these compounds in
Table 23.

N^6-Isopentenyladenosine [N^6-(γ,γ-dimethylallyl)adenosine] (Fig. 58) was
first identified in tRNA^{Ser} from *S. cerevisiae* (Feldman et al., 1966); later,
it was found by Madison et al. (1967a) in tRNA^{Tyr} from *S. cerevisiae*, by
Staehelin et al. (1968) in tRNA^{Ser} from rat liver, by Hashimoto et al. (1969)
in tRNA^{Tyr} from *T. utilis* and by Hecht et al. (1969) in yeast tRNA^{Cys}. In
those cases where its position could be determined, ipeA is the 3′-neighbor
of the anticodon.

Quantitative determination of ipeA in the tRNA of yeast and of mammalian (Hall et al., 1966; Robins et al., 1967) and plant (Hall et al., 1967b)
tissues showed that it is present in very small quantities, one residue to every

twenty tRNAs. Distribution of ipeA in tRNA from *Lactobacillus acidophilus* was studied by Peterkofsky and Jesensky (1969), who determined the radio-activity in the tRNA from this microorganism after cultivation on medium containing labeled mevalonic acid. So far ipeA has been found in leucine, tyrosine, cysteine, serine, and tryptophan tRNAs; however, not all isoacceptor tRNAs (isoleucine and serine) contain ipeA. Peterkofsky and Jesensky concluded that the enzyme modifying A into ipeA recognizes the sequence A-A-A-ψ-C in which the 5'-terminal A is the last letter of the anticodon; we have already discussed this matter in connection with the biosynthesis of ipeA.

The structure of ipeA (see Fig. 58) was established by physical methods (Biemann et al., 1966) and subsequently confirmed by chemical synthesis (Robins et al., 1967). Despite the fact that a 1→6-reorganization of ipeA was shown to be possible (Leonard et al., 1966; Grim and Leonard, 1967), Hall (1967) concludes that the C^6 isomer is probably present in tRNA. The spectrophotometric parameters of ipeAde and ipeA are as follows (Robins et al., 1967):

Spectrophotometric parameters	Isopentenyladenine	Isopentenyladenosine
pH 1		
λ_{max}, nm	273	265
$\varepsilon_{max} \times 10^{-3}$	18.6	20.4
pH 7		
λ_{max}, nm	269	269
$\varepsilon_{max} \times 10^{-3}$	19.4	20.0
pH 12		
λ_{max}, nm	275	269
$\varepsilon_{max} \times 10^{-3}$	18.1	20.0

The nucleoside ipeA possesses cytokinin activity, i.e., it stimulates growth and differentiation of cultures of plant cells (Skoog et al., 1966; Hall et al., 1967a). The double bond in the allyl group and the position of the isopentenyl radical in the heterocyclic ring are responsible for the high reactivity of ipeA. For example, ipeA reacts with aqueous solutions of iodine under conditions when no other nucleotides react with it (except the thiopyrimidines). This fact has been used for the selective modification of ipeA in tRNA^{Ser} (Fittler and Hall, 1966). On treatment of tRNA^{Ser} with iodine the ipeA disappears; this has no effect on acceptor activity, but it inhibits the binding of seryl-tRNA^{Ser} with the mRNA-ribosome complex.

The treatment of ipeA with a dilute solution of permanganate leads to loss of the isopentenyl group and to the formation of adenosine. This reaction has been used to produce a tRNA deficient in the isopentenyl group, and this has been used to study the biosynthesis of ipeA (Kline et al., 1969).

Besides ipeA, two other analogous compounds have so far been found in tRNAs from different sources (Hall et al., 1967b; Robins et al., 1967; Harada et al., 1968; Burrows et al., 1968). This suggests that a series of isoprenoid nucleosides is present in tRNA (Hall, 1970). A detailed survey of the chemical properties, biosynthesis, metabolism, and role of ipeA has been published by Hall (1970).

The nucleoside on the 3'-side of the anticodon has been identified in tRNA^{Tyr} from *E. coli* (Goodman et al., 1968). As had been supposed, it does contain a methylthio-group. Chromatographic, spectrophotometric, and mass-spectrometric investigations have shown that the nucleoside is 2-methylthio-N^6-(Δ^2-isopentenyl)adenosine (Harada et al., 1968). This minor nucleoside was identified in tRNA^{Ser} and tRNA^{Phe} from *E. coli* by Nishimura et al. (1969). It was isolated from a total preparation of tRNA from *E. coli* by Burrows et al. (1968). Since under certain conditions the A residue on the 3'-side of the anticodon of tRNA^{Tyr} from *E. coli* can remain unmodified, while in the yeast tRNAs the corresponding position is occupied by ipeA, these workers suggest that 2-methylthio-N^6-(Δ^2-isopentenyl)adenosine is formed from A at the polynucleotide level through the consecutive action of several enzymes: adenosine→6-isopentenyladenosine→2-thio-6-isopentenyladenosine→2-methylthio-6-isopentenyladenosine.

Gefter (1969) showed that the source of the CH$_3$-group of mtpA is S-adenosylmethionine, while Gefter and Russell (1969) isolated three forms of suppressor tRNA^{Tyr} from *E. coli* Su$_{III}^+$, with the same primary structure but differing in the degree of modification of the minor component on the 3'-side of the anticodon (mtpA). A study of the properties of these tRNAs showed that they are indistinguishable in the velocity of their aminoacylation, but in protein synthesis *in vitro* they function differently. Complete modification of A into mtpA was found to be essential for normal binding of the aminoacyl-tRNA with ribosomes.

The same minor component has recently been found during the study of the primary structure of tRNA^{Phe} (Barrell and Sanger, 1969a) and tRNA^{Trp} from *E. coli* (Hirsh, 1970), in which it also lies next to the anticodon on the 3'-side. Its absorption spectra at neutral, acid, and alkaline pH values are illustrated in Fig. 59.

In 1964, Hall (1964a) reported the presence of compounds with the structure of N^6-(aminoacyl)adenosines in yeast tRNA. These compounds were found to contain adenosine and one of the amino acids (predominantly alanine, aspartic acid, glycine, threonine, and valine), or even a polypeptide chain. The stability of the compounds during acid hydrolysis (0.5 N HCl, 100°C, 2 h), in alkaline buffer, pH 10.5, and in 2 M hydroxylamine (pH 7.0, 3 h, room temperature) led to the conclusion that the amino acid is not linked to the adenosine by phosphoanhydride, ester, or phosphoamino-bonds. The

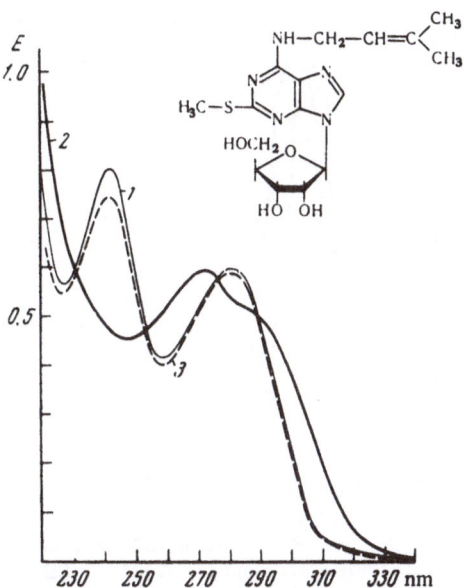

Fig. 59. Absorption spectrum of 2-methyl-thio-N^6-(Δ^2-isopentenyl)adenosine: (1) at pH 7.0; (2) at pH 1.0; (3) at pH 13.

spectrum of the compound and certain other properties indicated that the most probable site of addition was N^6. Hall and Chheda (1965) later reported that a formyl residue is attached to the α-amino-group of these compounds, and that they are thus N^6-(N-formyl-α-aminoacyl)adenosines (but see below).

A compound N-(purinyl-6-carbamoyl)threonine (thrA) was originally isolated from an acid digest of total tRNA (Chheda et al., 1969), and it was shown to be fairly widely distributed; its content in yeast tRNA is 0.28, in tRNA from *E. coli* 0.07, and in tRNA from bovine liver 0.19 mole %. Schweizer et al. (1969) identified thrA and confirmed its structure by chemical synthesis. It was first discovered in an individual tRNA by Takemura et al. (1969a); during determination of the primary structure of tRNA^{Ile} from *T. utilis*, thrA was found in the 3'-position next to its anticodon. The distribution of thrA is evidently fairly wide; it has now been identified in tRNA$_3^{\text{Ser}}$, tRNA^{Met}, and tRNA^{Lys} from *E. coli* (Ishikura et al., 1969). Later, after tRNA$_3^{\text{Ser}}$ had been fractionated into 3A and 3B, it was found that thrA is in fact present in tRNA$_{3A}^{\text{Ser}}$, where it lies on the 3'-side of the anticodon; a compound similar to, but not identical with thrA was found in tRNA$_{3B}^{\text{Ser}}$ (Ishikura et al., 1971). The significance of the position of this hypermodified nucleotide in the recognition of codons in general, and of codons beginning

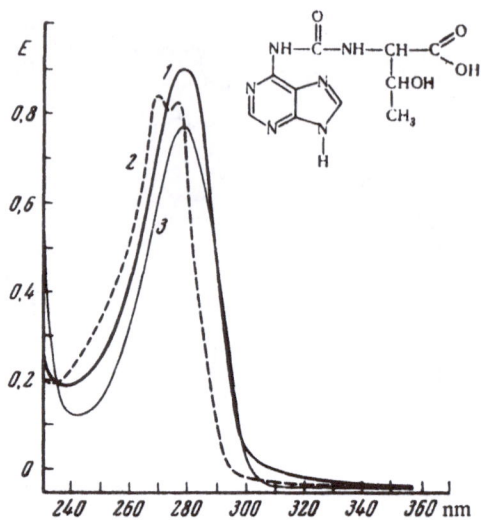

Fig. 60. Absorption spectrum of *N*-(purinyl-6-carbamoyl)threonine: (1) at pH 1.5; (2) at pH 7.0; (3) at pH 11.5.

with A in particular, has been discussed (Nishimura et al., 1969). The spectrum of *N*-(purinyl-6-carbamoyl)threonine is shown in Fig. 60.

In connection with the identification of this minor component, Hall reexamined some of his earlier findings. The compound previously identified as N^6-(aminoacyl)adenosine (Hall, 1964a) was converted, after removal of the carbohydrate residue, into a product identical with *N*-(purinyl-6-carbamoyl)threonine, and unlike aminoacyladenosines studied previously, it contained only one amino acid, threonine. However, the possibility is not ruled out that this is only one representative of a group of compounds with the general structure of an *N*-(nebularinyl-6-carbamoyl)amino acid. It is very likely that this is true, for compounds related to but not identical with thrA have been found in a number of isoacceptor tRNAs (Yamada et al., 1969; Ishikura et al., 1971).

INOSINE

The presence of I and 1meI in yeast tRNA was established as long ago as in 1963, when Hall (1963d) isolated these minor components from enzymic hydrolysates of yeast tRNA and investigated their chromatographic and spectrophotometric properties. However, special attention was paid to I after it had been identified in the anticodons of a number of tRNAs (alanine, serine, and valine tRNAs from *S. cerevisiae,* serine tRNA from rat liver,

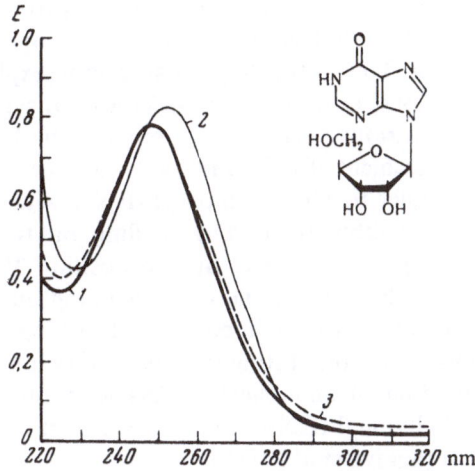

Fig. 61. Absorption spectrum of inosine: (1) in 0.1 N HCl; (2) in 0.1 N KOH; (3) after bromination.

valine and isoleucine tRNAs from *T. utilis)*. In all these cases inosine was found to be the first letter of the anticodon. According to Crick's wobble hypothesis, the I of the anticodon can pair with U, C, and A; the presence of I in the anticodon is thus one cause of degeneracy of the code. In tRNA^Ala, 1meI occupies the first position on the right of the anticodon, which in most tRNAs is occupied by A or by one of its derivatives. It is postulated that 1meI is formed in tRNA^Ala from A as the result of the modifying action of two enzymes: deaminase and methylase.

Inosine possesses clearly defined spectral characteristics, for its absorption maximum lies in an unusual region, namely at 250 nm. In addition, as a deaminated derivative of A, inosine shares with it a convenient property for its identification: the spectrum of inosine is not altered by bromination (Fig. 61). The spectrophotometric and chromatographic parameters for I and 1meI are to be found in a number of publications (Beaven et al., 1955; Hall, 1963d). Some of them are shown in Table 24.

TABLE 24. Spectrophotometric Parameters of Hypoxanthine and Its Derivatives

Compound	Acid medium			Alkaline medium		
	λ_{max}, nm	ε_{max} $\times 10^{-3}$	$\varepsilon_{260} \times 10^{-3}$	λ_{max}, nm	$\varepsilon_{max} \times 10^{-3}$	$\varepsilon_{260} \times 10^{-3}$
Hypoxanthine	248	10.8	7.4	259	11.1	11
1-Methylhypoxanthine	249	—	—	260	—	—
Inosine	248–249	12.2	9.0	253	13.1	11.7
1-Methylinosine	250	—	—	249	—	—

The similarity between I and G in relation to hydrogen bonding (Davies and Rich, 1958; Davies, 1960) led to the view that I can replace G in codon–anticodon interaction. To test this hypothesis and to explain the ability of I to pair with other bases, Grünberger et al. (1967a,b) synthesized trinucleotides in which G in various positions was replaced by I. Their experiments showed that on replacement of G by I in the 5'- and in the middle positions, the ability of the triplet to bind aminoacyl-tRNA with ribosomes is completely lost or considerably reduced depending on the character of the aminoacyl-tRNA. Replacement of G by I in position 3' of the triplets has virtually no effect on binding of the corresponding aminoacyl-tRNAs, in accordance with Crick's wobble hypothesis. Blocking the proton in the N^1-position of inosine, it is interesting to note, makes it no longer capable of stimulating the binding of aminoacyl-tRNAs with ribosomes; G-A-I, for instance, binds ^{14}C-glutamyl-tRNA, but as a result of cyanoethylation of I, this property is lost (Sekiya et al., 1967).

Investigation by the nmr method has demonstrated the ability of I to pair with A, U, and 4thioU (Scheit, 1967a). Chemical modifications of I have been discussed above.

DERIVATIVES OF GUANOSINE

The following methylated derivatives of guanosine are known to be present in tRNA (18): 1-methylguanosine, N^2-methylguanosine, N^2-dimethylguanosine, and 7-methylguanosine.

Methylated derivatives of guanosine are much more widely distributed in tRNAs than derivatives of adenosine; 1meG, 2meG, and 2me$_2$G are found mainly in yeast tRNAs and are absent from tRNAs from *E. coli*. The dis-

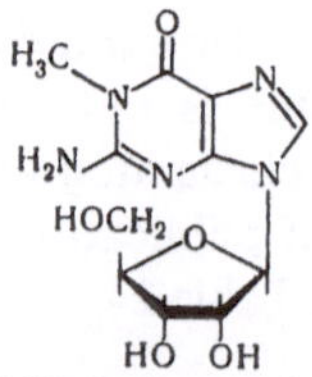

1-Methylguanosine

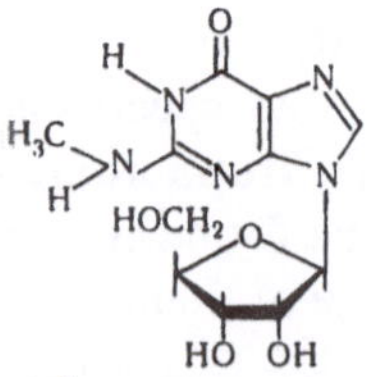

N^2-Methylguanosine

18   N^2-Dimethylguanosine

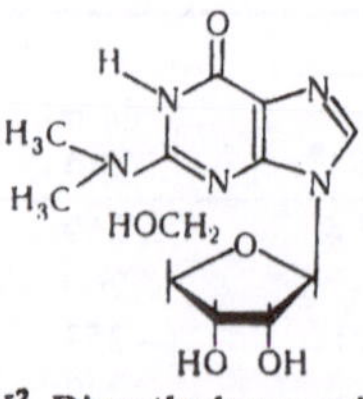

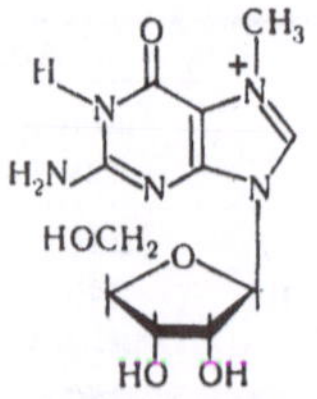

7-Methylguanosine

tribution of modified guanosines along the polynucleotide chain is interesting, and as in the case of other minor components, it exhibits surprisingly regular patterns; 1meG is always found in positions 9 or 10 from the 5'-end of the molecule in the single-stranded segment. In yeast tRNA^{Asp}, 1meG lies next to the anticodon on the 3'-side. In tRNA^{Leu} from yeast and *E. coli,* 1meG also, evidently, is present in this position, although in these two cases the identification is not yet complete; 2meG occupies position 10 from the 5'-end of the molecule and is a component of the first pair of the double-helical segment of the dihydrouridyl branch. In every case 2me$_2$G occupies a completely analogous position, namely position 9 from the first letter of the anticodon toward the 5'-end of the molecule, so that in the clover leaf it is in the bend between the anticodon and H$_2$U-branches. 7meGua (Fig. 62), most commonly found in tRNA from *E. coli,* but also present in tRNAs from other sources (tRNA^{Phe} from *S. cerevisiae* and wheat germ, and also in tRNA$_1^{Val}$ from *S. cerevisiae*) lies in the extra branch in every case, where it invariably occupies position 31 counting from the 3'-end of the molecule.

The most complete descriptions of 1meG, 2meG, and 2me$_2$G are given by Smith and Dunn (1959b). The spectrophotometric and chromatographic parameters for G derivatives have been surveyed by Venkstern (1964). Data for 7meG are given in several papers (Reiner and Zamenhof, 1957; Brookes and Lawley, 1961; Magee and Farber, 1962; Dunn, 1963). 7meG is the main product formed in tRNA by the action of various methylating agents. Large quantitites of 7meG are found in tRNA after the action of dimethylsulfate (Zakharyan et al., 1967, 1968), dimethylnitrosamine (Magee and Farber,

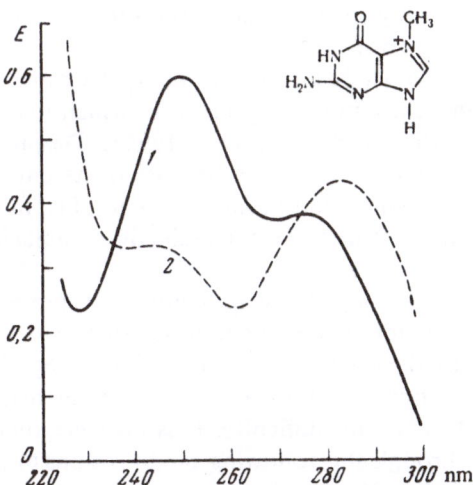

Fig. 62. Absorption spectrum of 7-methylguanine: (1) in acid; (2) alkaline medium.

1962), etc. Since tumor tissues have a high content of 7meG, the carcino-genic action of dimethylnitrosamine has been attributed to its ability to form this compound.

Methylation ta N^7 makes the imidazole ring labile, so that it is opened under comparatively mild conditions (Townsend and Robins, 1963), and 7meG is converted into a pyrimidine derivative. This property of 7meG has been used in the development of a method of purely chemical degradation of the tRNA molecule (Wintermeyer and Zachau, 1970). An important feature of 7meG for structural research is that the phosphodiester bonds formed by it are not ruptured by guanyl RNases. This is used in the study of nucleotide sequences in RNA molecules (for example, the 5S rRNA from *E. coli,* Brownlee et al., 1968). The RNA is partially methylated by dimethylsulfate and then completely hydrolyzed by guanyl RNase. The single-stranded seg-ments, which are methylated first, are split by the RNase only at the unmodi-fied G residues, and they give long sequences providing the essential overlaps.

Igo-Kemenes and Zachau (1971), using the reaction of reduction with NaBH$_4$, showed that 7meGua lies in a screened segment of the molecule and they postulated that by interacting with negatively charged phosphate groups, this positively charged minor component stabilizes the tertiary structure of the tRNA.

A characteristic property of guanine and its derivatives, namely, violet fluorescence (Udenfriend and Zaltzman, 1962; Udenfriend et al., 1963), is widely employed for the qualitative detection of G and, in particular, its derivatives at both mononucleotide and oligonucleotide levels; so far, how-ever, the fluorescence has never been used in practice for the quantitative estimation of these components, despite the greater sensitivity of this test than that based on uv-absorption.

In 1966, during a study of the primary structure of tRNA^{Phe} from *S. cerevisiae,* an unusual minor component, to which the name Y has stuck, was discovered (RajBhandary and Stuart, 1966a). Despite the fact that this compound attracted general attention because of its unusual properties, its structure was not determined until four years later, i.e., in 1970. The reasons were the small quantity of material available, the complexity of its structure, and its lability.

Having isolated 300 μg of the individual substance, Nakanishi et al. (1970) identified this compound and confirmed that its structure as determin-ed in model compounds was correct. Y proved to be the most hypermodified nucleoside of all the minor components, in which the features of guanosine can be recognized only with difficulty. It is possible that the biosynthesis of Y takes place through the following modifications of guanine: C_1 and C_2 units are added to the N^3 and C^2 amino groups, respectively; a residue of glutaric acid is added to N^1, and cyclization takes place between C^{10} and C^{11}

CH₃OOC—NH
CH₃OOC—CH
CH₂
CH₂   O----H
CH₃

CH₃

to form the third ring. The amino-group of glutaric acid is converted into carbamate at some stage (Nakanishi et al., 1970).The presence of dimethyl-*N*-carbamoyl-2-aminobutyric acid as a side chain was proved by a combination of mass-spectrometric and nmr methods; the composition of the ring was determined from the results of mass spectrometry with a high level of resolution, and the uv-spectrum characterizes the nature of the chromophore. Finally, the location of the substituents was demonstrated by nmr shifts.

The biosynthesis of Y is still a field for hypothesis and speculation. The spectrum of Y is shown in Fig. 63.

The distinguishing feature of Y is its intensive violet fluorescence and extreme lability in an acid medium, as a result of which not only its spectral properties but all its other properties are modified: its susceptibility to enzyme attack, its fluorescence, and its chromatographic mobility (RajBhandary et al., 1968a). The fluorescence exhibited by Y was used to study the conformation of tRNA^{Phe} (Beardsley and Cantor, 1970; Eisinger et al., 1970). The nucleoside Y, or compounds related to it, are evidently present in tRNA^{Phe} from many sources (RajBhandary et al., 1967; Yoshikami et al., 1968; Fink

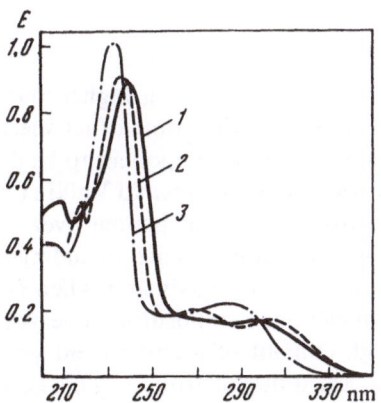

Fig. 63. Absorption spectrum of nucleoside Y from yeast tRNA^{Phe}: (1) at pH 7.0; (2) at pH 13.0; (3) at pH 1.0.

et al., 1968; Dudock et al., 1969), as shown by their intense violet fluorescence. The differences between the Y residues in yeast and wheat germ tRNA^{Phe} are probably concerned with the side groups, for the fluorescence emission spectra of the two Y components are virtually identical (Yoshikami et al., 1968).

The Y residue lies on the 3'-side of the anticodon in tRNA^{Phe}, and it has been suggested that its high level of modification reflects the need for stabilization of the comparatively weak codon–anticodon interaction which can be expected for tRNA^{Phe}. The important role of the base Y in codon–anticodon interaction has in fact been established (Thiebe and Zachau, 1968a). They succeeded in removing the Y base from yeast tRNA^{Phe} by mild acid treatment without disturbing the integrity of the polynucleotide chain (Thiebe and Zachau, 1968a). A study of the properties of the tRNA^{Phe} modified in this way showed that it still retains its ability to regenerate the CCA-end, and to undergo aminoacylation in a homologous system, but it is completely unable to recognize the codon. Most probably the removal of Y essentially changes the conformation of the anticodon loop, as has recently been confirmed experimentally (Ghosh and Ghosh, 1970).

When treated with acid under the degradation conditions described by Whitfeld and Markham (1953), tRNA^{Phe} could be divided into two halves (Philippsen et al., 1968); the functional activity of a mixture of homologous and heterologous halves could then be investigated, using for this purpose halves of tRNA^{Phe} from yeast and wheat germ (Thiebe and Zachau, 1969a, b; Zachau, 1969).

URIDINE DERIVATIVES

Pseudouridine

The relative content of ψ in tRNA is higher than that of other modified nucleotides. This fact, together with some of its special properties, were the reasons why ψ was the first minor component to be discovered in RNA.

As long ago as in 1951, when Cohn and Volkin (1951) were fractionating enzymic and alkaline hydrolysates of bovine liver RNA on Dowex-1 they noticed a number of unusual components, including a nucleotide that eluted immediately before uridylic acid. Davis and Allen (1957) observed that the chief property of the nucleotide composition of yeast ribonucleic acid soluble in 1 M NaCl is a high content of a component with properties similar to those of uridylic acid, yet different from it. This component, known as the "fifth nucleotide," was isolated from enzymic hydrolysates of RNA by two-dimensional paper chromatography and in preparative amounts on a column of cellulose powder; the spectrum of pseudouridylic acid was obtained for the first time. A series of brilliant investigations of pseudouridylic acid was carried out by Cohn (1957, 1958, 1959, 1960, 1961), who showed that the nucleotide which he had studied was identical with the fifth nucleotide of Davis and Allen, and was in fact a modified uridylic acid. The nucleotide was accordingly named pseudouridylic acid (ψp). In an important paper, Cohn (1960) described the properties and structure of ψ. From the results of chemical, chromatographic, and spectrophotometric analyses, Cohn concluded that ψ is 5-ribosyluracil, in which, instead of the N-glycosidic bond usually found in nucleosides, there is a C-glycosidic bond. The bathochromic shift of the absorption curve of ψ in the ultraviolet region at pH values over 8 compared with the curve in an acid medium indicates the presence of two ionizing tautomeric groups in the molecule. Hence it was concluded that the carbohydrate residue replaces the hydrogen atom at C^5 or C^6 rather than at N^1 or N^3 of uracil. On the other hand, the nmr spectrum and reactions with borate and periodate indicated the presence of a substituent containing glycolic groups in the C^5 position of the nitrogenous base. Methylated derivatives of ψ with two methyl groups at N^1 and N^3 were obtained; the study of their properties confirmed that the most probably site of addition of ribose is C^5 in the uracil moiety (Scannell et al., 1959). Four isomers of ψ, differing in the size of the ring and the configuration at C^1, are known (Cohn, 1960).

The structural peculiarity of ψ, namely the addition of ribose to uracil by means of a C–C bond, leaves its imprint on its chemical properties. All the methods used to remove ribose from the pyrimidine ring of nucleosides are unsuitable in the case of ψ. Vigorous treatment with acid leads to isomerization rather than degradation. Labilization of the glycosidic bond likewise could not be achieved through saturation of the 5,6-double bond by hydrogenation, bromination, or uv-irradiation. The reaction with orcinol, used to determine ribose, follows an unusual course, as also does periodate oxidation. The view was expressed that periodate oxidation cannot in general be used with ψ, because any absorption in the uv region disappears after this procedure (Dlugajczyk and Allen, 1961; Yu and Allen, 1959). The fact that ψ can be an obstacle in the way of determination of the primary structure of nucleic acids by stepwise degradation using periodate oxidation has also been described by Bitte and Lis (1967). Tomasz and Chambers (Tomasz and Chambers, 1965; Chambers, 1966) found that on oxidation with sodium periodate ψ is converted into 5-formyl- and 5-carboxyuracil, which can be detected with o-dianisidine. These workers point out that if the oxidation procedure is repeated, rupture of the polynucleotide chain through labilization of the 3'-phosphomonoester bond cannot be ruled out.

The most characteristic property which distinguishes ψ from any other nucleoside is the marked bathochromic shift of the absorption curve when

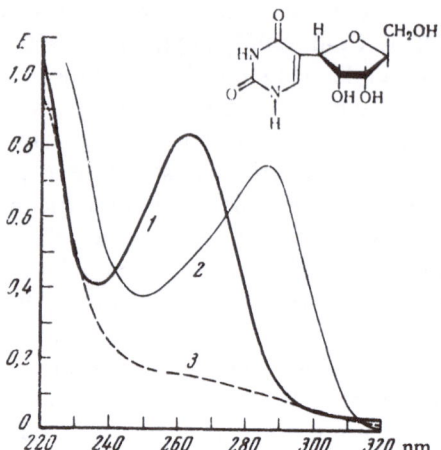

Fig. 64. Absorption spectrum of pseudouridine: (1) in 0.1 N HCl; (2) in 0.1 N KOH;
(3) after bromination.

the compound is transferred from an acid to an alkaline medium (Fig. 64).
This shift is characteristic not only of ψ and ψp; as a rule it is also exhibited
at the level of the lower oligonucleotides containing this component. The
appearance of a shoulder in the region of 280 nm or a bathochromic shift
of the entire spectrum when a solution of any oligonucleotide is made alkaline thus reliably indicates the presence of ψ in its molecule (Fig. 12A).
The spectrophotometric parameters of ψ and ψp are given in Table 25.

Because of the presence of two unsubstituted amido groups, ψ has more
opportunities for hydrogen bonding than does U (Pochon et al., 1964; 1965,)
this has served as the basis for many suggestions regarding the role of ψ in
tRNA. The higher reactivity of N^1 in the ψ residue compared with N^3 in other
nucleosides is used for the selective modification of ψ in tRNA. For example,
ψ reacts with acrylonitrile under conditions when C, G, and A are for practical purposes inert (Chambers, 1965, 1966; Ofengand, 1965, 1967). The
possibility of selective modification of ψ by this reagent has been demon-

TABLE 25. Spectrophotometric Parameters of Pseudouridine and
Pseudouridylic Acid

Compound	pH	λ_{max}, nm	$\varepsilon_{max} \times 10^{-3}$	λ_{min}, nm	$\varepsilon_{min} \times 10^{-3}$	$\varepsilon_{260} \times 10^{-3}$	$\dfrac{E_{250}}{E_{260}}$	$\dfrac{E_{280}}{E_{260}}$	$\dfrac{E_{290}}{E_{260}}$
ψ	2	263	7.5	233	2.0	7.4	0.74	0.42	0.06
ψp	2	263	8.4	233	2.3	8.3	0.76	0.38	0.06
ψ	12	286	7.3	245	1.7	3.2	0.60	2.11	2.44
ψp	12	286	8.4	246	2.1	3.7	0.66	2.04	2.10

strated independently in several laboratories (Chambers, 1965; Yoshida and Ukita, 1965a,b; Ofengand, 1965; Rake and Tener, 1966).

Like U, G, and I, ψ reacts with water-soluble carbodiimide (Gilham, 1962; Naylor et al., 1965). The carbodiimide can be removed under comparatively mild conditions. The reaction with carbodiimide depends on secondary structure (Augusti-Tocco and Brown, 1965; Knorre et al., 1966), a fact which is used when the conformation and functional topography of tRNA are studied (Brostoff and Ingram, 1967). The essence of the method is that, after modification, the phosphodiester bonds become resistant to pancreatic RNase and also to snake-venom and spleen PDEase (Naylor et al., 1965; Lee, J. C., et al., 1965).

One of the chief problems in determination of the primary structure of nucleic acids is how to obtain large overlapping fragments. The search for new methods for use for this purpose, in conjunction with partial hydrolysis by T_1- and pancreatic RNases, still continues; ψ has attracted attention in this connection because degradation of tRNA at only those phosphodiester bonds formed by it would yield a few large fragments, on account of the small number of ψ residues found in the molecules of individual tRNAs. Three methods have been suggested for degradation in this way. One is based on the removal of U by hydroxylamine, splitting of the cytidine by hydrazine, and hydrolysis of the RNA thus treated with pancreatic RNase at the unmodified T and ψ residues (Verwoerd and Zillig, 1963). The second method is based on periodate oxidation under vigorous conditions, leading to cleavage of the ribose moiety of ψ between C^1 and C^2, with the resulting formation at C^2 of an aldehyde, labilizing the phosphate bond at C^3 (Tomasz et al., 1965). Finally, the third method consists of photochemical degradation of the ψp residues with the formation of 5-formyluracil and certain other compounds (Tomasz and Chambers, 1964). The last reaction has been tested on the tetranucleotide T-ψ-C-G and also on tRNA^{Ala} (Tomasz and Chambers, 1966), and in the opinion of these workers the results are encouraging.

The chemical synthesis of ψ has been described (Shapiro and Chambers, 1961; Brown et al., 1965, 1968). However, the simplest method of obtaining ψ so far is its isolation from natural sources, namely, yeast RNA (Cohn, 1961) and human urine (Cohn et al., 1963), especially the urine of patients with leukemia (Adler and Gutman, 1959).

Because of the unusual chemical structure of ψ, its behavior with enzymes is particularly interesting. In principal, specific enzymes might exist which can distinguish ψ from other pyrimidines, but as yet such nucleases are unknown. On the contrary, the impression has been gained that most enzymes behave in the same way toward ψ as toward other pyrimidines. Pyrimidyl RNase removes pseudouridylic acid as the 3'-phosphate or leaves it at the ends of oligonucleotide fragments depending on whether its neighbor on the 5'-side is a purine or pyrimidine; ψp is dephosphorylated by PMEase from prostate and *E. coli;* snake-venom PDEase removes a nucleotide (ψ-5'-p) from RNA, and this is subsequently dephosphorylated by 5'-nucleotidase. However, the velocity of PDEase hydrolysis at ψ is much less than in the case of other nucleotides. The behavior of ψ in enzymic reactions has been examined in detail by Goldwasser and Heinrikson (1966).

Let us now turn to the content and localization of ψ in tRNA, a problem which has now been completely settled in connection with the decoding of a number of primary structures, so that it can be dealt with briefly.

It was pointed out above (Table 19) that the content of ψ in individual tRNAs varies from one to four residues, so that it corresponds approximately to that postulated from analysis of total preparations of tRNA.

All tRNAs have one ψ residue in position 22 counting from the 3'-end of the molecule, next to a T residue; this pair of minor components occur in the universal oligonucleotide and in fact constitute its universal part, because no exception to this rule has yet been discovered.* The remaining ψ residues are found in the anticodon branch, except in $tRNA_1^{Val}$ and $tRNA^{Asp}$ from *S. cerevisiae,* and also $tRNA_1^{Val}$ from *T. utilis,* in which one of the four ψ residues is situated in the H_2U-loop. In eleven tRNAs ψ occupies position 3 from the last letter of the anticodon and forms the first pair of the helical part of the anticodon branch. In rat-liver $tRNA^{Ser}$ this ψ residue is methylated at the 2'-OH-group of the ribose. In eleven tRNAs ψ is found in the unpaired segment of the anticodon branch, in which it occupies only three of the seven possible positions: in six tRNAs the extreme 5'-terminal, in three the 3'-terminal, and in the $tRNA^{Tyr}$ from *S. cerevisiae* and *T. utilis,* the central position, so that it is the second nucleotide in their anticodon.

It follows logically from the results for the quantitative content and distribution of ψ in the polynucleotide chain of different tRNAs that the various ψ residues cannot have identical roles. It is unanimously accepted that the universal oligonucleotide and, consequently, the ψ contained in it, participates in a function which is common to all tRNAs, such as binding with ribosomes. I have already mentioned the selective modification of this ψ residue which indicates its participation in supporting the native conformation of the molecule, an essential factor for acceptor activity (Siddiqui et al., 1970). The functional universality of ψ in tRNAs of different specificity and different origin, postulated by Zaitseva et al. (1962), probably applies fully to this nucleotide. In particular, this residue cannot play a role in the determination of species specificity, because its presence is independent of the

---

*The only exception is the $tRNA^{Gly}$ from *Staphylococcus epidermidis,* with only one minor component, namely thiouridine. This tRNA, however, is also an exception in that it participates in the synthesis of peptidoglucan only, and not in protein synthesis (Stewart et al., 1971).

source of the tRNA. The role of ψ in the anticodon of tRNA^{Tyr} is obvious. During codon–anticodon interaction, the extra facility for hydrogen bonding which ψ possesses because of the fact that both amino groups of the heterocyclic ring are free, could have a role to play. The role of the other ψ residues is less clear. However, elucidation of the primary structure of tRNA, and the development of methods of selective modification of nucleotides will now enable this problem to be solved on a solid experimental basis. This applies, in particular, to ψ because in this case it is possible to modify some residues selectively while leaving others intact (Yoshida and Ukita, 1965a,b; Rake and Tener, 1966).

Dihydrouridine

Dihydrouridine (Fig. 65), like the other dihydropyrimidines, has been known for a long time and its properties have been reasonably well studied. It was not until 1965 that H_2U was identified in the composition of nucleic acids (Madison and Holley, 1965), when oligonucleotides not containing terminal pyrimidines were found in a pancreatic RNase hydrolysate of tRNA^{Ala}; more detailed analysis revealed a pyrimidine nucleotide in these fragments: it does not absorb in the uv region in water or in acid medium, it gives a positive reaction when tested for dihydropyrimidines (Fink et al., 1956), and it contains phosphorus. These results, together with those of determination of the

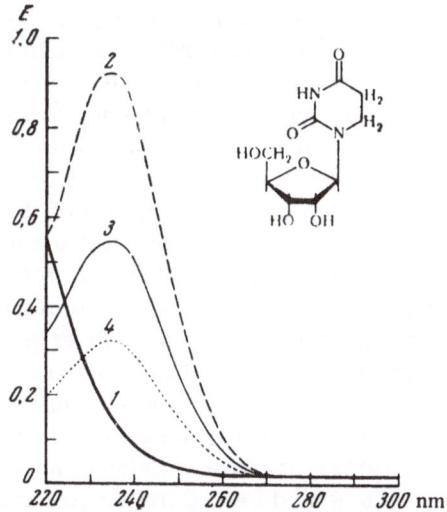

Fig. 65. Absorption spectrum of 5,6-dihydrouridine: (1) in water; (2, 3, 4) in 0.1 N KOH 2, 4, and 6 min, respectively, after the addition of alkali to an aqueous solution of dihydrouridine.

mobility of the compound during chromatography, led to the conclusion that the "invisible nucleotide" is 5,6-dihydrouridylic acid. During 1965–1969, H_2Up was identified in all tRNAs whose primary structure had been investigated, with the exception of tRNA[Tyr] from *E. coli*. In addition, Huang and Bonner (1965) identified H_2Up in a special type of RNA consisting of only 38 nucleotides and bound with histones; its content in this case was extremely high, namely 27.5 mole %.

As was stated above, H_2Up does not absorb in the uv region in neutral and acid solutions; the spectrum appears when the solution is made alkaline, but the absorption at once begins to diminish (Fig. 65). The chemical mechanism of these conversions is as follows: hydrogenation of Ura(I), U, or Up, usually carried out on a rhodium catalyst (Cohn and Doherty, 1956), leads to the addition of hydrogen at the $C^5–C^6$ double bond of the pyrimidine ring:

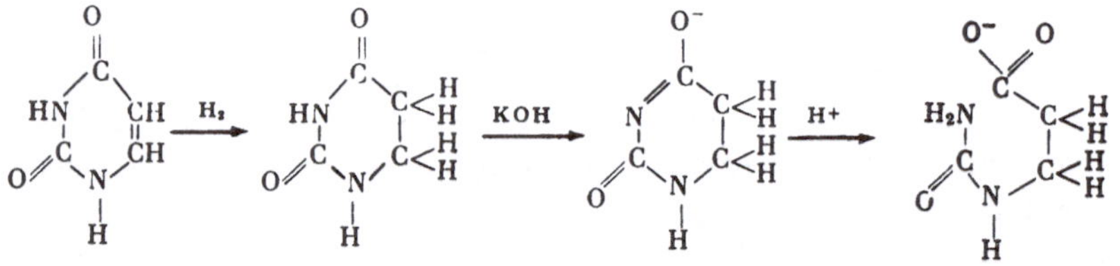

The absence of absorption bands in the uv region is explained by the fact that in neutral solutions dihydrouracil exists in the keto-form (II), and conjugated double bonds are absent in the pyrimidine ring reduced at the 5,6-double bond. At alkaline pH values the compound is converted into the enolic form (III); under these circumstances one hydroxyl group dissociates, and the resulting 3,4-double bond is conjugated with the double bond of the carbonyl group at C^2. This causes the appearance of the new maximum at 230 nm; the subsequent disappearance of the spectrum is explained by rupture of the double bond between N^3 and C^4 of the heterocyclic ring (McLaren and Shugar, 1964) and by the formation of β-ureidopropionic acid (IV). The true value of the absorption of the dihydro-derivatives in an alkaline medium is established by extrapolation of the spectrum observed a certain time interval after the solution is made alkaline to the zero mark (Batt et al., 1954). H_2Up is determined quantitatively as phosphorus (Madison and Holley, 1965; Venkstern et al., 1968a); as ribose (Neelon et al., 1967); or as ribosyl phosphate of ureidopropionic acid formed from H_2Up as a result of alkaline treatment (Sanger et al., 1965). Magrath and Shaw (1967) proposed determining H_2Up as β-alanine, into which they converted the ureidopropionic acid by alkaline hydrolysis. However, this method is not quantitative, for the yield of β-alanine is only 70%. Recently, Jacobs and Hedgcoth (1970) suggested two methods for determining H_2U in RNA; one is colorimetric, requires 25–100 μg tRNA, and has an accuracy of $\pm 5\%$. The second method is based on fractionation by two-dimensional thin-layer chromatography of an enzymic hydrolysate of the RNA.

Heating β-ureidopropionic acid in 0.1 N HCl closes the ring. However, this method can only be used with the base, for in nucleotides and nucleosides the glycosidic bond is ruptured at the same time. According to Janion and Shugar (1960), H_2Ura has the highest lability in an alkaline medium, for H_2U is slightly more resistant, and H_2Up even more resistant; in the composition of the oligonucleotide, a further increase in stability is observed. This is shown, e.g., by the fact that the decrease in absorption of the dinucleotide A-H_2Up at 230 nm after the solution is made alkaline takes place much more slowly than in the case of free H_2Up. The half-decay times of dihydrouracil, dihydrouridine, and dihydrouridylic acid (0.1 N NaOH, 22°C) are:

H_2Ura	3.7 min
H_2U	4.8 min
H_2Up (free)	6.5 min
H_2Up (as a component of A-H_2Up)	11.5 min

A method of cleavage of the polynucleotide chain of tRNA at H_2U, based on the reduction of H_2U by $NaBH_4$ into ureidopropanol-N-riboside (Igo-Kemenes and Zachau, 1969) and subsequent removal of the latter with hydrochloric acid (Beltchev and Grunberg-Manago, 1970a), has been described. The method has been used to obtain fragments of yeast tRNA[Phe] required for structural and functional investigations (Beltchev and Grunberg-Manago, 1970b).

The study of the properties of polymers containing H_2U is of great interest. Since H_2U is not a substrate for polynucleotide phosphorylase (Ochoa and Heppel, 1957; Szer and Shugar, 1961; Grunberg-Manago, 1963), the method of photochemical reduction of polyU (Cerutti et al., 1965) was used to obtain these polymers. The activity of poly(U, H_2U) was tested for the binding of aminoacyl-tRNA with ribosomes, and also as regards incorporation of amino acids into protein. A positive result was obtained only with phenylalanine, and only if the percentage content of H_2U in the polymer was low (Rottman and Cerutti, 1966). It was shown (Cerutti et al., 1966) that the presence of H_2U in a polymer lowers the melting temperature; H_2U does not pair with A and weakens interaction with neighboring nucleotides. It is postulated that the formation of H_2U in a polynucleotide chain may deprive the corresponding fractions of their template function.

Smrt et al. (Smrt, 1966; Smrt et al., 1966) showed that replacement of U by H_2U in the triplet G-U-U leads to complete loss of its ability to bind valyl-tRNA[Val] in the system of Nirenberg and Leder (1964). Hence, this would seem to suggest that H_2U cannot replace U functionally. Smrt et al. consider that H_2U in tRNA can also act as an uncoupling unit or as an element determining secondary structure. According to Huang and Bonner (1965), the role of H_2U may be to act as the link between the RNA which they described and protein; they point out that the RNA-histone complex which they discovered gives a yellow color with p-dimethylaminobenzaldehyde without preliminary treatment with alkali. This means that H_2U is in an open form and the complex can be represented as follows:

$$
\begin{array}{c}
\text{O} \\
\parallel \\
\text{C} - \text{O} - \text{histone} \\
\text{H}_2\text{N} \quad \text{CH}_2 \\
| \quad | \\
\text{O}{=}\text{C} \quad \text{CH}_2 \\
\diagdown \text{N} \diagup \\
| \\
\text{ribose}
\end{array}
$$

These workers suggest that the H_2U in tRNA, which also interacts with proteins, may serve the same purpose, such as enabling interaction to take place with aminoacyl-tRNA synthetases. Other workers are inclined to ascribe a role to H_2U in interaction with ribosomes. I have already mentioned the contradictory data obtained by the study of the role of H_2U in the acceptor and transfer function of tRNA with the aid of $NaBH_4$.

H_2U has been found in all tRNAs so far studied except tRNA^{Tyr} from *E. coli*. There are only two segments in which H_2U is found in the polynucleotide chain of tRNA. These are the H_2U-loop in the 5'-half and the extra loop in the 3'-half of the molecule. Sequences containing H_2U in the 5'-half of the molecule occur in all tRNAs except tRNA^{Tyr} from *E. coli* (Table 26); in the 3'-half of the molecule, H_2U is found in only certain tRNAs.

Hydrogenation of the pyrimidine ring does not cause qualitative changes in the action of enzymes acting on phosphate and phosphodiester bonds–RNase, snake-venom PDEase, and 5'-nucleotidase (Janion and Shugar, 1960). However, the lability of H_2U in an alkaline medium, in which some of these reactions are carried out, adds considerably to the difficulty of determination of the structure both of tRNA and of oligonucleotides containing H_2U.

Spectrophotometric parameters of H_2Ura, H_2U, and H_2Up in 0.1 N NaOH are given below:

Compound	λ_{max}, nm	$\varepsilon_{max} \times 10^{-3}$
H_2Ura	230	8.3
H_2U	235	10.1
H_2Up	235	10.1

TABLE 26. H_2U Content of Individual tRNAs and Its Distribution between 3'- and 5'-Halves of the Molecules

tRNA specific to	Source	Total content moles/mole of tRNA	5'-half, H_2U-branch	3'-half, extra branch
Ala	S. cerevisiae	2–3	G-H_2U-C-G-G-H_2U-A-G	A-G-H_2U-C
Ser	''	3	G-H_2U-OmeG-G-(H_2U)$_2$-A-A-G	—
Tyr	''	6	G-(H_2U)$_2$-OmeG-G-(H_2U)$_3$-A-A-G	A-G-A-H_2U-5meC
Val	''	4	G-H_2U-C-G-G-(H_2U)$_2$-A-U-G	A-C-7meG-H_2U-5meC
Phe	''	2	G-(H_2U)$_2$-G-G	
Asp	''	2	A-A-H_2U-G-G-H_2U-C-A	—
Leu$_3$	''	2	G-C-OmeG-G-(H_2U)$_2$-C-A	—
Tyr	E. coli	0	—	—
fMet	''	1	G-H_2U-A-G	—
Met	''	3–4	G-(H_2U)$_2$-OmeG-G-(H_2U)$_2$-A-G	—
Val	''	1	G-C-H_2U-G-G-G	—
Phe	''	2	G-H_2U-C-G-G-H_2U-A-G	—
Gln	''	1	G-C-OmeG-G-H_2U-A-A-G	—
Gly	''	3	G-(H_2U)$_2$-G-G-H_2U-A	—
Leu	''	3	A-A-(H_2U)$_2$-OmeG-G-H_2U-A-G	—
Trp	''	3	A-A-(H_2U)$_2$-G-G-H_2U-A	—
Val$_2$	''	4	A-G-(H_2U)$_2$-G-G-(H_2U)$_2$-A	—
Val	T. utilis	3	G-(H_2U)$_2$-G-G-H_2U-C-A-U-G	—
Ile	''	5	G-(H_2U)$_2$-G-G-(H_2U)$_2$-A-A-G	A-G-A-H_2U-5meC
Tyr	''	6	G-(H_2U)$_2$-OmeG-G-(H_2U)$_3$-A-A-G	A-C-A-H_2U-5meC
Ser	Rat liver	3	G-H_2U-OmeG-G-(H_2U)$_2$-A-A-G	—
Phe	Wheat germ	3	G-(H_2U)$_2$-G-G-G	A-G-7meG-H_2U-C

Other Uridine Derivatives

Thymine is a uracil derivative and is formed like other methylated derivatives of RNA; unlike the thymine of DNA it is formed by the transfer of a methyl group from S-adenosylmethionine to uracil (Mandel and Borek, 1961).

The presence of thymine (5meUra) in specimens of RNA was disputed for a long time, although it was first found in RNA in 1958 (Littlefield and Dunn, 1958b); the suggestion was made that it might be formed by enzymic deamination of 5meCyt. As the result of numerous investigations conducted on total samples of tRNA (Dunn, 1960; Price et al., 1963) this problem has now been unequivocally solved; moreover, T has been identified in all individ-

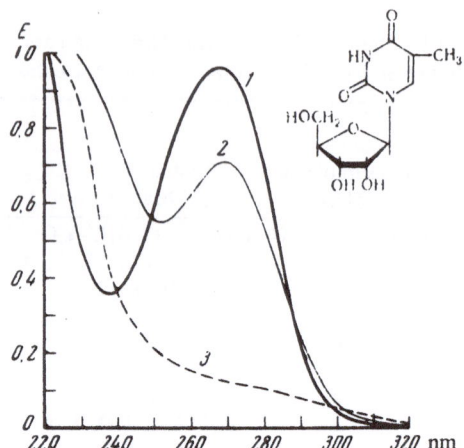

Fig. 66. Absorption spectrum of ribothymidine: (1) in 0.1 N HCl; (2) in 0.1 N KOH; (3) after bromination.

ual tRNAs so far studied, and it occurs in the proportion of 1 mole per mole of tRNA. The position of this minor component is also surprisingly constant: T occupies position 23 from the 3'-end of the molecule, which lies in the universal oligonucleotide. The properties of T have been studied in detail by Littlefield and Dunn (1958b). The spectrum of T is illustrated in Fig. 66.

Hall (1963a) found a derivative of U in enzymic digests of tRNA that corresponded to synthetic 3meU (Miles, 1956). The presence of 3meU in tRNA, judging from the findings of Szer and Shugar (1961) obtained with poly(N^3-methyluridylic acid), has a marked effect on the physical properties of the polymer. Lis and Passarge (1966) reported the identification of 5-hydroxyuridine in RNA from *Torulopsis utilis.*

Iwanami and Brown (1968) identified 5-hydroxymethyluridine in tRNA from L cells, and Gray and Lane (1967, 1968) found 5-carboxymethyluridine in tRNA from yeast and wheat germ. Later, in connection with the discovery of the methylester of 2-thio-5-carboxymethyluridine (Baczinskyj et al., 1968, 1969), tests were carried out to ascertain whether this minor component also exists in tRNA in a blocked form. The methylester of 5-carboxymethyluridine was actually identified, and results were obtained to show that the esterification reaction takes place at the polynucleotide level, S-adenosylmethionine acting as donor of the CH_3-group (Tumaitis and Lane, 1970). The derivative 5-carboxymethyluridine has been synthesized (Fissekis and Sweet, 1970).

Since the time of the determination of the structure of $tRNA_1^{Val}$ from *E. coli* it has been known that position 1 in its anticodon is occupied by a minor component, evidently a derivative of uridine (Yaniv and Barrell, 1969). The nature of this minor component was determined later: it was found to be uridine-5-oxyacetic acid (Harada et al., 1969b). Chemical synthesis confirmed that this identification was correct (Murao et al., 1970). Uridine-5-oxyacetic acid has recently been found in the composition of an isoacceptor tRNA^{Ser} from *E. coli;* judging from its coding properties, in this case also the minor component is located in the anticodon (Ishikura et al., 1971). The formulas of the last five compounds (**19**) are given below.

3-Methyluridine   5-Hydroxyuridine   Uridine-5-oxyacetic acid

19   5-Hydroxymethyluridine   5-Carboxymethyluridine

The spectrophotometric parameters of methylated derivatives of uracil are given in Table 27.

Examination of this group of minor components shows strikingly the wide variety of modifications occurring in nature: in position 5 of uridine

TABLE 27. Spectrophotometric Parameters of Uracil and Some of Its Derivatives

Compound	Acid medium		Alkaline medium	
	λ_{max}, nm	ε_{max} $\times 10^{-3}$	λ_{max}, nm	ε_{max} $\times 10^{-3}$
Uracil	259	8.2	284	6.1
Uridine	262	10.1	262	8.5
Uridine-3-phosphate	262	9.9	261	7.3
Thymine	265	7.9	290	5.4
Ribothymidine	267	9.6	268	7.4
Ribothymidine-3-phosphate	267	9.8	268	—
3-Methyluridine	260	—	260	—
5-Hydroxyuridine	280	—	306	—

alone, methyl, hydroxy, hydroxymethyl, oxyacetyl, and carboxymethyl groups, the latter in either the free form or esterified by a methyl radical, have been found; minor components formed as a result of modification of uridine in other positions (e.g., 3meU, 2thioU, etc.) or in two positions at once (e.g., the methyl ester of 2-thio-5-uridineacetic acid) also are known. The role of these modifications is not yet quite clear, and indeed the precise position of the corresponding minor components in the tRNA molecule is known in only a few cases. However, there can be no doubt that the introduction of different, although similar, substituents into the same position of the molecule or of identical substituents into different positions in the molecule can alter the fine details of tRNA structure and thus perform an important, possibly regulatory, function.

CYTIDINE DERIVATIVES

Only three C derivatives have been identified in individual tRNAs, namely, 5-methylcytidine, 3-methylcytidine, and 4-acetylcytidine. The presence of 5-hydroxymethyl- and N^4-methylcytidine in tRNA has also been reported (Cantoni et al., 1962), but these reports have not subsequently been confirmed. The formulas of these compounds (**20**) are given below.

5-Methylcytidine was discovered as a component of RNA by Amos and Korn (1958). Its presence in RNA of animal, plant, and bacterial origin was subsequently confirmed by Dunn et al. (Dunn and Smith, 1959; Dunn, 1960). The spectra of the compound were obtained and its chromatographic mobility studied.

The distribution of 5meC has been studied by several workers (Dunn, 1959; Dunn et al., 1960; Sluyser and Bosch, 1962; Zaitseva et al., 1962). Its highest content has been found in liver (10 moles/100 moles uridine); there is significantly less in various species of bacteria (0.15–0.2 mole/100 moles uridine), and no 5meC whatever has been found in *E. coli*. In fact, 5meC has not been found in any tRNA from this source so far studied. One residue of 5meC has been identified in all tRNAs from *S. cerevisiae* and *T. utilis*, except tRNA^{Ala}, which contains none, and tRNA^{Phe}, which contains two residues. As a rule this minor component is found in the extra branch,

5-Methylcytidine   5-Hydroxymethylcytidine N^4-Methylcytidine   3-Methylcytidine

TABLE 28. Spectrophotometric Parameters and Chromatographic Mobility
on Paper of Cytosine and Some of Its Derivatives

Compound	Acid medium			Alkaline medium			R_f in solvent[a]			
	λ_{max}, nm	λ_{min}, nm	ε_{max} $\times 10^{-3}$	λ_{max}, nm	λ_{min}, nm	ε_{max} $\times 10^{-3}$	1	2	3	4
Cytosine	274	238	10.2	267	247	6.13	—	—	—	—
5-Methylcytosine	283	242	9.8	287	253	6.9	0.60	0.59	0.25	0.30
Cytidine	280	242	13.4	270	250	9.1	0.55	0.54	0.10	0.10
5-Methylcytidine	287	245	11.6	277	255	8.8	0.57	0.57	0.12	0.12

[a]For solvents see Table 23, page 229.

where in some cases it is next to H_2U. In tRNA[Phe] and tRNA[Asp], 5meC is a
member of the first pair in the duplex segment of the Tψ-branch, while the
second 5meC residue in tRNA[Phe] lies in the helical segment of the anticodon
branch. The spectrophotometric parameters and some R_f values of 5meCyt
and 5meC are given in Table 28.

Hall (1963a) found 3meC in enzymic hydrolysates of tRNA and studied
its properties. On analysis of tRNA[Ser] from rat liver, Staehelin et al. (1968)
found that, besides 5meC, it contains two 3meC residues. Considering the
results obtained with a total sample of tRNA, it may be that the other tRNAs
obtained from liver also have a high content of methylated C derivatives.

Zachau et al. (1966a) identified a new minor component in yeast
tRNA[Ser]; its properties are exactly the same as the properties of cytidine
acetylated chemically at the NH_2-group (Watanabe and Fox, 1966). This
N^4-acetylcytidine is labile: in an alkaline medium the acetyl group is removed,
with the resulting formation of C. The spectrum of acC (Feldmann et
al., 1966) is shown in Fig. 67. The spectrophotometric parameters of N^4-

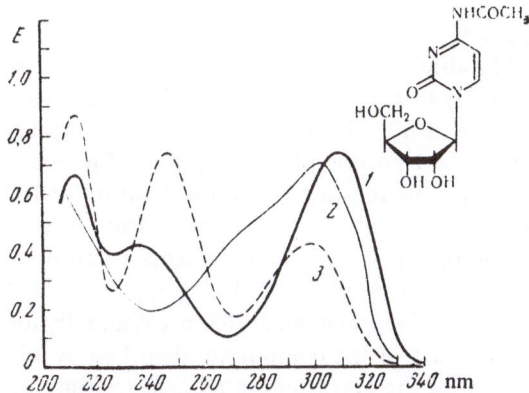

Fig. 67. Absorption spectrum of N^4-acetyl-
cytidine: (1) in acid medium; (2) in alkaline
medium; (3) in water.

acetylcytidine in water are given below (Watanabe and Fox, 1966):

λ_{max}, nm	$\varepsilon_{max} \times 10^{-3}$	λ_{min}, nm	$\varepsilon_{max} \times 10^{-3}$
294	8.6	268	4.4
245	15.1	225	6.1
212	18.2		

Determination of the acC content indicates that this compound occurs in only a few individual tRNAs. Cerutti et al. (1968), for instance, when investigating tRNAs from different sources by a combination of ion-exchange and thin-layer chromatography, found no perceptible quantities of acC. More recently, however, acC has been found in tRNA^{Ser} from rat liver, where it occupies precisely the same position as in yeast tRNA^{Ser} (Staehelin et al., 1968).

When investigating the effect of NaBH$_4$ on nucleotides and polymers, Miller and Cerutti (1967) found that this reagent, at pH 10, destroys not only H$_2$U but also acC, and the reaction takes place more readily with the latter. These workers concluded from their experimental results that acC has no role in the acceptor and transfer functions of tRNA, because yeast tRNA^{Ser}, which contains this minor component, behaves exactly like tRNA^{Val} and tRNA^{Ala} from the same source, i.e., its functional activity is unchanged after treatment with NaBH$_4$ (Cerutti and Miller, 1968).

In tRNA^{Ser} from *S. cerevisiae* and rat liver, acC occurs in the helical part of the H$_2$U-branch and occupies position 12 from the 5'-terminal pGp. The minor component found in this position in yeast tRNA$_3^{Leu}$ is also acetylcytidine. In most other tRNAs, this position is occupied by an unmodified C residue (Table 18).

THIOPYRIMIDINES

If tRNA from *E. coli* or rabbit liver (but not from yeast) is treated with dilute solutions of iodine, part of the acceptor activity relative to a number of amino acids is quickly lost. It was noted that the residual acceptor activity is insensitive to iodine, while the lost activity can be restored by reducing agents. Meanwhile, two groups of investigators (Lipsett, 1965; Carbon et al., 1965) showed that the reason for the inactivation of tRNA by iodine is oxidation of the thiopyrimidines which are normal structural elements of tRNA. Thiopyrimidines have in fact been isolated from alkaline digest of tRNA, and the principal member of this group was identified by Lipsett (1965, 1966) as 4thioU. Carbon et al. (1965) provisionally identified a 2thioU derivative. The spectrum of the compound identified by Lipsett is shown in Fig. 68; the position of the maximum in an acid medium at 331 nm and the characteristic hypochromic shift of the spectrum to 317 nm if the medium

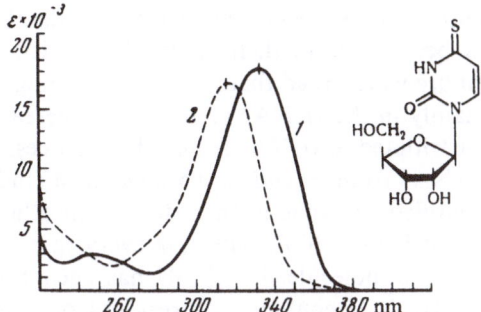

Fig. 68. Absorption spectrum of 4-thiouri-
dine: (1) at pH 2; (2) at pH 12.

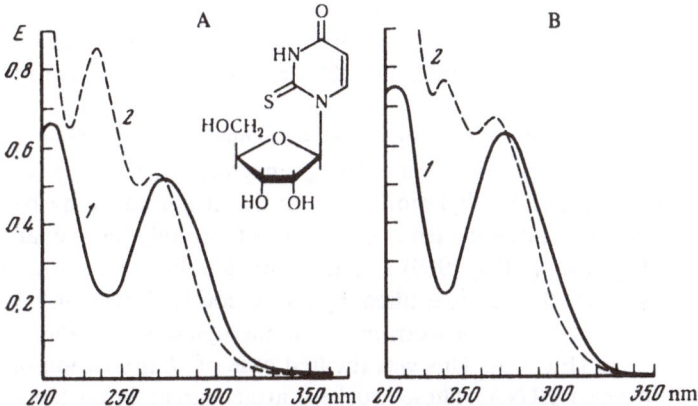

Fig. 69. Absorption spectrum of 2-thiouridine (A) and of the compound isolated by Carbon
et al. (1965) (B): (1) at pH 1; (2) at pH 13.

is made alkaline are evidence that the compound is in fact 4thioU. The spectrum of the compound isolated by Carbon et al. is very close, but not completely identical with the spectrum of 2thioU (Shugar and Fox, 1952) (Fig.
69). 4thioU has been found mainly in tRNA from *E. coli*, while 2thioU has
also been found in tRNA from rat liver (Carbon et al., 1965). Analysis of
tRNA from *E. coli*, grown on medium containing ^{35}S-sulfate, showed that
the total content of thionucleotides is about one per 60, and of 4thioU one
per 120 nucleotides. The use of various methods, including countercurrent
distribution (Schleich and Goldstein, 1965) has shown that the thiopyrimidines are unevenly distributed between the individual polynucleotide chains
of tRNA from *E. coli*.

The thionucleotides have recently attracted considerable attention and much progress has been made in their study. Their distribution has been found to be much more widespread than was hitherto supposed. Thiopyrimidines are found mainly in the tRNAs of bacteria, whereas thiopurines are also fairly widely distributed in tRNAs from other sources.

In nearly all tRNAs from *E. coli* which have been studied, thiopyrimidines have been identified: in some of them finally, in others still only provisionally. One of these is the 4thioU which, in every case, occupies position 8 from the 5′-end of the molecule. No thioU has been found in tRNA^{Leu} from *E. coli*, and its tRNA^{Tyr} contains two residues of thioU (in positions 8 and 9 from the 5′-end of the molecule). In addition, in the tyrosine, phenylalanine, and tryptophan, and also in one of the serine (Ishikura et al., 1971) tRNAs from *E. coli*, a hypermodified nucleoside is found on the 3′-side of the anticodon: 2-thiomethyl-N^6-isopentenyladenosine (Fig. 59). Two other thionucleosides (21) have been identified in tRNA from *E. coli*: 2-thiocytidine and 5-methylaminomethyl-2-thiouridine (Carbon et al.,1968). 2-Thiocytidine has now been identified in tRNA$^{Ser}_{3A}$ and tRNA$^{Ser}_{3B}$ from *E. coli* (Yamada et al., 1970).

5-Methylaminomethyl-2-thiouridine is formed from uridine evidently through the action of a series of modifying enzymes, performing thiolation, aminomethylation, and methylation respectively; it is interesting because it is the first minor component with a side chain of strongly basic character.

Baczynskyj et al. (1968, 1969) isolated a nucleotide from a total sample of yeast tRNA which was identified by mass spectrometry and from its chemical properties and uv-spectrum as the methylester of 2-thio-5-(or 6)-uridineacetic acid (Fig. 70). This was the first case of identification of a thiopyrimidine in yeast tRNA. These workers attach great importance to the presence of an acetic acid ester residue in this compound and they consider that enzymic esterification could be an important mechanism of modification of the properties of the corresponding tRNA.

It was later shown that 2-methylaminomethyl-2-thiouridine and the methylester of 2-thio-5-uridineacetic acid occupied position 1 in the anticodons of tRNA^{Glu} from *E. coli* (Ohashi et al., 1970) and yeast (Yoshida et al., 1970,

21   2-Thiocytidine   5-Methylaminomethyl-2-thiouridine

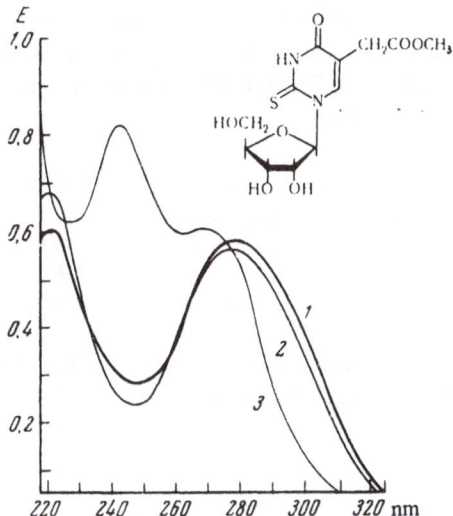

Fig. 70. Absorption spectrum of methylester
of 2-thio-5(6)-uridineacetic acid: (1) at pH
1.0; (2) at pH 7.0; (3) at pH 12.5.

1971) respectively, and that they are responsible for the unusual coding properties of these tRNAs (Sekiya et al., 1969), for they recognize only one
codon (G-A-A). The suggestion has been made that the minor component in
position 1 of the anticodon of these tRNAs can pair only with A, and not
with G.

Investigation of tRNAs from other sources led to the discovery of yet
another thiopyrimidine; the compound 5-methyl-2-thiouridine was identified
in tRNA^{Glu} and tRNA^{Lys} from rat liver (Kimura-Harada et al., 1971). It is
postulated that this minor component also occupies position 1 of the anticodon. Thiopyrimidines thus exist, evidently, in tRNAs from mammalian
tissues also.

Thiopyrimidines are unstable in neutral and acid solutions, in which
they lose sulfur and are converted into the corresponding pyrimidines. This
property precludes work with thiopyrimidines at acid and neutral pH values.
Oxidation of 4thioU with periodate also causes the removal of sulfur and
liberation of uracil and the ribose residue. Thionucleosides can be detected
on paper by the iodine–azide reaction, specific for sulfide and sulfhydryl
derivatives (Lipsett, 1965). The most characteristic property of the thiopyrimidines is their ability to be oxidized by iodine with the formation of disulfide
derivatives. Oxidation is accompanied by a decrease in absorption and a shift
of the maximum to 311 nm. As a component of tRNA, 4thioU has a separate
absorption band with a maximum at 336 nm, the intensity of which is 1.5%

of E_{260} (Fig. 11B). For this reason, the behavior of 4thioU can be studied spectrophotometrically, without the need for its preliminary isolation (Lipsett, 1966; Scheit, 1967b). The spectrophotometric parameters of 4thioU (Lipsett, 1965) are given below:

	pH 2	pH 12
λ_{max}, nm	331; 245	317
$\varepsilon_{max} \times 10^{-3}$	17.0; 5.2	14.9
λ_{min}, nm	285	260
E_{320}/E_{260}	2.09	3.58

The mobility of 4-thiouridine and of some related compounds during chromatography and electrophoresis is illustrated by the following data (Lipsett, 1965):

Compound	Butanol–H$_2$O (86 : 14)	Electrophoresis, pH 8.6, cm/h
4-Thiouridine disulfide	0.06	0.9
Uridine	0.15	2.7
4-Thiouridine	0.35	7.0

Little is yet known about the role of the thio-derivatives in tRNA. It is postulated that the easy and reversible oxidation of S may play an important role in the determination of tRNA conformation. Lipsett (1966) suggested that the formation of disulfide bonds between thionucleotides in the same polynucleotide chain, which probably must be accompanied by rupture of certain hydrogen bonds (Fig. 71), may be an effective method of modifying secondary structure. However, it is now known that the two 4thioUs in *E. coli* tRNA^{Tyr} are next to each other. Some interesting work in this connection was carried out by Doi and Goehler (Doi and Goehler, 1966; Goehler

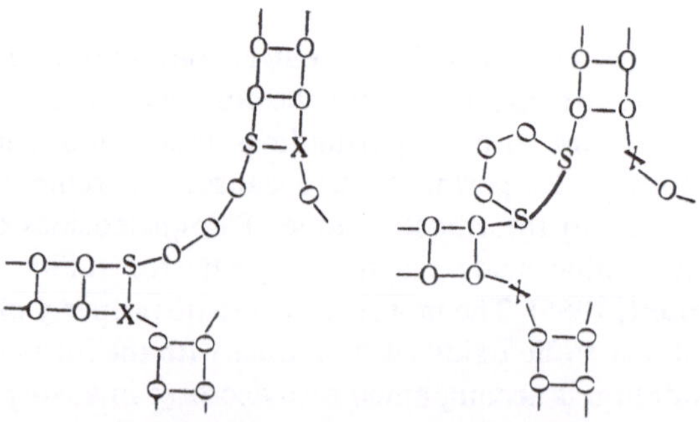

Fig. 71. Hypothetical scheme of intramolecular formation of disulfide bridges in tRNA.

et al., 1966), who showed that oxidation of tRNA from *B. subtilis* with iodine solution reversibly changes the elution profile from MAK columns, and depresses the acceptor function relative to lysine and the binding of lysyl-tRNA with the poly(A)–ribosome complex. These workers consider that the action of iodine is indirect, and is exerted primarily on the conformation of the tRNA, on which the binding of lysyl-tRNA with ribosomes depends.

It has been shown by the nmr method that 4thioU can form hydrogen bonds with A and I, but not with C and G (Scheit, 1967a). In their investigations of polymers containing 4thioU, Scheit and co-workers (Scheit and Gaertner, 1969a; Faerber and Scheit, 1970) showed that they can form complexes of the [poly(U,4thioU)]₂·[poly(A)] type, but they are much less stable than [poly(U)·poly(A)] complexes. Longer polymers of the poly(U,4thioU) type were subsequently obtained with the aid of polynucleotide-phosphorylase from *Micrococcus lysodeikticus* (Scheit and Gaertner, 1969a) and it was shown that the stability of the [poly(U, 4thioU)·poly(A)] complexes is indistinguishable from that of [poly(U)·poly(A)] complexes and also is independent of the Up : 4thioU ratio (Scheit and Gaertner, 1969b). Scheit cites the uv-spectra of polymers containing U and 4thioU, from which it is possible to determine the percentage content of 4-thiouridine directly (Fig. 11B) in a sample without its isolation (Scheit, 1967b).

Because of the wide distribution of thionucleotides, established by recent work, and their increasing importance in the study of tRNA, technical aspects of their investigation are being intensively developed. Several chemical modifications of the thionucleotides have been suggested (Saneyoshi, 1970; Irie, 1970; Sato and Kanaoko, 1971; Hecht et al., 1971); methods of identification have been developed, e.g., by means of phosphorescence spectroscopy (Hélène and Yaniv, 1970); techniques have been perfected for the conversion of thiouridine into uridine (Pleiss et al., 1969; Walker and RajBhandary, 1970) and for the selective labeling of thionucleotides (Burton, 1970; Ziff and Fresco, 1970; Hara et al., 1970; Kwong and Lane, 1970); the behavior of thionucleotides in enzymic reactions is being investigated (Sawada and Ishii, 1969); and so on.

The hypothesis has been put forward that some types of tRNA can act as modulators of protein synthesis at the translation level (Ames and Hartman, 1963; Stent, 1964). This hypothesis of adaptor modification stipulates the existence of a mechanism which could activate and inactivate particular tRNA molecules rapidly. Conversion of thionucleotides is regarded as one such case. The presence of thioU in tRNA may confer unique properties on the molecule. I have already mentioned changes in the coding properties of tRNAs containing thiopyrimidines in their anticodon. The formation of disulfide bridges, as has already been said, may play an important role in changes in the conformation of the molecule.

2'-O-METHYLRIBOSIDES

The presence of a methyl group at the 2'-OH group of ribose gives these compounds a number of distinctive chemical properties. It has long been known that total hydrolysates of tRNA contain, besides mononucleotides, a small quantity (3–5%) of alkali-resistant dinucleotides (Magasanik and Chargaff, 1951; Smith and Allen, 1953; Crestfield et al., 1955; Potter and Dounce, 1956; Tamaoki and Lane, 1967) and even of trinucleotides (Singh and Lane, 1964). It was suggested that the resistance of the small number of phosphodiester bonds is dependent on the presence of a substituent at the 2'-OH group of ribose, so that cyclization is impossible and rupture of the bond is prevented. In fact, the presence of an unusual carbohydrate, identified as 2-O-methylribose (22), has been demonstrated in specimens of RNA of animal and plant origin and also in RNA from *Euglena gracilis* (Brawerman and Chargaff, 1959; Smith and Dunn, 1959a):

22   2-O-Methylribose

Morisawa and Chargaff (1963) isolated an alkali-resistant fraction of dinucleotides from rat-liver RNA and from yeast tRNA; phosphodiesterase digests of this fraction contained OmeG and OmeC in addition to the four principal nucleotides. Later, 2'-O-methyl-derivatives of the four principal nucleosides and ψ were identified in tRNAs isolated from yeast and from other sources (Hall, 1964b, 1965; Hudson et al., 1965; Gray and Lane, 1967). Hall has investigated the 2'-O-methylnucleosides (Hall, 1963b,c; 1964b).

Three methods of determining 2'-O-methylnucleosides are known. One is to degrade the RNA by the combined action of snake-venom PDEase and alkaline phosphatase to nucleosides, and then to separate the OmeN together with other minor components painstakingly by column and paper chromatography (Hall, 1963c, 1964a, 1965). In this way 1.2 mg of 2'-O-methyl-derivatives was isolated from 1 g of yeast tRNA; calculations show that for every 600 nucleotides in tRNA there is one nucleoside methylated in the ribose residue. Judging by results obtained with individual tRNAs, this figure is too low. The second method is based on the isolation and identification of alkali-resistant dinucleotides (Singh and Lane, 1964; Hudson et al., 1965; Gray and Lane, 1967). By this method, 1.3 mole % of alkali-stable dinucleotides was found in wheat germ tRNA and 0.8 mole % in yeast tRNA. The third method of determining OmeN uses the lability of their

TABLE 29. Mobility of Nucleosides (R_f) during Paper Chromatography
(Hall, 1965)

Com-pound	Solvents[a]					Com-pound	Solvents[a]				
	1	2	3	4	5		1	2	3	4	5
A	0.27	0.58	0.34	0.54	0.35	2meG	—	—	0.45	0.39	0.21
OmeA	0.49	0.67	0.53	0.71	0.57	2me$_2$G	0.29	0.59	0.42	0.45	0.19
6meA	0.47	0.73	0.51	0.72	0.57	I	0.05	0.53	0.33	0.37	0.14
A[b]	0.19	0.64	0.34	0.61	0.11	1meI	0.21	0.61	0.48	0.56	0.22
C	0.15	0.49	0.47	0.44	0.10	U	0.10	0.62	0.64	0.38	0.27
OmeC	0.30	0.67	0.64	0.62	0.19	OmeU	0.30	0.73	0.86	0.52	0.58
3meC	0.34	0.62	0.54	0.62	0.04	3meU	0.45	0.75	0.88	0.70	0.59
5meC	0.20	0.51	0.48	0.44	0.07	5meU	0.27	0.68	0.74	0.54	0.44
G	0.05	0.47	0.30	0.26	0.13	ψ	0.04	0.50	0.52	0.29	0.11
OmeG	0.21	0.64	0.42	0.43	0.30	Omeψ	—	—	0.56	0.75	0.18
1meG	0.17	0.54	0.34	0.47	0.18						

[a]Solvents: (1) n-butanol–H_2O–conc. NH_4OH (86:14:5); (2) isopropanol–1% aqueous $(NH_4)_2SO_4$ (2:1); (3) isopropanol–conc. HCl–H_2O (68:17:14.4); (4) isopropanol–H_2O–conc. NH_4OH (7:2:1); (5) ethyl acetate–n-propanol–H_2O (4:1:2).
[b]A = N^6-(aminoacyl)adenosine (thrA).

ether bond on hydrolysis with 60% $HClO_4$ and the resulting formation of methanol (Baskin and Dekker, 1967).

Nucleosides methylated in the ribose moiety are indistinguishable in their spectrophotometric properties from nonmethylated nucleosides, but they can easily be separated from them by chromatography on paper, because the methyl group gives them great mobility in a number of systems of chromatographic solvents (Table 29). One result of the absence of the two vicinal hydroxyl groups in OmeN is that cyclic phosphates cannot be formed; this, as has already been said, means that the corresponding phosphodiester bonds are not ruptured by alkali. An important property of OmeN for structural research is that the bonds formed by them are not ruptured by the corresponding RNases, because in this case also the reaction proceeds through the intermediate formation of cyclic phosphates, and this is impossible. The OmeN do not form borate complexes and do not interact with sodium periodate; the electrophoretic mobility of these compounds and that of the corresponding deoxyribonucleosides at pH 9.2 are identical.

The distribution of the OmeN group of compounds in individual tRNAs has proved to be wider than was supposed on the basis of analysis of the total preparations: one tRNA (serine, from rat liver) contains three OmeN residues; four tRNAs contain two residues each, and eight tRNAs contain one OmeN residue each; no nucleosides methylated in the ribose residue have yet been found in the other tRNAs. It is interesting to note that no OmeN is found in any of the valine-specific tRNAs so far decoded. Nucleosides methylated in the ribose moiety are located in the H_2U-loop, in which only OmeG is found, and the anticodon loop. OmeG occurs most widely: Omeψ has been identified only in tRNA^{Ser} from rat liver; OmeA has not yet been found in individual tRNAs, in agreement with its comparatively low content in total tRNA from yeast (Gray and Lane, 1967).

UNIDENTIFIED MINOR COMPONENTS

Work is still in progress on a number of modified nucleosides present in individual tRNAs.

Most of the unidentified minor components belong to the tRNAs from *E. coli,* which is not surprising because these have been studied by Sanger's micromethod, which is based entirely on the determination of radio-activity.

Position 1 of the anticodon in tRNA^{Tyr} from *E. coli* is occupied by a minor component which, judging from its spectrum, is a guanine derivative, which differs from guanine in having a basic substituent (pK5–6) in the imidazole part of the purine ring. Nucleosides occupying position 1 in the anticodons of tRNA^{Met} and tRNA$_1^{Gln}$ likewise have not been identified; the nature of the nucleoside in the extra branch of tRNA^{Met} and tRNA^{Phe}, as well as the nucleoside N in tRNA$_2^{Val}$ has not yet been established; the nature of the nucleosides on the 3′-side of the anticodon in tRNA^{Met} and tRNA^{Gln} likewise is still unknown. The minor components in yeast tRNA$_3^{Leu}$ have not yet been identified.

The material described shows that minor components form an extremely varied and rapidly growing group of compounds, whose only connecting link as yet is their location in tRNA. The study of the chemical nature of the modified nucleotides has gone far ahead of the study of their functional role, for we do not yet know what role is played by any single minor component, nor do we know why modification has occurred at a particular site of the polynucleotide chain and not at any other.

The material which has been collected in recent years shows more clearly than ever that the functional properties of the tRNAs are determined, not by their individual nucleotides, or even, probably, by their nucleotide sequences, but by the conformation of particular three-dimensional centers of the tRNA. It can be conjectured that, because of their modified reactivity with other nucleotides, the minor components have a special role in this matter. The rapid progress in research into the primary structure of tRNA, the development of methods of selective chemical modification in general, and of the minor components in particular, and above all the crystallization of tRNA which is the essential preliminary to its x-ray structural investigation, lead to the hope that these problems will find concrete solutions in the very near future.

CONCLUSION

The structure of approximately twenty tRNAs has now been studied; examples are known of the determination of the primary structure of tRNAs of different amino-acid specificity from the same source, of the same amino-acid specificity from different sources and, finally, of isoacceptor tRNAs. Some general principles and conclusions can be drawn from the wealth of evidence that has been accumulated. The first striking feature is the homology in the structure of the tRNA molecules, manifested at all levels of their organization: the covalent structures of the tRNAs, although individual in each particular case, are built in accordance with the same common plan; the arrangement of the Watson–Crick pairs, i.e., the secondary structure of the tRNA molecules, is the same, and this is reflected in the clover-leaf model, which is common to all structures so far identified; finally, we now have evidence that this class of nucleic acids possesses the same three-dimensional organization (one fact which points in this direction, e.g., is that a specimen of total tRNA can be crystallized).

The unity of structure of the transfer RNAs is evidently a reflection of their functional needs. Integration of a host of diverse functions in one molecule places strict limitations on structure; for this reason, in the course of evolution, as the result of mutations and the rigid application of natural selection, molecules with a sharply defined structure, not permitting substantial deviations, have developed from a single common prototype tRNA. The unity of the common structural plan of the tRNAs is thus the result of their multiplicity of function as well as the result of their common origin.

On the other hand, we cannot fail to notice the multiplicity of the tRNAs, and judging from the results yielded by more sophisticated analytical methods, this multiplicity is much greater than was hitherto supposed. Slight changes in structure, consisting of the replacement of single nucleotides or of a modification to a single component, have been detected. The significance of these changes is not yet clear; however, considering the central role of tRNA in protein synthesis and the rigid demands on structure, which I mentioned above, it can be postulated that even the slightest modification to a tRNA will significantly affect protein synthesis. This suggests that variability of structure, leading to a change in the relative proportion of isoacceptor fractions of tRNA is an effective means of regulating protein synthesis at the translation level.

The study of primary structure has led to an understanding of the general structural plan of the transfer RNAs. These investigations are also important because they have prepared the ground for the study of the three-dimensional structure of these biopolymers, which is an essential preliminary to the study of the finer details of the protein–nucleic-acid interactions in which the transfer RNAs participate.

APPENDIX

Comparison of the primary structures of transfer RNAs are offered with reference to the separate branches or arms of the cloverleaf.* The nucleotides have been arranged according to the various arms. Underlined bases indicate those base-paired in the cloverleaf model. Yeast (B) = bakers' yeast, yeast (T) =Torula yeast. For the abbreviations of minor nucleotides the system using small letters for substitutions has been used (cf. *Handbook of Biochemistry,* H.A. Sober, ed., The Chemical Rubber Company, Cleveland (1968), pp. H64–H65), i.e., hU = dihydrouridine, m^1G = 1-methylguanosine, m^2G = N^2-methylguanosine, m_2^2G = N^2-dimethylguanosine; m^7G = 7-methylguanosine; m^1I = 1-methylinosine; m^1A = 1-methyladenosine; m^6A = N^6-methyl adenosine; T = ribothymidine; m^5C = 5-methylcytidine; m^3C = 3-methylcytidine; Um = 2'-O-methylguanosine; I = inosine; s^4U =4-thiouridine; i^6A = N^6-isopentenyladenosine; ms^2-i^6A = 2-methylthio-N^6-isopentenyladenosine; ac^4C = N^4-acetylcytidine; tcP = N-(purin-6-yl-carbamoyl)-threonine; Y = fluorescent base in phenylalanine tRNA; X = unknown derivatives of parent bases. Pseudouridine is indicated as Ʋ.

---

*The papers are taken from the paper by Staehelin (1971) and supplemented by the author's data for glutamine, glycine, and valine 2 tRNAs from *E. coli,* and also for leucine tRNA 3 from bakers' yeast.

	p	1	2	3	4	5	6	7	...	81	82	83	84	85	86	87	88	89	90	91
Tyr$_{yeast(B)}$	p	C	U	C /U/	C	G	G		......	C	C	G /G/	G	A	G	A	C	C	A	
Tyr$_{yeast(T)}$	p	C	U	C	U	C	G	G	......	C	C	G	A	G	A	G	A	C	C	A
Val$_{yeast(B)}$	p	G /G/	U	U	U	C	G		......	C	G	A	A	A /U/	C	A	C	C	A	
Val$_{yeast(T)}$	p	G /G/	U	U	U	C	G		......	C	G	A	A	A /U/	C	A	C	C	A	
Ile$_{yeast(T)}$	p	G	G	U	C	C	C	U	......	A	G	G	G	A	C	C	A	C	C	A
Gln$_{2E.coli}$	p	U	G	G	G	G	U	A	......	U	A	C	C	C	C	A	G	C	C	A
Ser$_{yeast(B)}$	p	G /G/	C	A	A	C	U		......	A	G	U	U	G /U/G	G	C	C	A	
Ser$_{rat}$	p	G	U	A	G	U	C	G	......	C	G	A	C	U	A	C	G	C	C	A
Tyr$_{E.coli}$	p	G	G	U	G	G	G	G	......	C	C	C	C	A	C	C	A	C	C	A
Leu$_{E.coli}$	p	G	C	G	A /A/	G	G		......	C	C /C/	U	C	G	C	A	C	C	A	
Leu$_{3yeast(B)}$	p	G	G	U	U	G /U/	U		......	A	G	C	A	A	C	C	A	C	C	A	
Val$_{E.coli}$	p	G	G	G	U	G	A	U	......	A	U	C	A	C	C	C	A	C	C	A
Phe$_{yeast(B)}$	p	G	C	G /G/	A	U	U		......	A	A	U /U/	C	G	C	A	C	C	A	
Phe$_{wheat}$	p	G	C	G	G /G/	G	A		......	U	C /A/	C	C	G	C	A	C	C	A	
Phe$_{E.coli}$	p	G	C	C	C	G	G	A	......	U	C	C	G	G	G	C	A	C	C	A
Ala$_{yeast(B)}$	p	G	G /G/	C	G /U/	G			......	C /U/	C	G /U/	C	C	A	C	C	A		
Asp$_{yeast}$	p	U	C	C	G /U/	G	A		......	U /C/	G	C	G	G	A	G	C	C	A	
Trp$_{E.coli}$	p	A	G	G	G	G	C	G	......	C	G	C	C	C	C	U	G	C	C	A
Met$_{E.coli}$	p	G	G	C	U	A	C	G	......	C	G	U	A	G	C	C	A	C	C	A
Gly$_{E.coli}$	p	G	C	G	G	G	A	A	......	U	U	C	C	C	G	C	U	C	C	A
Val$_{2E.coli}$	p	G	C	G	U	C	C	G	......	C	G	G	A	C	G	C	A	C	C	A
			...U...		A...	U			...A...												
fMet$_{E.coli}$	p	C	G	C	G	G	G	G	......	C	C	C	C	G	C	A	A	C	C	A
		1	2	3	4	5	6	7	81	82	83	84	85	86	87	88	89	90	91	

hU Arm

	8	9	10	11	12	13	14	15	16	17	18	19	20	21	22	23	24	25	26	27	28	29	30
Tyr$_{yeast(B)}$	U	A	m²G	C	C	A	–	–	A	G	hU	hU	–	Gm	G	hU	hU	hU	A	A	G	G	C
Tyr$_{yeast(T)}$	U	m¹G	G	C	C	A	–	–	A	G	hU	hU	–	Gm	G	hU	hU	hU	A	A	G	C	C
Val$_{yeast(B)}$	U	m¹G	G	U	C	Ψ	–	–	A	G	hU	C	–	G	G	hU	hU	–	A	U	G	G	C
Val$_{yeast(T)}$	U	m¹G	G	U	C	Ψ	–	–	A	G	hU	hU	–	G	G	hU	C	–	A	U	G	G	C
Ile$_{yeast(T)}$	U	G	G	C	C	C	–	–	A	G	hU	hU	–	G	G	hU	hU	–	A	A	G	G	C
Gln$_2$ $_{E.coli}$	sU	C	G	C	C	A	A	G	C	Gm	G	hU	A	A	G	G	C	–	–	–	G	G	C
Ser$_{yeast(B)}$	U	G	G	C	acC	G	–	–	A	G	hU	–	–	Gm	G	hU	hU	–	A	A	G	G	C
Ser$_{rat}$	U	G	G	C	acC	G	–	–	A	G	hU	–	–	Gm	G	hU	hU	–	A	A	G	G	C
Tyr$_{E.coli}$	sU	sU	C	C	C	G	–	–	A	G	C	–	–	Gm	G	C	C	A	A	A	G	G	G
Leu$_{E.coli}$	U	G	G	C	G	G	–	–	A	A	hU	hU	–	G	G	hU	A	–	G	A	C	G	C
Leu$_3$ $_{yeast(B)}$	U	G	G*	C	C*	G	–	–	A	G	C	hU	–	Gm	G	–	hU	C	A	A	G	G	C
Val$_{E.coli}$	sU	A	G	C	U	C	–	–	A	G	C	hU	–	G	G	G	–	–	A	G	A	G	C
Phe$_{yeast(B)}$	U	A	m²G	C	U	C	–	–	A	G	hU	hU	–	G	G	G	–	–	A	G	A	G	C
Phe$_{wheat}$	U	A	m²G	C	U	C	–	–	A	G	hU	hU	–	G	G	G	–	–	A	G	A	G	C
Phe$_{E.coli}$	U	A	G	C	U	C	–	–	A	G	hU	C	–	G	G	hU	–	–	A	G	A	G	C
Ala$_{yeast(B)}$	U	m¹G	G	C	G	C	G	U	A	G	hU	C	–	G	G	hU	–	–	A	G	C	G	C
Asp$_{yeast}$	U	A	G	U	U	Ψ	–	–	A	A	hU	–	–	G	G	hU	C	–	A	G	A	A	U
Trp$_{E.coli}$	sU	A	G	U	U	C	–	–	A	A	hU	hU	–	G	G	hU	–	–	A	G	A	G	C
Met$_{E.coli}$	sU	A	G	C	U	C	–	–	A	G	hU	hU	–	Gm	G	hU	hU	–	A	G	A	G	C
Gly$_{E.coli}$	sU	A	G	C	U	C	–	–	A	G	hU	hU	–	G	G	hU	–	–	A	G	A	G	C
Val$_2$ $_{E.coli}$	sU	A	G	C	U	C			A	G	hU	hU	–	G	G	hU	hU	–	A	G	A	G	C
fMet$_{E.coli}$	sU	G	G	A	G	C	–	–	A	G	C	C	U	G	G	hU	–	–	A	G	C	U	C

...U... (Met$_{E.coli}$)

Anticodon Arm

	31	32	33	34	35	36	37	38	39	40	41	42	43	44	45	46	47	48
Tyr$_{yeast(B)}$	m_2^2G	C	A	A	G	A	C	U	G	Ψ	A	i^6A	A	Ψ	C	U	U	G
Tyr$_{yeast(T)}$	m_2^2G	Ψ	C	A	G	A	C	U	G	Ψ	A	i^6A	A	Ψ	C	U	G	A
Val$_{yeast(B)}$	A	Ψ	C	U	G	C	Ψ	U	I	A	C	A	C	G	C	A	G	A
Val$_{yeast(T)}$	A	Ψ	C	U	G	C	Ψ	U	I	A	C	A	C	G	C	A	G	A
Ile$_{yeast(T)}$	m_2^2G	Ψ	G	G	U	G	C	U	I	A	U	tcP	A	C	G	C	C	A
Gln$_{2\ E.coli}$	A	C	C	G	G	A	Um	U	C	U	G	R	Ψ	Ψ	C	C	G	G
					...U...								...A...					
Ser$_{yeast(B)}$	m_2^2G	A	A	A	G	A	Ψ	U	I	G	A	i^6A	A	Ψ	C	U	U	U
Ser$_{rat}$	m_2^2G	A	Ψ	G	G	A	m^3C	U	I	G	A	i^6A	A	Ψm	C	C	A	U
Tyr$_{E.coli}$	A	G	C	A	G	A	C	U	G^*	U	A	s^iA	A	Ψ	C	U	G	C
Leu$_{E.coli}$	G	C	U	A	/G/	C	U	U	C	A	G	G^*	Ψ	G	/Ψ/	U	A	G
Leu$_{3yeast(B)}$	G^*	C	C	U	G	A	Ψ	U	C	A	A	G^+	C	Ψ	C	A	G	G
Val$_{E.coli}$	A	C	C	U	C	C	C	U	X	A	C	m^6A	A	G	G	A	G	G
Phe$_{yeast(B)}$	m_2^2G	C	C	A	G	A	Cm	U	Gm	A	A	Y	A	Ψ	m^5C	U	G	G
Phe$_{wheat}$	m_2^2G	Ψ	C	A	G	A	C	U	Gm	A	A	Y	A	Ψ	C	U	G	A
Phe$_{E.coli}$	A	G	G	G	G	A	Ψ	U	G	A	A	s^iA	A	Ψ	C	C	C	C
Ala$_{yeast(B)}$	m_2^2G	C	U	C	C	C	U	U	I	G	C	mI	Ψ	G	G	G	A	G
Asp$_{yeast}$	G	G	G	C	/G/	C	Ψ	U	G	U	C	m^1G	C	G	/U/	G	C	C
Trp$_{E.coli}$	A	C	C	G	G	U	Cm	U	C	C	A	A^*	A	A	C	C	G	G
Met$_{E.coli}$	A	C	A	U	C	A	C	U	C^*	A	U	A^*	A	Ψ	G	A	U	G
Gly$_{E.coli}$	A	C	G	A	C	C	U	U	G	C	C	A	A	G	G	U	C	G
Val$_{2\ E.coli}$	A	C	C	A	C	C	U	U	G	A	C	A	U	G	G	U	G	G
fMet$_{E.coli}$	G	U	C	G	G	G	Cm	U	C	A	U	A	A	C	C	C	G	A

Extra Arm

	49	50	51	52	53	54	55	56	57	58	59	60	61	62	63
Tyr$_{yeast(B)}$	A	—	—	—	-	G	A	hU	—	—	—	—	—	—	m⁵C
Tyr$_{yeast(T)}$	A	—	—	—	—	C	A	hU	—	—	—	—	—	—	m⁵C
Val$_{yeast(B)}$	A	—	—	—	—	C	m⁷G	hU	—	—	—	—	—	—	m⁵C
Val$_{yeast(T)}$	A	—	—	—	—	C	—	—	—	—	—	—	—	—	m⁵C
Ile$_{yeast(T)}$	A	—	—	—	—	G	A	hU	—	—	—	—	—	—	m⁵C
Gln$_2$ $_{E.coli}$	C	—	—	—	—	A	U	U	—	—	—	—	—	—	C
Ser$_{yeast(B)}$	Um	G	G	G	C	U	C	U	—	G	C	C	C	G	m⁵C
							...U...								
Ser$_{rat}$	Um	G	G	G	G	Um³	C	U	—	C	C	C	C	G	m⁵C
Tyr$_{E.coli}$	C	—	G	U	C	A	U	C	—	G	A	C	U	U	C
						...C	A...								
Leu$_{E.coli}$	U	G	U	C	C	U	U	A	C	G	G	A	C	G	U
Leu$_3$ $_{yeast(B)}$		U	A	U	C	G	U	A	A	G	A	U	G	—	m⁵C
Val$_{E.coli}$	G	—	—	—	—	G	m⁷G	U	—	—	—	—	—	—	C
Phe$_{yeast(B)}$	A	—	—	—	—	G	m⁷G	U	—	—	—	—	—	—	C
Phe$_{wheat}$	A	—	—	—	—	G	m⁷G	hU	—	—	—	—	—	—	C
Phe$_{E.coli}$	G	—	—	—	—	U	m⁷G	X	—	—	—	—	—	—	C
Ala$_{yeast(B)}$	A	—	—	—	—	G	—	U	—	—	—	—	—	—	C
							...hU...								
Asp$_{yeast}$	A	—	—	—	—	G	—	A	—	—	—	—	—	—	U
Trp$_{E.coli}$	G	—	—	—	—	U	m⁷G	U	—	—	—	—	—	—	U
Met$_{E.coli}$	G	—	—	—	—	G	m⁷G	X	—	—	—	—	—	—	C
Gly$_{E.coli}$	G	—	—	—	—	G	m⁷G	U	—	—	—	—	—	—	C
Val$_{2E.coli}$	G	—	—	—	—	G	m⁷G	N	—	—	—	—	—	—	C
fMet$_{E.coli}$	A	—	—	—	—	G	A	U	—	—	—	—	—	—	C
							...m⁷G...								

TψC Arm

	64	65	66	67	68	69	70	71	72	73	74	75	76	77	78	79	80
Tyr$_{yeast(B)}$	G	G	G	C	G	T	ψ	C	G	m¹A	C	U	C	G	C	C	C
Tyr$_{yeast(T)}$	G	G	G	C	G	T	ψ	C	G	m¹A	A	U	C	G	C	C	C
Val$_{yeast(B)}$	C	C	C	A	G	T	ψ	C	G	m¹A	U	C	C	U	G	G	G
Val$_{yeast(T)}$	C	C	C	A	G	T	ψ	C	G	m¹A	U	C	C	U	G	G	G
Ile$_{yeast(T)}$	A	G	C	A	G	T	ψ	C	G	m¹A	U	C	C	U	G	C	U
Gln$_{2E.coli}$	C	C	U	G	G	T	ψ	C	G	A	A	U	C	C	A	G	G
	...G...A...														U...C...		

	64	65	66	67	68	69	70	71	72	73	74	75	76	77	78	79	80
Ser$_{yeast(B)}$	G	C	A	G	G	T	ψ	C	A	A	A	U	C	C	U	G	C
									...G...		...G...						
Ser$_{rat}$	G	C	A	G	G	T	ψ	C	G	m¹A	A	U	C	C	U	G	C
Tyr$_{E.coli}$	G	A	A	G	G	T	ψ	C	G	A	A	U	C	C	U	U	C
Leu$_{E.coli}$	G	G	G	G	G	T	ψ	C	A	A	G	U	C	C	C	C	C
Leu$_{3yeast(B)}$	A	A	G	A	G	T	ψ	C	G	A	A	U	C	U	C	U	U

	64	65	66	67	68	69	70	71	72	73	74	75	76	77	78	79	80
Val$_{E.coli}$	G	/G/	C	G	G	T	ψ	C	G	A	U	C	C	C	G	/U/	C
Phe$_{yeast(B)}$	m⁵C	U	G	U	G	T	ψ	C	G	m¹A	U	C	C	A	C	A	G
Phe$_{wheat}$	G	C	G	U	G	T	ψ	C	G	m¹A	U	C	C	A	C	G	C
Phe$_{E.coli}$	C	U	/U/	G	G	T	ψ	C	G	A	U	U	C	C	/G/	A	G
Ala$_{yeast(B)}$	U	C	C	G	G	T	ψ	C	G	A	U	U	C	C	G	G	A
Asp$_{yeast}$	m⁵C	G	G	C₍	G	T	ψ	C	A	A	U	U	C	C	C	C	G

	64	65	66	67	68	69	70	71	72	73	74	75	76	77	78	79	80
Trp$_{E.coli}$	G	/G/	G	A	G	T	ψ	C	G	A	G	U	C	U	C	/U/	C
Met$_{E.coli}$	A	C	/A/	G	G	T	ψ	C	G	A	A	U	C	C	/C/	G	U
Gly$_{E.coli}$	/G/	C	G	A	G	T	ψ	C	G	A	G	U	C	U	C	G	/U/
Val$_{2E.coli}$	G	G	U	G	G	T	ψ	C	G	A	G	U	C	C	A	C	/U/
	...U...														...A...		
fMet$_{E.coli}$	G	/U//C/	G	G	T	ψ	C	A	A	U	C	C	G	/G/	C		

    64   65   66   67   68   69   70   71   72   73   74   75   76   77   78   79   80

BIBLIOGRAPHY

Abelson, J., Barnett, L., Brenner, S., Gefter, M., Landy, A., Russell, R., and Smith, J. D., 1969, *FEBS Letters*, **3**:1.

Abelson, J. N., Gefter, M. L., Barnett, L., Landy, A., Russell, R. L., and Smith, J. D., 1970, *J. Mol. Biol.*, **47**:15.

Abrell, J. W., Kaufman, E. E., and Lipsett, M. N., 1971, *J. Biol. Chem.*, **246**:294.

Abrosimova-Amel'yanchik, N. M., 1968, Dissertation for degree of Candidate, Moscow.

Abrosimova-Amel'yanchik, N. M., Tamarskaya, R. I., Venkstern, T. V., Aksel'rod, V. D., and Baev, A. A., 1965, *Biokhimiya*, **30**:1269.

Adams, A., and Zachau, H. G., 1968, *Europ. J. Biochem.*, **5**:556.

Adams, J. M., and Capecchi, M. R., 1966, *Proc. Nat. Acad. Sci. (Washington)*, **55**:147.

Adams, J. M., and Cory, S., 1970, *Nature*, **227**:570.

Adams, J. M., Jeppesen, P. G. N., Sanger, F., and Barrell, B. G., 1969, *Nature*, **223**:1009.

Adler, M., and Gutman, A. B., 1969, *Science*, **130**:862.

Adler, M., Weissmann, B., and Gutman, A. B., 1958, *J. Biol. Chem.*, **230**:717.

Aksel'rod, V. D., Venkstern, T. V., and Baev, A. A., 1965, *Biokhimiya*, **30**:999.

Aksel'rod, V.D., Fodor, I., and Baev, A. A., 1967, *Dokl. Akad. Nauk SSSR*, **174**:707.

Alexander, M., Heppel, L. A., and Hurwitz, J., 1961, *J. Biol. Chem.*, **236**:3014.

Alvino, C. G., and Clarke, H. T., 1968, *Federat. Proc.*, **27**:342.

Alvino, C. G., Remington, L., and Ingram, V. M., 1969, *Biochemistry*, **8**:282.

Ames, B. N., and Hartman, P. E., 1963, *Cold Spring Harbor Symp. Quant. Biol.*, **28**:349.

Amos, H., and Korn, M., 1958, *Biochim. Biophys. Acta*, **29**:444.

Anderson, F., 1969, *Biochemistry*, **8**:3687.

Aoyagi, S., and Inoue, Y., 1968, *J. Biol. Chem.*, **243**:514.

Apgar, J., and Holley, R. W., 1964, *Biochem. Biophys. Res. Commun.*, **16**:21.

Apgar, J., Holley, R. M., and Merrill, S. H., 1961, *Biochim. Biophys. Acta*, **53**:220.

Apgar, J., Holley, R. W., and Merrill, S. H., 1962, *J. Biol. Chem.*, **237**:796.

Apgar, J., Everett, G. A., and Holley, R. W., 1965, *Proc. Nat. Acad. Sci. (Washington)*, **53**:546.

Apgar, J., Everett, G. A., and Holley, R. W., 1966, *J. Biol. Chem.*, **241**:1206.

Arima, T., Uchida, T., and Egami, F., 1969, *J. Biochem.*, **106**:609.

Armstrong, A., Hagopian, H., Ingram, V. M., and Wagner, E. K., 1966, *Biochemistry*, **5**:3027.

Asano, K., 1961, *J. Biochem. (Tokyo)*, **50**:544.

Aubert, M., Reynier, M., and Monier, R., 1967, *Bull. Soc. Chim. Biol.*, **49**:1191.

Augusti-Tocco, G., and Brown, G. L., 1965, *Nature*, **206**:683.

Azegami, M., and Iwai, K., 1964, *J. Biochem. (Tokyo)*, **55**:346.

Baczynskyj, L., Biemann, K., Fleysher, M. N., and Hall, R. H., 1969, *Can. J. Biochem.*, **47**:1202.

Baczynskyj, L., Biemann, E., and Hall, R. H., 1968, *Science*, **159**:1481.

Baev (Bajew), A. A., Venkstern (Wenkstern), T. V., and Mirzabekov, A. D., 1963, *Jenaer Symp. Physikalische Chem. Biogen. Makromol.*, Akademie-Verlag, Berlin, p. 423.

Baev, A. A., Venkstern, T. V., Mirzabekov, A. D., and Tatarskaya, R. I., 1963a, *Biokhimiya*, **28**:931.

Baev, A. A., Mirzabekov, A. D., Gorshkova, V. I., and Venkstern, T. V., 1963b, *Dokl. Akad. Nauk SSSR*, **152**:331.

Baev (Bayev), A. A., Venkstern, T. V., Mirzabekov, A. D., Krutilina, A. I., Li, L., and Aksel'rod (Axelrod), V. D., 1966a, *Biochim. Biophys. Acta*, **108**:162.

Baev (Bayev), A. A., Venkstern, T. D., Mirzabekov, A. D., Krutilina, A. I., Li, L., and Aksel'rod (Axelrod), V. D., 1966b, *Abstr. 3rd Meeting FEBS*, Warsaw, p. 13.

Baev (Bayev), A. A., Venkstern, T. V., Mirzabekov, A. D., Krutilina, A. I., Aksel'rod (Axelrod), V. D., Li, L., and Engelhardt, V. A., 1966c, in: *Genetic Elements*, Academic Press, London, p. 287.

Baev, A. A., Venkstern, T. V., Mirzabekov, A. D., Krutilina, A. I., Li, L., and Aksel'rod (Axelrod), V. D., 1967a, *Molekul. Biol.*, **1**:754.

Baev, A. A., Mirzabekov, A. D., Aksel'rod, V. D., Venkstern, T. V., Li, L., Krutilina, A. I., Fodor, I., and Kazarinova, L. Ya., 1967b, *Dokl. Akad. Nauk SSSR*, **173**:204.

Baev, A. A., Fodor, I., Mirzabekov, A. D., Aksel'rod, V. D., and Kazarinova, L. Ya., 1967c, *Molekul. Biol.*, **1**:859.

Baev (Bayev), A. A., Venkstern, T. V., Mirzabekov, A. D., Aksel'rod (Axelrod), V. D., Li, L., Fodor, I., Kasarinova, Ya., and Engelhardt, V. A., 1968, in: *Structure and Function of Transfer RNA and 5S-RNA*, Universitetsforlaget, Oslo, p. 17.

Baev (Bayev), A. A., Mirzabekov, A. D., Aksel'rod (Axelrod), V. D., and Chuguev., I. I., 1969, *Abstr. 6th Meeting FEBS*, Madrid, p. 61.

Baguley, B. C., and Staehelin, M., 1968a, *Biochemistry*, **7**:45.

Baguley, B. C., and Staehelin, M., 1968b, *Europ. J. Biochem.*, **6**:1.

Baguley, B. C., and Staehelin, M., 1969, *Biochemistry*, **8**:257.

Baguley, B. C., Bergquist, P. L., and Ralph, R. K., 1965a, *Biochim. Biophys. Acta*, **95**:510.

Baguley, B. C., Bergquist, P. L., and Ralph, R. K., 1965b, *Biochim. Biophys. Acta*, **108**:139.

Baguley, B. C., Wehrli, W., and Staehelin, M., 1970, *Biochemistry*, **9**:1645.

Baliga, B. S., Srinivasan, P. R., and Borek, E., 1965, *Nature*, **208**:555.

Baliga, B. S., Borek, E., Weinstein, I. B., and Srinivasan, P. R., 1969, *Proc. Nat. Acad. Sci. (Washington)*, **62**:899.

Barnett, W. E., and Brown, D. H., 1967, *Proc. Nat. Acad. Sci. (Washington)*, **57**:452.

Barrell, B. G., and Sanger, F., 1969, *FEBS Letters*, **3**:275.

Bartz, J. K., Kline, L. K., and Söll, D., 1970, *Biochem. Biophys. Res. Commun.*, **40**:1481.

Baskin, F., and Dekker, C. A., 1967, *J. Biol. Chem.*, **242**:5447.

Batt, D. B., Martin, J. K., Ploeser, J. McT., and Murray, J., 1954, *J. Amer. Chem. Soc.*, **76**:3663.

Beardsley, K., and Cantor, C. R., 1970, *Proc. Nat. Acad. Sci. (Washington)*, **65**:39.

Beaven, G. H., Holiday, E. R., and Johnson, E. A., 1955, in: *The Nucleic Acids* (Chargaff and Davidson, eds.), Vol. 1, Academic Press, New York, p. 502.

Beck, G., Fellner, P., Ebel, J.-P., and Blasi, O., 1970, *FEBS Letters*, **7**:41.

Beer, M., and Moudrianakis, E. M., 1962, *Proc. Nat. Acad. Sci. (Washington)*, **48**:409.

Beers, R. F., Jr., 1960, *J. Biol. Chem.*, **235**:2393.

Bekker, Zh. M., Molin, Yu. N., Girshovich, A. S., Grachev, M. A., and Knorre, D. G., 1969, *Molekul. Biol.*, **3**:366.

Belikova, A. M., Zarytova, V. F., and Grineva, N. I., 1967, *Tetrahedron Letters*, p. 3557.

Bell, D., and Russell, G. J., 1967, *Biochemistry*, **6**:3363.

Bell, D., Tomlinson, R. V., and Tener, G. M., 1964, *Biochemistry*, **3**:317.

Belozersky, A. N., and Spirin, A. S., 1960, in: *The Nucleic Acids* (Chargaff and Davidson, eds.), Vol. 3, Academic Press, New York, p. 147.

Beltchev, B., and Grunberg-Manago, M., 1970a, *FEBS Letters*, **12**:24.

Beltchev, B., and Grunberg-Manago, M., 1970b, *FEBS Letters*, **12**:27.

Bennett, T. P., Goldstein, J., and Lipmann, F., 1963, *Proc. Nat. Acad. Sci. (Washington)*, **49**:850.
Bennett, T. P., Goldstein, J., and Lipmann, F., 1965, *Proc. Nat Acad. Sci. (Washington)*, **53**:385.
Berg, P., Bergmann, F. H., Ofengand, E. J., and Dieckmann, M., 1961, *J. Biol. Chem.*, **236**: 1726.
Berg, P., Lagerkvist, U., and Dieckmann, M., 1962, *J. Mol. Biol.*, **5**:159.
Bergquist, P. L., 1966a, *Biochim. Biophys. Acta*, **103**:347.
Bergquist, P. L., 1966b, *Cold Spring Harbor Symp. Quant. Biol.*, **31**:435.
Bergquist, P. L., and Matthews, R. E. F., 1962, *Biochem. J.*, **85**:305.
Bergquist, P. L., and Robertson, J. M., 1965, *Biochim. Biophys. Acta*, **103**:579.
Bergquist, P. L., Baguley, B. C., Robertson, J. M., and Ralph, R. K., 1965, *Biochim. Biophys. Acta*, **108**:531.
Bernardi, A., and Bernardi, G., 1966, *Biochim. Biophys. Acta*, **129**:23.
Bernardi, A., and Bernardi, G., 1968, *Biochim. Biophys. Acta*, **155**:360.
Bernardi, A., and Cantoni, G. L., 1969, *J. Biol. Chem.*, **244**:1468.
Bernhardt, D., and Darnell, J. E., 1969, *J. Mol. Biol.*, **42**:43.
Biemann, K., Tsunakawa, S., Sonnenbichler, J., Feldmann, H., Dütting, D., and Zachau, H. G., 1966, *Angew. Chem.*, **78**:600.
Billeter, M. A., Dahlberg, J. E., Goodman, H. M., Hindley, J., and Weissmann, C., 1969, *Nature*, **224**:1083.
Bishop, D. H. L., Mills, D. R., and Spiegelman, S., *Biochemistry*, **7**:3744.
Bitte, L. F., and Lis, A. W., 1967, *7th Intern. Congr. Biochem., Tokyo, Abstr. IV*, p. 655.
Björk, G. R., and Svensson, I., 1967, *Biochim. Biophys. Acta*, **138**:430.
Björk, G. R., Svensson, I., and Johansson, K. E., 1968, in: *Studies on Aminoacylation and Methylation of Transfer Ribonucleic Acids*, Upsala.
Blake, R. D., Fresco, J. R., and Langridge, R., 1969, *Federat. Proc.*, **28**:409.
Blake, R. D., Fresco, J. R., and Langridge, R., 1970, *Nature*, **225**:32.
Bollack, C., Dirheimer, G., and Ebel, J. P., 1969, *Abstr. 6th Meeting FEBS*, Madrid, p. 59.
Bonnet, J., Ebel, J. P., and Dirheimer, G., 1971, *FEBS Letters*, **15**:286.
Borek, E., and Christman, J., 1965, *Federat. Proc.*, **24**:292.
Borek, E., Ryan, A., and Rockenbach, J., 1955, *J. Bacteriol.*, **69**:460.
Brawerman, G., and Chargaff, E., 1959, *Biochim. Biophys. Acta*, **31**:172.
Brimacombe, B. L. C., Griffin, B. E., Haines, J. A., Haslam, W. J., and Reese, C. M., 1965, *Biochemistry*, **4**:2542.
Brookes, S. W., and Lawley, P. D., 1961, *J. Chem. Soc.*, p. 3923.
Brostoff, S. W., and Ingram, V. M., 1967, *Science*, **158**:666.
Brown, D. M., Fried, M., and Todd, A. R., 1955, *J. Chem. Soc.*, p. 808.
Brown, D. M., Burdon, M. G., and Slatcher, R. P., 1965, *Chem. Commun.*, p. 77.
Brown, D. M., Burdon, M. G., and Slatcher, R. P., 1968 *J. Chem. Soc. (C)*, p. 1051.
Brown, G., and Attardi, G., 1965, *Biochem. Biophys. Res. Commun.*, **20**:298.
Brownlee, G. G., 1971, *Nature New Biology*, **229**:148.
Brownlee, G. G., and Sanger, F., 1967, *J. Mol. Biol.*, **23**:337.
Brownlee, G. G., and Sanger, F., 1969, *Europ. J. Biochem.*, **11**:395.
Brownlee, G. G., Sanger, F., and Barrell, B. G., 1967, *Nature*, **215**:735.
Brownlee, G. G., Sanger, F., and Barrell, B. G., 1968, *J. Mol. Biol.*, **34**:379.
Bruton, C. J., and Hartley, B. S., 1968, *Biochem. J.* **108**:281.
Buck, C., and Nass, M. K., 1968, *Federat. Proc.*, **27**:342.
Budovskii, É. I., 1968, Doctoral Dissertation, Moscow.
Budovskii, É. I., and Demushkin, V. P., 1964, *Biokhimiya*, **29**:1063.
Budovskii, É. I., and Klebanova, L. M., 1967, *Vopr. Med. Khim.* **13**:299.
Burdon, R. H., 1970, *Abstr. 8th Intern. Congr. Biochem.*, Switzerland, p. 184.
Burdon, R. H., and Clason, A. E., 1969, *J. Mol. Biol.*, **39**:113.
Burdon, R. H., Martin, B. T., and Lal, B. M., 1967, *J. Mol. Biol.*, **28**:357.
Burrows, W. J., Armstrong, D. J., Skoog, F., Hecht, S. M., Boyle, J. T. A., Leonard, N. J., and Occolowitz, J., 1968, *Science*, **161**:691.

Burrows, W. J., Armstrong, D. J., Kaminek, M., Skoog, F., Bock, R. M., Hecht, S. M., Cammann, L. G., Leonard, N. J., and Occolowitz, J., 1970, *Biochemistry*, 9:1867.
Burton, K., 1969, *FEBS Letters*, 6:77.
Cantoni, G. L., 1951, *J. Biol. Chem.*, 189:745.
Cantoni, G. L., Gelboin, H. V., Luborsky, S. W., and Richards, H. H., 1962, *Biochim. Biophys. Acta*, 61:354.
Cantor, C. R., and Tinoco, I., 1965, *J. Mol. Biol.*, 13:65.
Cantor, C. R., and Tinoco, I., 1967, *Biopolymers*, 9:821.
Cantor, C. R., Jaskunas, S. R., and Tinoco, J., Jr., 1966, *J. Mol. Biol.*, 20:39.
Capra, J. D., and Peterkofsky, A., 1966, *Federat. Proc.*, 25:403.
Capra, J. D., and Peterkofsky, A., 1968, *J. Mol. Biol.*, 33:591.
Carbon, J. A., Hung, L., and Jones, D. S., 1965, *Proc. Nat. Acad. Sci. (Washington)*, 53:979.
Carbon, J., David, H., and Studier, M. H., 1968, *Science*, 161:1146.
Cashmore, A., 1971, *Nature New Biology*, 230:236.
Černa, J., Rychlik, I., and Šorm, F., 1964, *Coll. Czech. Chem. Commun.*, 29:2832.
Cerutti, P., 1968, *Biochem. Biophys. Res. Commun.*, 30:434.
Cerutti, P., and Miller, N., 1967, *J. Mol. Biol.*, 26:55.
Cerutti, P., and Miller, N., 1968, *Biochem. Biophys. Res. Commun.*, 30:434.
Cerutti, P., Ikeda, K., and Witkop, B., 1965, *J. Amer. Chem. Soc.*, 87:2505.
Cerutti, P., Miles, H. T., and Frazier, J., 1966, *Biochem. Biophys. Res. Commun.*, 22:466.
Cerutti, P., Holt, J. W., and Miller, N., 1968, *J. Mol. Biol.*, 34:505.
Chambers, R. W., 1965, *Biochemistry*, 4:219.
Chambers, R. W., 1966, *Progr. Nucleic Acid Res. Mol. Biol.*, 5:349.
Chang, S. H., 1968, *Federat. Proc.*, 27:767.
Chang, S. H., and RajBhandary, U. L., 1968, *J. Biol. Chem.*, 243:592.
Chen, C. M., and Hall, R. H., 1969, *Phytochemistry*, 8:1687.
Cherayil, J. D., and Bock, R. M., 1965, *Biochemistry*, 4:1174.
Chheda, G. B., Hall, R. H., Magrath, D. I., Mozejko, J., Schweizer, M. P., Stasiuk, L., and Taylor, P. R., 1969, *Biochemistry*, 8:3278.
Christman, J., and Borek, E., 1967, *Federat. Proc.*, 26:867.
Chuguev, I. I., Aksel'rod (Axelrod), V. D., and Baev (Bayev), A. A., 1969, *Biochem. Biophys. Res. Commun.*, 34:348.
Chuguev, I. I., Aksel'rod (Axelrod), V. D., and Baev (Bayev), A. A., 1970, *Biochem. Biophys. Res. Commun.*, 41:108.
Clark, B. F. C., and Marcker, K. A., 1965, *Nature*, 207:1038.
Clark, B. F. C., and Marcker, K. A., 1966, *J. Mol. Biol.*, 17:394.
Clark, B. F. C., Doctor, B. P., Holmes, K. C., Klug, A., Marcker, K. A., Morris, S. J., and Paradies, H. H., 1968a, *Nature*, 219:1222.
Clark, B. F. C., Dube, S. K., and Marcker, K. A., 1968b, *Nature*, 219:484.
Cohn, W. E., 1949, *J. Amer. Chem. Soc.*, 71:2275.
Cohn, W. E., 1951, *J. Cell. Comp. Physiol.*, 38 Suppl. 1:21.
Cohn, W. E., 1957, *Federat. Proc.*, 16:166.
Cohn, W. E., 1958, *Federat. Proc.*, 17:203.
Cohn, W. E., 1959, *Biochim. Biophys. Acta*, 32:569.
Cohn, W. E., 1960, *J. Biol. Chem.*, 235:1488.
Cohn, W. E., 1961, *Biochem. Preparations*, 8:116.
Cohn, W. E., and Doherty, D. G., 1956, *J. Amer. Chem. Soc.*, 78:2863.
Cohn, W. E., and Volkin, E., 1951, *Nature*, 167:483.
Cohn, W. E., Kurkov, V., and Chambers, R. W., 1963, *Biochem. Preparations*, 10:135.
Connors, P. G., Labanauskas, M., and Beeman, W. W., 1969, *Science*, 166:1528.
Cory, S., and Marcker, K. A., 1970, *Europ. J. Biochem.*, 12:177.
Cory, S., Marcker, K. A., Dube, S. K., and Clark, B. F. C., 1968, *Nature*, 220:1039.
Cousin, M., 1963, *Bull. Soc. Chim. Biol.*, 45:1363.
Craddock, C. G., and Nakai, G. S., 1962, *J. Clin. Invest.*, 41:306.
Cramer, F., 1967, *Angew. Chem.*, 79:653.
Cramer, F., 1971, *Progr. Nucleic Acid Res. Mol. Biol.*, 11:391.

Cramer, F., Doepner, H., Haar, F., Schlimme, E., and Seidel, H., 1968, *Proc. Nat. Acad. Sci. (Washington)*, **61**:1384.
Crestfield, A. M., Smith, K. C., and Allen, F. W., 1955, *J. Biol. Chem.*, **216**:185.
Crick, F. H. C., 1966, *J. Mol. Biol.*, **19**:548.
Culp, L. A., and Brown, G. M., 1968, *Arch Biochem. Biophys.*, **124**:483.
Dahlberg, J. E., 1968, *Nature*, **220**:548.
Davies, D. R., 1960, *Nature*, **186**:1030.
Davies, D. R., and Rich, A., 1958, *J. Amer. Chem. Soc.*, **80**:1003.
Davis, B. J., 1964, *Ann. New York Acad. Sci.*, **12**:404.
Davis, F. F., and Allen, F. W., 1957, *J. Biol. Chem.*, **227**:907.
Delihas, N., 1967, *Biochemistry*, **6**:3356.
Delihas, N., and Bertman, J., 1966, *J. Mol. Biol.*, **21**:391.
De Wachter, R., and Fiers, W., 1967, *J. Mol. Biol.*, **30**:507.
De Wachter, R., and Fiers, W., 1969, *Nature*, **221**:233.
De Wachter, R., Verhassel, J.-P., and Fiers, W., 1968a, *Biochim. Biophys. Acta*, **157**:195.
De Wachter, R., Verhassel, J.-P., and Fiers, W., 1968b, *FEBS Letters*, **1**:93.
Dirheimer, G., and Ebel, J.-P., 1967, *Bull. Soc. Chim. Biol.*, **49**:1679.
Dirheimer, G., Sabeur, G., and Ebel, J.-P., 1967, *Biochim. Biophys. Acta*, **149**:587.
Dlugajczyk, A., and Allen, F. W., 1961, *Biochim. Biophys. Acta*, **51**:215.
Dlugajczyk, A., and Eiler, J. J., 1966, *Biochim. Biophys. Acta*, **119**:11.
Doctor, B. P., 1967, *Abstr. 7th Intern. Congr. Biochem.*, Tokyo, IV, p. 657.
Doctor, B. P., and Mudd, J. A., 1963, *J. Biol. Chem.*, **238**:3677.
Doctor, B. P., Apgar, J., and Holley, R. W., 1961, *J. Biol. Chem.*, **236**:1117.
Doctor, B. P., Loebel, J. E., and Kellogg, D. A., 1966, *Cold Spring Harbor Symp. Quant. Biol.*, **31**:543.
Doctor, B. P., Loebel, J. E., and Winters, D. B., 1967, *Federat. Proc.*, **26**:871.
Doctor, B. P., Fuller, W., and Webb, N. L., 1969, *Nature*, **221**:58.
Doctor, B. P., Loebel, J. E., Sodd, M. A., and Winters, D. B., 1969, *Science*, **163**:693.
Doi, R. H., and Goehler, B., 1966, *Cold Spring Harbor Symp. Quant. Biol.*, **31**:457.
Dube, S. K., and Marcker, K. A., 1969, *Europ. J. Biochem.*, **8**:256.
Dube, S. K., Marcker, K. A., Clark, B. F. C., and Cory, S., 1968, *Nature*, **218**:232.
Dube, S. K., Marcker, K. A., Clark, B. F. C., and Cory, S., 1969a, *Europ. J. Biochem.*, **8**:244.
Dube, S. K., Rudland, P. S., Marcker, K. A., and Clark, B. F. C., 1969b, *Abstr. 6th Meeting FEBS*, Madrid, p. 60.
Dube, S. K., Marcker, K. A., and Yudelevich, A., 1970, *FEBS Letters*, **9**:168.
Dubin, D. T., and Günalp, A., 1967, *Biochim. Biophys. Acta*, **134**:106.
Dudock, B. S., and Katz, G., 1969, *J. Biol. Chem.*, **244**:3069.
Dudock, B. S., Katz, G., Taylor, E. K., and Holley, R. W., 1968, *Federat. Proc.*, **27**:342.
Dudock, B. S., Katz, G., Taylor, E. K., and Holley, R. W., 1969, *Proc. Nat. Acad. Sci. (Washington)*, **62**:941.
Dulbecco, R., and Smith, J. D., 1960, *Biochim. Biophys. Acta*, **39**:550.
Dunn, D. B., 1959, *Biochim. Biophys. Acta*, **34**:286.
Dunn, D. B., 1960, *Biochim. Biophys. Acta*, **38**:176.
Dunn, D. B., 1961, *Biochim. Biophys. Acta*, **46**:198.
Dunn, D. B., 1963, *Biochem. J.*, **86**:14P.
Dunn, D. B., and Smith, D., 1959, *Proc. 4th Intern. Congr. Biochem.*, Vienna, Vol. 7, p. 72.
Dunn, D. B., Smith, J. D., and Spahr, P. F., 1960, *J. Mol. Biol.*, **2**:113.
Dütting, D., and Zachau, H. G., 1964a, *Biochim. Biophys. Acta*, **91**:573.
Dütting, D., and Zachau, H. G., 1964b, *Z. Physiol. Chem.*, **336**:132.
Dütting, D., Karau, W., Melchers, F., and Zachau, H. G., 1965, *Biochim. Biophys. Acta*, **108**:194.
Dütting, D., Feldmann, H., and Zachau, H. G., 1966, *Z. Physiol. Chem.*, **347**:249.
Ebel, J. P., 1968, *Bull. Soc. Chim. Biol.*, **50**:2255.
Egami, F., Takahashi, K., and Uchida, T., 1964, *Progr. Nucleic Acid Res. Mol. Biol.*, **3**:59.
Ehrenstein, G. von, and Dais, O., 1963, *Proc. Nat. Acad. Sci. (Washington)*, **50**:81.

Ehresmann, C., and Ebel, J. P., 1970, *Europ. J. Biochem.*, **13**:577.

Ehresmann, C., Fellner, P., and Ebel, J. P., 1971, *FEBS Letters*, **13**:32.

Eisinger, Y., Fener, B., and Yamane, T., 1970, *Proc. Nat. Acad. Sci. (Washington)*, **65**:638.

Elion, G. B., 1962, *J. Org. Chem.*, **27**:2478.

Epler, J. L., 1969, *Biochemistry*, **8**:2285.

Erdmann, V. A., Haar, F., Schlimme, E., and Cramer, F., 1969, *Abstr. 6th Meeting FEBS*, Madrid, p. 61.

Everett, G. A., Merrill, S. H., and Holley, R. W., 1960, *J. Amer. Chem. Soc.*, **82**:5757.

Faerber, P., and Scheit, K. H., 1970, *FEBS Letters*, **11**:11.

Favorova, O. O., Frolova, L. Yu., Kharatyan, S. G., and Kiselev, L. L., 1968, *Molekul. Biol.*, **2**:455.

Feldmann, H., 1967a, *Abstr. 4th Meeting FEBS*, Oslo, p. 173.

Feldmann, H., 1967b, *Europ. J. Biochem.*, **2**:102.

Feldmann, H., Dütting, D., and Zachau, H. G., 1966, *Z. Physiol. Chem.*, **347**:236.

Fellner, P., Ebel, J.-P., and Blasi, O., 1970a, *FEBS Letters*, **6**:102.

Fellner, P., Ehresmann, C., and Ebel, J.-P., 1970b, *Nature*, **225**:26.

Fellner, P., Ehresmann, C., and Ebel, J.-P., 1970c, *Europ. J. Biochem.*, **13**:583.

Felsenfeld, G., and Cantoni, G. L., 1964, *Proc. Nat. Acad. Sci. (Washington)*, **51**:818.

Felsenfeld, G., and Miles, H. T., 1967, *Ann. Rev. Biochem.*, **36**:407.

Fiers, W., 1967, *Virology*, **33**:413.

Fiers, W., De Wachter, R., Lepoutre, L., and Vandendriessche, J., 1965, *J. Mol. Biol.*, **13**:451.

Fink, L. M., Cline, R. E., McGaughey, C., and Fink, K., 1956, *Anal. Chem.*, **28**:4.

Fink, L. M., Goto, T., Frankel, F., and Weinstein, I. B., 1968, *Biochem. Biophys. Res. Commun.*, **32**:963.

Fink, L. M., Goto, T., and Weintsein, I. B., 1969, *Federat. Proc.*, **28**:409.

Fiskin, A. M., and Beer, M., 1965, *Biochim. Biophys. Acta*, **108**:159.

Fissekis, J. D., and Sweet, F., 1970, *Biochemistry*, **9**:3136.

Fittler, F., and Hall, R. H., 1966, *Biochem. Biophys. Res. Commun.*, **25**:441.

Fittler, F., Kline, L. K., and Hall, R. H., 1968a, *Biochemistry*, **7**:940.

Fittler, F., Kline, L. K., and Hall, R. H., 1968b, *Biochem. Biophys. Res. Commun.*, **31**:571.

Fleissner, E., 1967, *Biochemistry*, **6**:621.

Fleissner, E., and Borek, E., 1962, *Proc Nat. Acad. Sci. (Washington)*, **48**:1199.

Folk, W. R., and Yaniv, M., 1972, *Nature New Biol.*, **237**:165.

Forget, B. G., and Weissman, S. M., 1967, *Science*, **158**:1695.

Forget, B. G., and Weissman, S. M., 1969, *J. Biol. Chem.*, **244**:3148.

Fresco, J. R., Alberts, B. M., and Doty, P., 1960, *Nature*, **188**:98.

Fresco, J. R., Klotz, L. C., and Richards, E. G., 1963, *Cold Spring Harbor Symp. Quant. Biol.*, **28**:83.

Fresco, J. R., Adams, A., Ascione, R., Henley, D., and Lindahl, T., 1966, *Cold Spring Harbor Symp. Quant. Biol.*, **31**:527.

Fresco, J. R., Lindahl, T., and Henley, D., 1968a, *Federat. Proc.*, **27**:796.

Fresco, J. R., Blake, R. D., and Langridge, R., 1968b, *Nature*, **220**:1285.

Frifz, H. G., and Röttger, B., 1963, *Z. Naturforsch.*, **18b**:124.

Fröholm, L. P., and Olsen, B. R., 1969, *Abstr. 6th Meeting FEBS*, Madrid, p. 61.

Frolova, L. Yu., Sandakhchiev, L. S., Knorre, D. G., and Kiselev, L. L., 1964, *Dokl. Adad. Nauk SSSR*, **158**:235.

Fuller, W., and Hodgson, A., 1967, *Nature*, **215**:817.

Fuller, W., Arnott, S., and Creek, J., 1969, *Biochem. J.*, **114**:26p.

Furuichi, Y., Wataya, Y., Hayatsu, H., and Ukita, T., 1970, *Biochem. Biophys. Res. Commun.*, **41**:1185.

Gaberman, V., 1963, *Uspekhi Sovrem. Biol.*, **55**:9.

Galizzi, A., 1967, *J. Mol. Biol.*, **27**:619.

Gal-Or, L., Mellema, Z. E., Moudrianakis, E. N., and Beer, M., 1967, *Biochemistry*, **6**:1909.

Gangloff, J., Keith, G., and Dirheimer, G., 1970, *Bull. Soc. Chim. Biol.*, **52**:125.

Gangloff, J., Keith, G., Ebel. J.-P., and Dirheimer, G., 1971, *Nature New Biology*, **230**:125.
Gartland, W. J., and Sueoka, N., 1966, *Proc. Nat. Acad. Sci. (Washington)*, **55**:948.
Gassen, H. G., 1969, *J. Chromatog.*, **39**: 147.
Gassen, H. G., and Uziel, M., 1969, *Abstr. 6th Meeting FEBS*, Madrid, p. 50.
Gefter, M. L., 1969, *Biochem. Biopyhs. Res. Commun.*, **36**:435.
Gefter, M. L., and Russell, R. L., 1969, *J. Mol. Biol.*, **39**:145.
Gefter, M. L., Smith, J. D., Abelson, J., and Russell, R., 1969, *Abstr. 6th Meeting FEBS*, Madrid, p. 60.
Ghosh, H. P., Söll, D., and Khorana, H. G., 1967 *J. Mol. Biol.*, **25**:275.
Ghosh, K., and Ghosh, H. P., 1970, *Biochem. Biophys. Res. Commun.*, **40**:135.
Gilham, P. T., 1962, *J. Amer. Chem. Soc.*, **84**:687.
Gillam, I., Millward, S., Blew, D., von Tigerstrom, M., Wimmer, E., and Tener, G. M., 1967, *Biochemistry*, **6**: 3043.
Gillam, I., Blew, D., Warrington, R. C., von Tigerstrom, M., and Tener, G. M., 1968, *Biochemistry*, **7**:3459.
Ginsberg, T., and Davis, F. F., 1968, *J. Biol. Chem.*, **243**:6300.
Girshovich, A. S., and Shubina, T. N., 1969, *Molekul. Biol.*, **3**:235.
Girshovich, A. S., Knorre, D. G., Nelidova, O. D., and Ovander, M. N. 1966, *Biochim. Biophys. Acta*, **119**:216.
Girshovich, A. S., Grachev, M. A., and Obukhova, L. V., 1968, *Molekul. Biol.*, **2**:351.
Glasky, A. J., and Simon, L. N., 1966, *Science*, **151**:207.
Glebov, R. N., Zaitseva, G. N., and Belozerskii, A. N., 1965, *Biokhimiya*, **30**:586.
Glitz, D. G., and Dekker, C. A., 1964, *Biochemistry*, **3**:1391.
Glitz, D. G., and Sigman, D. S., 1970, *Biochemistry*, **9**:3408.
Glitz, D. G., Bradley, A., and Fraenkel-Conrat, H., 1968, *Biochim. Biophys. Acta*, **161**:1.
Goehler, B., Kaneko, I., and Doi, R. H., 1966, *Biochem. Biophys. Res. Commun.*, **24**:466.
Gold, M., Hurwitz, J., and Anders, M., 1963a, *Biochem. Biophys. Res. Commun.*, **11**:107.
Gold, M., Hurwitz, J., and Anders, M., 1963b, *Proc. Nat. Acad. Sci. (Washington)*, **50**:164.
Goldstein, J., 1967, *J. Mol. Biol.*, **25**:123.
Goldstein, J., Bennett, T. P., and Craig, L. C., 1964, *Proc. Nat. Acad. Sci. (Washington)*, **51**:119.
Goldthwait, D. A., and Kerr, D. S., 1962, *Biochim. Biophys. Acta.* **61**:930.
Goldwasser, E., and Heinrikson, R. L., 1966, *Progr. Nucleic Acid Res. Mol. Biol.*, **5**:399.
Goodman, H. M., Abelson, J., Landy, A., Brenner, S., and Smith J. D., 1968, *Nature*, **217**: 1019.
Goodman, H. M., Abelson, J. N., Landy A., Zadrazil, S., and Smith, J. D., 1970, *Europ. J. Biochem.*, **13**:461.
Gould, H., 1966, *Biochemistry*, **5**:1103.
Gould, H. J., 1967, *J. Mol. Biol.*, **29**:307.
Gould, H. J., Pinder, J. C., and Matthews, H. R., 1969, *Anal. Biochem.*, **29**:1.
Grachev, M. A., Budovskii (Budowsky), E. I., Mirzabekov, A. D., Krutilina, A. I., and Sandakhchiev (Sandakchiev), L. S., 1965, *Biochim. Biophys. Acta*, **108**:506.
Grachev, M. A., Menzorov, N. I., Sandakhchiev, L. S., Budovskii, É. I., and Knorre, D. G., 1966, *Biokhimiya*, **31**:840.
Gray, M. W., and Lane, B. G., 1967, *Biochim. Biophys. Acta*, **134**:243.
Gray, M. W., and Lane, B. G., 1968, *Biochemistry*, **7**:3441.
Grimm, W. A. H., and Leonard, N. J., 1967, *Biochemistry*, **6**:3625.
Grineva, N. I., Zarytova, V. F., and Knorre, D. G., 1968, *Iz. Sibirsk. Otdel. Akad. Nauk SSSR, Ser. Khim.*, No. 5, 118.
Grunberg-Manago, M., 1962, *Ann. Rev. Biochem.*, **31**:301.
Grunberg-Manago, M., 1963, *Progr. Nucleic Acid Res. Mol. Biol.*, **1**:93.
Günberger, D., Holy, A., and Šorm, F., 1967a, *Biochim. Biophys. Acta*, **134**:484.
Grünberger, D., Holy, A., and Šorm, F., 1967b, *Biochim. Biophys. Acta*, **149**:246.
Gupta, N. K., Ohtsuka, E., Sgaramella, V., Buchi, H., Kumar, A., Weber. H., and Khorana, H. G. 1968, *Proc. Nat. Acad. Sci. (Washington)*, **60**:1338.

Hadjiolov, A. A., Venkov, P. V., Dolapchiev, L. B., and Genchev, D. D., 1967a, *Biochim. Biophys. Acta*, **142**:111.

Hadjiolov, A. A., Venkov, P. V., and Dolapchiev, L. B., 1967b, *Biochem. Biophys. Res. Commun.*, **28**:166.

Haines, J. A., Reese, C. B., and Todd, L., 1962, *J. Chem. Soc.*, p. 5281.

Hall, R. H., 1963a, *Biochem. Biophys. Res. Commun.*, **12**:361.

Hall, R. H., 1963b, *Biochem. Biophys. Res. Commun.*, **12**:429.

Hall, R. H., 1963c, *Biochim. Biophys. Acta*, **68**:278.

Hall, R. H., 1963d, *Biochem. Biophys. Res. Commun.*, **13**:394.

Hall, R. H., 1964a, *Biochemistry*, **3**:769.

Hall, R. H., 1964b, *Biochemistry*, **3**:876.

Hall, R. H., 1965, *Biochemistry*, **4**:661.

Hall, R. H., 1967, *Abstr. 7th Internat. Congr. Biochem.*, Tokyo, Vol. 1, p. 43.

Hall, R. H., 1970, *Progr. Nucleic Acid Res. Mol. Biol.*, **10**:57.

Hall, R. H., 1971, *The Modified Nucleosides in Nucleic Acids*, Columbia Univ. Press, New York.

Hall, R. H., and Allen, F. W., 1960, *Biochim. Biophys. Acta*, **45**:163.

Hall, R. H., and Chheda, G. B., 1965, *J. Biol. Chem.*, **240**:PC2754.

Hall, R. H., Robins, M. J., Stasiuk, L., and Thedford, R., 1966, *J. Amer. Chem. Soc.*, **88**:2614.

Hall, R. H., Csonka, L., David, H., and McLennan, B., 1967a, *Science*, **156**:69.

Hall, R. H., Hacker, B., and Kline, L. K., 1967b, *Federat. Proc.*, **26**:733.

Hampel, A., Labanauskas, M., Connors, P. G., Kirkegard, L., RajBhandary, U. L., Sigler, P. B., and Bock, R. M., 1968, *Science*, **162**:1384.

Hancock, R. L., 1967a, *Cancer Res.*, **27**:646.

Hancock, R. L., 1967b, *Canad. J. Biochem.*, **45**:1513.

Hancock, R. L., 1968, *Biochem. Biophys. Res. Commun.*, **31**:77.

Hara, H., Horiuchi, T., Saneyoshi, M., and Nishimura, S., 1970, *Biochem. Biophys. Res. Commun.*, **38**:305.

Harada, F., Gross, H. J., Kimura, F., Chang, S. H., Nishimura, S., and RajBhandary, U. L., 1968, *Biochem. Biophys. Res. Commun.*, **33**:299.

Harada, F., Kimura, F., and Nishimura, S., 1969a, *Biochim. Biophys. Acta*, **182**:590.

Harada, F., Kimura, F., and Nishimura, S., 1969b, *Biochim. Biophys. Acta*, **195**:590.

Hashimoto, S., Miyazaki, M., and Takemura, S., 1969, *J. Biochem. (Tokyo)*, **65**:659.

Hayashi, H., and Miura, K., 1966, *Cold Spring Harbor Symp. Quant. Biol.*, **31**:63.

Hayashi, H., and Ukita, T., 1964, *Biochem. Biophys. Res. Commun.*, **14**:198.

Hayward, R. S., and Weiss, S. B., 1966, *Proc. Nat. Acad. Sci. (Washington)*, **55**:1161.

Hayward, R. S., Eliceiri, G. L., and Weiss, S. B., 1966, *Cold Spring Harbor Symp. Quant. Biol.*, **31**:459.

Hecht, S. M., Kirkegaard, L. H., and Bock, R. M., 1971 *Proc. Nat. Acad. Sci. (Washington)*, **68**:48.

Hecht, S. M., Leonard, N. J., Occolowitz, J., Burrows, W. J., Armstrong, D. J., Skoog, F., Bock, R. M., Gillam, I., and Tener, G. M., 1969, *Biochem. Biophys. Res. Commun.*, **35**:205.

Heinrikson, R. L., and Hartley, B. S., 1968, *Biochem. J.*, **105**:17.

Hélène, C., and Yaniv, M., 1970, *Europ. J. Biochem.*, **15**:500.

Hemmes, W. F., 1964, *Biochim. Biophys. Acta*, **91**:332.

Henes, C., Krauskopf, M., and Ofengand, J., 1969, *Biochemistry*, **8**:3024.

Henley, D. D., Lindahl, T., and Fresco, J. R., 1966, *Proc. Nat. Acad. Sci. (Washington)*, **55**:191.

Herbert, E., and Wilson, C. W., 1962a, *Biochim. Biophys. Acta*, **61**:750.

Herbert, E., and Wilson, C. W., 1962b, *Biochim. Biophys. Acta*, **61**:762.

Herbert, E., Smith, C. J., and Wilson, C. W., 1964, *J. Mol. Biol.*, **9**:376.

Highton, P. J., Murr, B. L., Shafa, F., and Beer, M., 1968, *Biochemistry*, **7**:825.

Hindley, J., 1967, *J. Mol. Biol.*, **30**:125.

Hindley, J., and Staples, D. H., 1969, *Nature*, **224**:964.

Hirsh, D., 1970, *Nature*, **228**:57.

Hoagland, M. B., Zamecnik, P. C., and Stephenson, M. L., 1957, *Biochim. Biophys. Acta*, **24**:215.

Holley, R. W., 1957, *J. Amer. Chem. Soc.*, **79**:658.

Holley, R. W., 1963, *Biochem. Biophys. Res. Commun.*, **10**:186.

Holley, R. W., 1966, *Sci. Amer.*, **214(2)**:30.

Holley, R. W., 1968, *Progr. Nucleic Acid Res. Mol. Biol.*, **8**:37.

Holley, R. W., and Doctor, B. P., 1960, *Federat. Proc.*, **19**:348.

Holley, R. W., and Merrill, S. H., 1959, *J. Amer. Chem. Soc.*, **81**:753.

Holley, R. W., Doctor, B. P., Merrill, S. H., and Saad, F. M., 1959, *Biochim. Biophys. Acta*. **35**:272.

Holley, R. W., Apgar, J., Doctor, B. P., Farrow, J., Marini, M. A., and Merrill, S. H., 1961a, *J. Biol. Chem.*, **236**:200.

Holley, R. W., Apgar, J., and Merrill, S. H., 1961b, *J. Biol. Chem.*, **236**:PC42.

Holley, R. W., Apgar, J., Merrill, S. H., and Zubkoff, P. L., 1961c, *J. Amer. Chem. Soc.*, **83**:4861.

Holley, R. W., Apgar, J., Everett, G. A., Madison, J. T., Merrill, S. H., and Zamir, A., 1963, *Cold Spring Harbor Symp. Quant. Biol.*, **28**:117.

Holley, R. W., Everett, G. A., Madison, J. T., Marquisee, M., and Zamir, A., 1964a, *Abstr. 6th Intern. Congr. Biochem.*, New York, p. 9.

Holley, R. W., Madison, J. T., and Zamir, A., 1964b, *Biochem. Biophys. Res. Commun.*, **17**:389.

Holley, R. W., Apgar, J., Everett, G. A., Madison, J. T., Marquisee, M., Merrill, S. H., Penswick, J. R., and Zamir, A., 1965a, *Science*, **147**:1462.

Holley, R. W., Everett, G. A., Madison, J. T., and Zamir, A., 1965b, *J. Biol. Chem.*, **240**: 2122.

Hoskinson, R. M., and Khorana, H. G., 1965, *J. Biol. Chem.*, **240**:2129.

Huang, R. C., and Bonner, J., 1965, *Proc. Nat. Acad. Sci. (Washington)*, **54**:960.

Hudson, L., Gray, M., and Lane, B. G., 1965, *Biochemistry*, **4**:2009.

Hunt, J. A., 1965, *Biochem. J.*, **95**:541.

Hunt, J. A., 1970, *Biochem. J.*, **120**:353.

Hurwitz, J., Gold, M., and Anders, M., 1964a, *J. Biol. Chem.*, **239**:3462.

Hurwitz, J., Gold, M., and Anders, M., 1964b, *J. Biol. Chem.*, **239**:3474.

Igo-Kemenes, T., and Zachau, H. G., 1969, *Europ. J. Biochem.*, **10**:549.

Igo-Kemenes, T., and Zachau, H. G., 1971, *Europ. J. Biochem.*, **18**:292.

Imura, N., Weiss, G. B., and Chambers, R. W., 1969, *Nature*, **222**:1147.

Ingram, V. M., and Pierce, J. G., (1962), *Biochemistry*, **1**:580.

Ingram, V. M., and Sjöquist, J. A., 1963, *Cold Spring Harbor Symp. Quant. Biol.*, **28**:133.

Inoue, Y., Aoyagi, S., and Nakanishi, K., 1967, *J. Amer. Chem. Soc.*, **89**:5701.

Irie, M., 1965 *J. Biochem. (Tokyo)*, **58**:599.

Irie, S., 1970, *J. Biochem. (Tokyo)*, **68**:129.

Ishida, T., and Miura, K., 1963, *J. Biochem. (Tokyo)*, **54**:378.

Ishida, T., and Miura, K., 1965, *J. Mol. Biol.*, **11**:341.

Ishikura, H., and Nishimura, S., 1967, *Abstr. 7th Internat. Congr. Biochem.*, Tokyo, p. 668.

Ishikura, H., and Nishimura, S., 1968, *Biochim. Biophys. Acta*, **155**:72.

Ishikura, H., Yamada, Y., Munro, K., Saneyoshi, M., and Nishimura, S., 1969, *Biochem. Biophys. Res. Commun.*, **37**:990.

Ishikura, H., Yamada, Y., and Nishimura, S., 1971, *Biochim. Biophys. Acta*, **228**:471.

Iwanami, G., and Brown, G. M., 1968, *Arch. Biochem. Biophys.*, **124**:472.

Jacobs, M., and Hedgcoth, C., 1960, *Anal. Biochem.*, **34**:459.

Jacobson, K. B., and Nishimura, S., 1963, *Biochim. Biophys. Acta*, **68**:490.

Jacobson, K. B., and Nishimura, S., 1964, *J. Chromatog.* **14**:46.

Janion, C., and Shugar, D., 1960, *Acta Biochem. Polon.*, **7**:309.

Jones, A. S., and Woodhouse, D. L., 1959, *Nature*, **183**:1603.

Jones, J. W., and Robins, R. K., 1963, *J. Amer. Chem. Soc.*, **85**:193.

Jukes, T. H., 1966, *Biochem. Biophys. Res. Commun.*, **24**:744.

Kahan, F. M., and Hurwitz, J., 1962, *J. Biol. Chem.*, **237**:3778.

Kamen, H. O., 1969, *Federat. Proc.*, **28**:865.

Kamen, H. O., and Spengler, S. J., 1970, *Biochim. Biophys. Acta*, **213**:332.

Karau, W., and Zachau, H. G., 1964, *Biochim. Biophys. Acta*, **91**:549.

Katchalsky, E., Yanofsky, S., Novogorodsky, A., Galenter, Y., and Littauer, U. Z., 1966, *Biochim. Biophys. Acta*, **123**:641.

Katz, G., and Dudock, B. S., 1969, *J. Biol. Chem.*, **244**:3062.

Kawade, Y., Okamoto, T., and Yamamoto, Y., 1963, *Biochem. Biophys. Res. Commun.*, **10**:200.

Keith, G., Gangloff, J., Ebel, J. -P., and Dirheimer, G., 1970, *Compt. Rend. Acad. Sci. (Paris) Ser. D*, **271**:613.

Keller, E. B., 1964, *Biochem. Biophys. Res. Commun.*, **17**:412.

Kellog, D. A., Doctor, B. P., Loebel, J. E., and Nirenberg, M. W., 1966, *Proc. Nat. Acad. Sci. (Washington)*, **55**:912.

Kelmers, A. D., 1966a, *J. Biol. Chem.*, **241**:3540.

Kelmers, A. D., 1966b, *Biochem. Biophys. Res. Commun.*, **25**:562.

Kelmers, A. D., Novelli, G. D., and Stulberg, M. P., 1965, *J. Biol. Chem.*, **240**:3979.

Kerr, S. J., 1970, *Biochemistry*, **9**:690.

Kerr, S. J., 1971, *Proc. Nat. Acad. Sci. (Washington)*, **68**:406.

Khorana, H. G., 1971, *Zh. Vsesoyuz. Khim. Obshch. im. D. I. Mendeleeva*, **16**:145.

Khym, J. X., and Cohn, W. E., 1961, *J. Biol. Chem.*, **236**:PC9.

Khym, J. X., and Uziel, M., 1968, *Biochemistry*, **7**:422.

Kikugawa, K., Hayatsu, H., and Ukita, T., 1967a, *Biochim. Biophys. Acta*, **134**:224.

Kikugawa, K., Muto, H., Hayatsu, K. I., and Ukita, T., 1967b, *Biochim. Biophys. Acta*, **134**:232.

Kim, S. -H., and Rich, A., 1968, *Science*, **162**:1381.

Kimura-Harada, F., Saneyoshi, M., and Nishimura, S., 1971, *FEBS Letters*, **13**:335.

Kirby, K. S., 1960, *Biochim. Biophys. Acta*, **41**:338.

Kiselev, L. A., 1971, in: *Molecular Bases of Protein Biosynthesis*, Nauka, Moscow (in Russian).

Kline, L., Fittler, F., and Hall, R. H., 1969, *Biochemistry*, **8**:4361.

Knorre, D. G., 1966, Doctoral Dissertation, Novosibirsk.

Knorre, D. G., Myzina, S. D., and Sandakhchiev, L. S., 1964, *Izvest. Sibirsk. Otdel. Akad. Nauk SSSR, Ser. Khim.*, **2**:135.

Knorre, D. G., Malygin, E. G., Mushinskaya, G. S., and Favorov, V. V., 1966, *Biokhimiya*, **31**:334.

Knorre, D. G., Korzhov, V. A., and Melamed, N. V., 1969, *Molekul. Biol.*, **3**:579.

Kochetkov, N. K., and Budovskii, E. I., 1964, in: *Molecular Biology, Problems and Perspectives*, Nauka, Moscow, p. 139 (in Russian).

Kochetkov, N. K., and Budovskii (Budowsky), E. I., 1969, *Progr. Nucleic Acid Res. Mol. Biol.*, **9**:403.

Kochetkov, N. K., Budovskii, E. I., and Simukova, N. A., 1962, *Biokhimiya*, **27**:519.

Kochetkov (Kotchetkov), N. K., Budovskii (Budowsky), E. I., and Shibaeva, R. P., 1963, *Biochim. Biophys. Acta*, **68**:493.

Kochetkov (Kotchetkov), N. K., Budovskii (Budowsky), E. I., Broude, N. E., and Klebanova, L. M.,1967, *Biochim. Biophys. Acta*, **134**:482.

Kowalski, S., and Fresco, J. R., 1971, *Science*, **172**:384.

Kowalski, S., Yamane, T., and Fresco, J. R., 1971, *Science*, **172**:385.

Krauskopf, M., and Ofengand, J., 1971, *FEBS Letters*, **15**:111.

Krutilina, A. I., Venkstern, T. V., and Baev, A. A., 1964, *Biokhimiya*, **29**:333.

Krutilina, A. I., Mirzabekov, A. D., Venkstern, T.V., and Baev, A. A., 1965, *Biokhimiya*, **30**:1225.

Krutilina, A. I., Mirzabekov, A. D., Venkstern, T. V., and Baev, A. A., 1970, *Molekul. Biol.*, **4**:97.
Kryukov, V. M., Isaenko, S. N., Aksel'rod, V. D., and Baev, A. A., 1971, *Molekul. Biol.*, (in press).
Kuchino, Y., and Nishimura, S., 1970, *Biochem. Biophys. Res. Commun.*, **40**:306.
Kuo, T. H., and Keller, E. B., 1968, *Federat. Proc.*, **27**:341.
Kwong, T. C., and Lane, B. G., 1970, *Biochim. Biophys. Acta*, **224**:405.
Lagerqvist, U., and Berg, P., 1962, *J. Mol. Biol.*, **5**:139.
Lake, J. A., and Beeman, W. W., 1967, *Science*, **156**:1371.
Landy, A., Abelson, J., Goodman, H. M., and Smith J. D., 1967, *J. Mol. Biol.*, **29**:457.
Lawley, P. D., and Brookes, P., 1963, *Biochem. J.*, **89**:127.
Lazzarini, R. A., and Peterkofsky, A., 1965, *Proc. Nat. Acad. Sci. (Washington)*, **53**:549.
Lebowitz, P., Ipata, P., Makman, M., Richards, H., and Cantoni, G. L., 1966, *Biochemistry*, **5**:3617.
Lebowitz, P., Dohan, F. C., Jr., Richards, H. H., Neelon, F. A., and Cantoni, G. L., 1967, *J. Biol. Chem.*, **242**:4523.
Leder, P., and Nirenberg, M., 1964, *Proc. Nat. Acad. Sci. (Washington)*, **52**:420.
Lee, J. C., and Gilham, P. T., 1965, *J. Amer. Chem. Soc.*, **87**:4000.
Lee, J. C., Ho, N. W. Y., and Gilham, P. T., 1965, *Biochim. Biophys. Acta.* **95**:503.
Lee, S., McMullen, D., and Brown, G. L., 1965, *Biochim. J.*, **94**:314.
Leonard, N. J., Achmatowicz, S., Loepky, R. N., Carraway, K. L., Grimm, W. A. H., Szwekkowska, A., Hamzi, H. Q., and Skoog, F., 1966, *Proc. Nat. Acad. Sci. (Washington)*, **56**:709.
Levitt, M., 1969, *Nature*, **224**:1759.
Li, L., Venkstern, T. V., Mirzabekov, A. D., Krutilina, A. I., and Baev, A. A., 1966, *Biokhimiya*, **31**:1158.
Lindahl, T., Adams, A., and Fresco, J. R., 1966 *Proc. Nat. Acad. Sci. (Washington)*, **55**:941.
Lindahl, T., Adams, A., and Fresco, J. R., 1967a, *J. Biol. Chem.*, **242**:3129.
Lindahl, T., Adams, A., Geroch, M., and Fresco, J. R., 1967b, *Proc. Nat. Acad. Sci. (Washington)*, **57**:178.
Lipsett, M. N., 1965, *J. Biol. Chem.*, **240**:3975.
Lipsett, M. N., 1966, *Cold Spring Harbor Symp. Quant. Biol.*, **31**:449.
Lipsett, M. N., and Doctor, B. P., 1967, *J. Biol. Chem.*, **242**:4072.
Lipsett, M. N., and Peterkofsky, A., 1966, *Proc. Nat. Acad. Sci. (Washington)*, **55**:1169.
Lipsett, M. N., Norton, J. S., and Peterkofsky, A., 1967, *Biochemistry*, **6**:855.
Lis, A. W., and Lis, E. W., 1962, *Biochim. Biophys. Acta*, **61**:799.
Lis, A. W., and Lis, E. W., 1964, *Federat. Proc.*, **23**:532.
Lis, A. W., and Passarge, W. E., 1966, *Arch. Biochem. Biophys.*, **114**:593.
Litt, M., 1969, *Biochemistry*, **8**:3249.
Litt, M., and Ingram, V. M., 1964, *Biochemistry*, **3**:560.
Littauer, U. Z., 1964, *Abstr. 6th Internat. Congr. Biochem.*, New York, p. 11.
Littauer, U. Z., Muench, K., Bern, P., Gilbert, W., and Spahr, P. F., 1963, *Cold Spring Harbor Symp. Quant. Biol.*, **28**:157.
Littauer, U. Z., Revel, M., and Stern, R., 1966, *Cold Spring Harbor Symp. Quant. Biol.*, **31**:501.
Littlefield, J. W., and Dunn, D. B., 1958a, *Nature*, **181**:254.
Littlefield, J. W., and Dunn, D. B., 1958b, *Biochem. J.*, **70**:642.
Loehr, J. S., and Keller, E. B., 1968, *Proc. Nat. Acad. Sci. (Washington)*, **61**:1115.
Ludlum, D. B., 1966, *Biochim. Biophys. Acta*, **119**:630.
Ludlum, D. B., Warner, R. C., and Wahba, A. J., 1964, *Science*, **145**:397.
McCully, K. S., and Cantoni, G. L., 1961, *Biochim. Biophys. Acta*, **51**:190.
McCully, K. S., and Cantoni, G. L., 1962a, *J. Biol. Chem.*, **237**:3760.
McCully, K. S., and Cantoni, G. L., 1962b, *J. Mol. Biol.*, **5**:497.
McLaren, A. D., and Shugar, D., 1964, in: *Photochemistry of Proteins and Nucleic Acids*, Pergamon Press, Oxford.

McLaughlin, C. S., and Ingram, V. M., 1965, *Biochim. Biophys. Acta,* **103**:344.
McLennan, B. D., and Lane, B. G., 1968, *Can. J. Biochem.,* **46**:81.
Macon, J. B., and Wolfenden, R., 1968, *Biochemistry,* **7**:3453.
Madison, J. T., 1968, *Ann. Rev. Biochem.,* **37**:131.
Madison, J. T., and Holley, R. W., 1965, *Biochem. Biophys. Res. Commun.,* **18**:153.
Madison, J. T., and Kung, H.-K., 1967, *J. Biol. Chem.,* **242**:1324.
Madison, J. T., Everett, G. A., and Kung, H. K., 1966a, *Science,* **153**:531.
Madison, J. T., Everett, G. A., and Kung, H. K., 1966b, *Cold Spring Harbor Symp. Quant. Biol.,* **31**:409.
Madison, J. T., Everett, G. A., and Kung, H.-K., 1967a, *J. Biol. Chem.,* **242**:1318.
Madison, J. T., Holley, R. W., Poucher, J. S., and Connett, P. H., 1967b, *Biochim. Biophys. Acta,* **145**:825.
Magasanik, B., and Chargaff, E., 1951, *Biochim. Biophys. Acta,* **7**:396.
Magee, P. N., and Farber, E., 1962, *J. Biochem.,* **83**:114.
Magrath, D. I., and Shaw, D. C., 1967, *Biochem. Biophys. Res. Commun.,* **26**:32.
Makman, M. H., and Cantoni, G. L., 1966, *Biochemistry,* **5**:2244.
Mandel, L. R., and Borek, E., 1961, *Biochem. Biophys. Res. Commun.,* **3**:138.
Mandels, S., 1967a, *Federat. Proc.,* **26**:865.
Mandels, S., 1967b, *J. Biol. Chem.,* **242**:3103.
Mandels, S., 1968, *J. Biol. Chem.,* **243**:3671.
Marcker, K. A., Dube, S. K., and Clark, B. F. C., 1968, *FEBS Symposium on Structure and Function of Transfer RNA and 5S RNA,* Oslo, 1967, Universitetsforlaget, Oslo, and Academic Press, New York, p. 54.
Mehler, A. H., and Bank, A., 1963, *J. Biol. Chem.,* **238**:PC2888.
Melcher, G., 1969, *FEBS Letters,* **3**:185.
Melchers, F., and Zachau, H. G., 1964, *Biochim. Biophys. Acta,* **91**:559.
Melchers, F., and Zachau, H. G., 1965, *Biochim. Biophys. Acta,* **95**:380.
Melchers, F., Dütting, D., and Zachau, H. G., 1965, *Biochim. Biophys. Acta,* **108**:182.
Merril, C. R., 1968, *Biopolymers,* **6**:1727.
Michelson, A. M., 1959, *J. Chem. Soc.,* p. 1371.
Michelson, A. M., 1966, *The Chemistry of Nucleosides and Nucleotides* (Russian translation), Mir, Moscow, p. 71. (Original version, Academic Press, New York and London, 1963).
Midgley, J. E. M., 1962, *Biochim. Biophys. Acta,* **61**:513.
Midgley, J. E. M., and McIlreavy, D. J., 1966, *Biochem. J.,* **101**:32P.
Midgley, J. E. M., and McIlreavy, D. J., 1967, *Biochem. J.,* **103**:50P.
Miles, H. T., 1956, *Biochim. Biophys. Acta,* **22**:247.
Millar, D. B., and Byrne, R. W., 1967, *Arch. Biochem. Biophys.,* **119**:398.
Miller, N., and Cerutti, P., 1967, *J. Amer. Chem. Soc.,* **89**:2767.
Min Jou, W., and Fiers, W., 1969, *J. Mol. Biol.,* **40**:187.
Min Jou, W., Fiers, W., Goodman, H., and Sphar, P., 1969, *J. Mol. Biol.,* **42**:143.
Min Jou, W., Contreras, R., and Fiers, W., 1970, *FEBS Letters,* **9**:222.
Min Jou, W., Haegeman, G., and Fiers, W., 1971, *FEBS Letters,* **13**:105.
Mirzabekov, A. D., Aksel'rod, V. D., Venkstern, T. V., Li, L., Krutilina, A. I., and Baev, A. A., 1970, *Molekul. Biol.,* **4**:76.
Mirzabekov, A. D., and Baev, A. A., 1965, *Izvest. Akad. Nauk SSSR, Ser. Biol.,* No. 2, 221.
Mirzabekov, A. D., Krutilina, A. I., Gorshkova, V. I., and Baev, A. A., 1964, *Biokhimiya,* **29**:1158.
Mirzabekov, A. D., Venkstern, T. V., and Baev, A. A., 1965a, *Biokhimiya,* **30**:825.
Mirzabekov, A. D., Krutilina, A. I., Reshetov, P. D., Sandakhchiev, L. S., Knorre, D. G., Khokhlov, A. S., and Baev, A. A., 1965b, *Dokl. Akad. Nauk SSSR,* **160**:1200.
Mirzabekov, A. D., Krutilina, A. I., and Baev, A. A., 1966, *Dokl. Akad. Nauk SSSR,* **169**:1199.
Mirzabekov, A. D., Grünberger, D., Holy, A., Baev (Bayev), A., and Šorm, F., 1967, *Biochim. Biophys. Acta,* **145**:845.
Mirzabekov, A. D., Kazarinova, L. Ya., and Baev, A. A., 1969a, *Molekul. Biol.,* **3**:879.

Mirzabekov, A. D., Kazarinova, L. Ya., Lastity, D., and Baev (Bayev), A. A., 1969b, *FEBS Letters*, 3:268.
Mirzabekov, A. D., Kazarinova, L. Ya., Lastity, D., and Baev, A. A., 1969c, *Molekul. Biol.*, 3:909.
Mirzabekov, A. D., Lastity, D., and Baev (Bayev), A. A., 1969d, *FEBS Letters*, 4:281.
Mizrabekov, A. D., Lastity, D., Levina, E. S., and Baev (Bayev), A. A., 1970, *FEBS Letters*, 7:95.
Mittelman, A., Hall, R. H., Yohn, D. S., and Grace, J. T., 1967, *Cancer Res.*, 27:1409.
Miura, K., 1964, *J. Mol. Biol.*, 8:371.
Miyazaki, M., and Takemura, S., 1966, *J. Biochem. (Tokyo)*, 60:526.
Miyazaki, M., Kawata, M., and Takemura, S., 1966, *J. Biochem. (Tokyo)*, 60:519.
Miyazaki, M., Kawata, M., Nakazawa, K., and Takemura, S., 1967, *J. Biochem. (Tokyo)*, 62:161.
Mizutani, T., Miyazaki, M., and Takamura, S., 1968, *J. Biochem. (Tokyo)*, 64:839.
Molinaro, M., Scheiner, L. B., Neelon, F. A., and Cantoni, G. L., 1968, *J. Biol. Chem.*, 243:1277.
Morisawa, A., and Chargaff, E., 1963, *Biochim. Biophys. Acta*, 68:147.
Moudrianakis, E. N., and Beer, M., 1965a, *Biochim. Biophys. Acta*, 95:23.
Moudrianakis, E. N., and Beer, M., 1965b, *Proc. Nat. Acad. Sci. (Washington)*, 53:564.
Muench, K. H., and Berg, P., 1966a, *Biochemistry*, 5:970.
Muench, K. H., and Berg, P., 1966b, *Biochemistry*, 5:982.
Müller, P., Wehrli, W., and Staehelin, M., 1971, Biochemistry, 10:1885.
Murao, K., Saneyoshi, M., Harada, F., and Nishimura, S., 1970, *Biochem. Biophys. Res. Commun.*, 38:657.
Muto, A., and Miura, K., cited by Miura, K., 1967, *Progr. Nucleic Acid Res. Mol. Biol.*, 6:39.
Nakada, D., 1965, *J. Mol. Biol.*, 12:695.
Nakanishi, K., Furutachi, N., Funamizu, M., Grünberger, D., and Weinstein, I. B., 1970, *J. Amer. Chem. Soc.*, 92:7617.
Naylor, R., Ho, N. W. Y., and Gilham, P. T., 1965, *J. Amer. Chem. Soc.*, 87:4209.
Neelon, F. A., Molinaro, M., Ishikura, H., Sheiner, L. B., and Cantoni, G. L., 1967, *J. Biol. Chem.*, 242:4515.
Nelson, J. A., Ristow, S. C., and Holley, R. W., 1967, *Biochim. Biophys. Acta*, 149:590.
Neu, H., and Heppel, L. A., 1964, *J. Biol. Chem.*, 239:2927.
Nichols, J. L., 1970, *Nature*, 225:147.
Nichols, J. L., and Lane, B. G., 1966, *Biochim. Biophys. Acta*, 119:649.
Nichols, J. L., and Lane, B. G., 1967, *J. Mol. Biol.*, 30:477.
Nichols, J. L., and Lane, B. G., 1968, *Can. J. Biochem.*, 46:109.
Nihei, T., and Cantoni, G. L., 1962, *Biochim. Biophys. Acta*, 61:463.
Nihei, T., and Cantoni, G. L., 1963, *J. Biol. Chem.*, 238:3991.
Ninio, J., Favre, A., and Yaniv, M., 1969, *Nature*, 223:1333.
Nirenberg, M. W., and Leder, P., 1964, *Science*, 145:1399.
Nirenberg, M., and Matthaei, G. H., 1961, *Proc. Nat. Acad. Sci. (Washington)*, 37:1588.
Nirenberg, M., Leder, P., Bernfield, M., Brimacombe, R., Trupin, J., Rottman, F., and O'Neal, C., 1965, *Proc. Nat. Acad. Sci. (Washington)*, 53:1161.
Nirenberg, M., Caskey, T., Marshall, R., Brimacombe, R., Kellog, R., Doctor, B. P., Hatfield, D., Levin, J., Rottman, F., Pestka, S., Wilcox, M., and Anderson, F., 1966, *Cold Spring Harbor Symp. Quant. Biol.*, 31:11.
Nishimura, S., and Novelli, G. D., 1963, *Biochem. Biophys. Res. Commun.*, 11:161.
Nishimura, S., and Novelli, G. D., 1965, *Proc. Nat. Acad. Sci. (Washington)*, 53:178.
Nishimura, S., Saneyoshi, M., and Harada, D., 1967a, *Abstr. 7th Internat. Congr. Biochem.*, Tokyo, p. 51.
Nishimura, S., Harada, F., Narushima, U., and Seno, T., 1967b, *Biochim. Biophys. Acta*, 142:133.
Nishimura, S., Yamada, Y., and Ishikura, H., 1969, *Biochim. Biophys. Acta*, 179:517.

Norton, J. S., and Lipsett, M. N., 1967, *Federat. Proc.*, **26**:868.
Novgorodsky, A., and Hurwitz, J., 1966, *J. Biol. Chem.*, **241**:2923.
Ochoa, S., and Heppel, L. A., 1957, in: *The Chemical Basis of Heredity* (McElroy and Glass, eds.), Baltimore, p. 631.
Ofengand, J., 1965, *Biochem. Biophys. Res. Commun.*, **18**:192.
Ofengand, J., 1967, *J. Biol. Chem.*, **242**:5034.
Ofengand, J., and Henes, C., 1969, *Federat. Proc.*, **28**:350.
Ofengand, J., Dieckmann, M., and Berg, P., 1961, *J. Biol. Chem.*, **236**:1741.
Ofengand, J., Chu, L., and Schaefer, H., 1966, *Federat. Proc.*, **25**:780.
Ogata, K., and Nohara, H., 1957, *Biochim. Biophys. Acta*, **25**:659.
Ogur, M., and Small J. D., 1960, *J. Biol. Chem.*, **235**:PC60.
Ohashi, Z., Saneyoshi, M., Harada, F., Hara, H., and Nishimura, S., 1970, *Biochem. Biophys. Res. Commun.*, **40**:866.
Ohsaka, A., Mukai, J. I., and Laskowski, M., Sr., 1964, *J. Biol. Chem.*, **239**:3498.
Okamoto, T., and Kawade, Y., 1963, *Biochem. Biophys. Res. Commun.*, **13**:324.
Ornstein, L., 1964, *Ann. New York Acad. Sci.*, **121**:321.
Pearson, R. L., Weiss, J. F., and Kelmers, A. D., 1971, *Biochim. Biophys. Acta*, **228**:770.
Penswick, J. R., and Holley, R. W., 1965, *Proc. Nat. Acad. Sci. (Washington)*, **53**:543.
Peterkofsky, A., 1964, *Proc. Nat. Acad. Sci. (Washington)*, **52**:1233.
Peterkofsky, A., 1966, *Cold Spring Harbor Symp. Quant. Biol.*, **31**:462.
Peterkofsky, A., 1968, *Biochemistry*, **7**:472.
Peterkofsky, A., and Jesensky, C., 1969, *Biochemistry*, **8**:3798.
Peterkofsky, A., and Lipsett, M. N., 1965, *Biochem. Biophys. Res. Commun.*, **20**:780.
Peterkofsky, A., Jesensky, C., Bank, A., and Mehler, A. H., 1964, *J. Biol. Chem.*, **239**:2918.
Peterkofsky, A., Jesensky, C., and Capra, J. D., 1966, *Cold Spring Harbor Symp. Quant. Biol.* **31**:515.
Peterson, E. A., and Sober, H. A., 1956, *J. Amer. Chem. Soc.*, **78**:751.
Phillips, J. H., and Kjellin-Stråby, K., 1967, *J. Mol.. Biol.*, **26**:509.
Phillipsen, P., and Zachau, H. G., 1971, *FEBS Letters*, **15**:69.
Phliipsen, P., Thiebe, R., Wintermeyer, W., and Zachau, H. G., 1968, *Biochem. Biophys. Res. Commun.*, **33**:922.
Pillinger, D., and Borek, E., 1969, *Proc. Nat. Acad. Sci. (Washington)*, **62**:1145.
Pleiss, M., Oghiui, H., and Cerutti, P. A., 1969, *Biochem. Biophys. Res. Commun.*, **34**:70.
Pochon, F., Michelson, A. M., Grunberg-Manago, M., Cohn, W. E., and Dondon, L., 1964, *Biochim. Biophys. Acta*, **80**:441.
Pochon, F., Michelson, A. M., Grunberg-Manago, M., Cohn, W. E., and Dondon, L., 1965, *Biochim. Biophys. Acta*, **108**:194.
Poland, D., Vournakis, J. N., and Scheraga, H. A., 1966, *Biopolymers*, **4**:273.
Portatius, H. von, Doty, P., and Stephenson, M. L., 1961, *J. Amer. Chem. Soc.*, **83**:3351.
Potter, J. L., and Dounce., A. J., 1956, *J. Amer. Chem. Soc.*, **78**:3078.
Pratt, A. W., Toal, J. M., Rushizky, G. W., and Sober, H. A., 1964, *Biochemistry*, **3**:1831.
Preiss, J., Dieckmann, M., and Berg, P., 1961, *J. Biol. Chem.*, **236**:1748.
Price, T. D., Hinds, H. A., and Brown, R. S., 1963, *J. Biol. Chem.*, **238**:311.
RajBhandary, U. L., 1968, *J. Biol. Chem.*, **243**:556.
RajBhandary, U. L., and Chang, S. H., 1968, *J. Biol. Chem.*, **243**:598.
RajBhandary, U. L., and Ghosh, H. P., 1969, *J. Biol. Chem.*, **244**:1104.
RajBhandary, U. L., and Kumar, A., 1970, *J. Mol. Biol.*, **50**:707.
RajBhandary, U. L., and Stuart, A., 1966a, *Ann. Rev. Biochem.*, **35**:759.
RajBhandary, U. L., and Stuart, A., 1966b, *Federat. Proc.*, **25**:520.
RajBhandary, U. L., Young, R. L., and Khorana, H. G., 1964, *J. Biol. Chem.*, **239**:3875.
RajBhandary, U. L., Stuart, A., Faulkner, R. D., Chang, S. H., and Khorana, H. G., 1966, *Cold Spring Harbor Symp. Quant. Biol.*, **31**:425.
RajBhandary, U. L., Chang, S. H., Staurt, A., Faulkner, R. D., Hoskinson, R. M., and Khorana, H. G., 1967, *Proc. Nat. Acad. Sci. (Washington)*, **57**:751.
RajBhandary, U. L., Faulkner, R. D., and Stuart, A., 1968a, *J. Biol. Chem.*, **243**:575.
RajBhandary, U. L., Stuart, A., and Chang, S. H., 1968b, *J. Biol. Chem.*, **243**:584.

RajBhandary, U. L., Stuart, A., Hoskinson, R. M., and Khorana, H. G., 1968c, *J. Biol. Chem,* **243**:565.

RajBhandary, U. L., Chang, S. H., Gross, H. J., Harada, F., Kimura, F., and Nishimura, S., 1969, *Federat. Proc.,* **28**:409.

Rake, A. V., and Tener, G. M., 1966, *Biochemistry,* **5**:3992.

Ralph, R. K., Young, R. J., and Khorana, H. G., 1963, *J. Amer. Chem. Soc.* **85**:2002.

Randerath, E., Ten Broeke, J. W., and Randerath, K., 1968, *FEBS Letters,* **2**:10.

Randerath, K., Flood, K. M., and Randerath, E., 1969, *FEBS Letters,* **5**:31.

Randerath, K., MacKinnon, S. K., and Randerath, E., 1971, *FEBS Letters,* **15**:81.

Raymond, S., and Wang, Y.-J., 1960, *Anal. Biochem.,* **1**:391.

Raymond, S., and Weintraub, L., 1969, *Science,* **130**:711.

Razzell, W. E., and Khorana, H. G., 1961, *J. Biol. Chem.,* **236**:1144.

Reid, J. C., and Pratt, A. W., 1960, *Biochem. Biophys. Res. Commun.,* **3**:337.

Reiner, B., and Zamenhof, S., 1957, *J. Biol. Chem.,* **228**:475.

Revel, M., and Littauer, U. Z., 1966, *J. Mol. Biol.,* **15**:389.

Rice, W. E., and Bock, R. M., 1963, *J. Theoret. Biol.* **4**:260.

Richards, E. G., Coll. J. A., and Gratzer, W. B., 1965, *Anal. Biochem.,* **12**:452

Richardson, C. C., 1965, *Proc. Nat. Acad. Sci. (Washington),* **54**:158.

Robbins, P. W., and Hammond, J. B., 1962, *J. Biol. Chem.,* **237**:PC1379.

Robins, M. J., Hall, R. H., and Thedford, R., 1967, *Biochemistry,* **6**:1837.

Robinson, W. E., Frist , R. H., and Kaesberg, P., 1969, *Science,* **166**:1291.

Rodeh, R., Feldman, M., and Littauer, U. Z., 1967, *Biochemistry,* **6**:451.

Röschenthaler, R., and Fromageot, P., 1965, *J. Mol. Biol.,* **11**:458.

Rottman, F., and Cerutti, P., 1966, *Proc. Nat. Acad. Sci. (Washington),* **55**:960.

Roy-Burman, P., Roy-Burman, S., and Visser, D. W., 1967, *Biochim. Biophys. Acta,* **142**:355.

Rushizky, G. W., and Knight, C. A., 1960, *Virology,* **11**:236.

Rushizky, G. W., and Sober, H. A., 1962, *Biochim. Biophys. Acta,* **55**:217.

Rushizky, G. W., and Sober, H. A., 1963, *J. Biol. Chem.,* **238**:371.

Rushizky, G. W., and Sober, H. A., 1964, *Biochem. Biophys. Res. Commun.,* **14**:276.

Rushizky, G. W., Greco, A. E., Hartley, R. W., and Sober, H. A., 1964, *J. Biol. Chem.,* **239**:2165.

Saneyoshi, M., 1970, *Biochem. Biophys. Res. Commun.,* **40**:1501.

Saneyoshi, M., and Nishimura, S., 1967, *Biochim. Biophys. Acta,* **145**:208.

Saneyoshi, M., Harada, F., and Nishimura, S., 1969, *Biochim. Biophys. Acta,* **190**:264.

Sanger, F., and Brownlee, G. G., 1967, *Methods Enzymol.,* **12**:361.

Sanger, F., Brownlee, G. G., and Barrell, B. G., 1965, *J. Mol. Biol.,* **13**:363.

Sarkar, N., and Comb, D., 1966, *J. Mol. Biol.,***17**:541.

Sato, E., and Kanaoko, Y., 1971, *Biochim. Biophys. Acta,* **232**:213.

Sato, K., and Egami, F., 1957, *J. Biochem. (Tokyo),* **44**:753.

Sawada, F., and Ishii, F., 1969, *J. Biochem. (Tokyo),* **64**:161.

Scanell, J. P., Crestfield, A. M., and Allen, F. W., 1959, *Biochim. Biophys. Acta,* **32**:406.

Scheit, K. H., 1967a, *Angew. Chem.,* **79**:190.

Scheit, K. H., 1967b, *Biochim. Biophys. Acta,* **145**:535.

Scheit, K. H., and Gaertner, E., 1969a, *Biochim. Biophys. Acta,* **182**:1.

Scheit, K. H., and Gaertner, E., 1969b, *Biochim. Biophys. Acta,* **182**:10.

Schleich, T., and Goldstein, J., 1964, *Proc. Nat. Acad. Sci.* (Washington), **52**:744.

Schleich, T., and Goldstein, J., 1965, *Science,* **150**:1168.

Schulman, L. H., and Chambers, R. W., 1968, *Proc. Nat. Acad. Sci. (Washington),* **61**:308.

Schweiger, M., and Zachau, H. G., 1965, *Z. Physiol. Chem.,* **342**:93.

Schweizer, M. P., Chheda, G. B., Baczynskyj, L., and Hall, R. H., 1969, *Biochemistry,* **8**:3283.

Sedat, J. W., and Sinsheimer, R. L., 1964, *J. Mol. Biol.,* **9**:489.

Seifert, W., and Zilling, W., 1967, *Z. Physiol. Chem.,* **348**:1017.

Sekiya, T., Yoshida, M., and Ukita, T., 1967, *Biochim. Biophys. Acta,* **149**:610.

Sekiya, T., Takeishi, K., and Ukita, T., 1969, *Biochim. Biophys. Acta,* **182**:411.

Seno, T., Kobayashi, M., and Nishimura, S., 1969, *Biochim. Biophys. Acta*, **174**:408.

Shakulov, R. S., Aitkhozhin, M. A., and Spirin, A. S., 1962, *Biokhimiya*, **27**:744.

Shapiro, H. S., and Chargaff, E., 1963, *Biochim. Biophys. Acta*, **76**:1.

Shapiro, R., and Chambers, R. W., 1961, *J. Amer. Chem. Soc.*, **83**:3920.

Sharma, O. K., and Borek, E., 1970, *J. Bacteriol.*, **101**:703.

Sheid, B., and Wilson, S. M., 1970a, *Federat. Proc.*, **29**:927.

Sheid, B., and Wilson, S. M., 1970b, *Biochim. Biophys. Acta*, **224**:382.

Shershneva, L. P., Venkstern, T. V., and Baev (Bayev), A. A. 1971, *FEBS Letters*, **14**:297.

Shugar, D., and Fox, J. J., 1952, *Bull Soc. Chim. Belges*, **67**:293.

Shugar, D., and Szer, W., 1962, *J. Mol. Biol.*, **5**:580.

Shugart, L., Novelli, G. D., and Stulberg, M. P., 1968a, *Biochim. Biophys. Acta*, **157**:83.

Shugart, L., Chastain, B. H., Novelli, G. D., and Stulberg, M. B., 1968b, *Biochem. Biophys. Res. Commun.*, **31**:404.

Siddiqui, M. A. Q., and Hosokawa, K., 1969, *Biochem. Biophys. Res. Commun.*, **36**:711.

Siddiqui, M. A. Q., and Ofengand, J., 1971, *FEBS Letters*, **15**:105.

Siddiqui, M. A. Q., Krauskopf, M., and Ofengand, J., 1970, *Biochem. Biophys. Res. Commun.*, **38**:156.

Silber, R., Berman, E., Goldstein, B., Stein, H., Farnham, G., and Bertino, J. R., 1966, *Biochim. Biophys. Acta*, **123**:638.

Silber, R., Goldstein, B., Berman, E., Decter, J., and Friend, C., 1967, *Cancer Res.*, **27**:1264.

Simon, L. N., Glasky, A. J., and Rejal, T.H ., 1967, *Biochim. Biophys. Acta*, **142**:99.

Simon, S., Littauer, U. Z., and Katchalski, E., 1964, *Biochim. Biophys. Acta*, **80**:169.

Simukova, N. A., and Budovskii, É. I., 1970, *Molekul. Biol.*, **4**:213.

Singer, B., and Fraenkel-Conrat, H., 1963, *Biochim. Biophys. Acta*, **72**:534.

Singer, M. F., 1958, *J. Biol. Chem.*, **232**:211.

Singer, M. F., Luborsky, S. W., Morrison, R. A., and Cantoni, G. L., 1960, *Biochim. Biophys. Acta*, **38**:568.

Singh, H., and Lane, B. G., 1964, *Canad. J. Biochem.*, **42**:1011.

Skoog, F., Armstrong, D. J., Cherayil, J. D., Hampel, A. E., and Bock, R. M., 1966, *Science*, **154**:1354.

Sluyser, M., and Bosch, L., 1962, *Biochim. Biophys. Acta*, **55**:479.

Smillie, E. J., and Burdon, R. H., 1970, *Biochim. Biophys. Acta*, **213**:248.

Smith, C. J., and Herbert, E., 1965, *Science*, **150**:384.

Smith, C. J., Herbert, E., and Wilson, C. W., 1964, *Biochim. Biophys. Acta*, **87**:341.

Smith, J. D., and Dunn, D. B., 1959a, *Biochim. Biophys. Acta* **31**:573.

Smith, J. D., and Dunn, D. B., 1959b, *Biochem. J.*, **72**:294.

Smith, J. D., Abelson, J. N., Clark, B. F. C., Goodman, H. M., and Brenner, S., 1966, *Cold Spring Harbor Symp. Quant. Biol.*, **31**:479.

Smith, J. D., Abelson, J. N., Barnett, L., Brenner, S., Gefter, M., Landy, A., and Russell, R., 1969, *Abstr. 6th Meeting FEBS*, Madrid, p. 60.

Smith, J. D., Barnett, L., Brenner, S., and Russell, R. L., 1970, *J. Mol. Biol.*, **54**:1.

Smith, J. D., Anderson, K., Cashmore, A., Hooper, M. L., and Russell, R. L., 1971, *Cold Spring Harbor Symp. Quant. Biol.*, **35**:21.

Smith, K. C., and Allen, F. W., 1953, *J. Amer. Chem. Soc.*. **75**:2131.

Smrt, J., 1966, *Coll. Czech. Chem. Commun.*, **32**:198.

Smrt, J., Skoda, J., Lisy, V., and Šorm, F., 1966, *Biochim. Biophys. Acta*, **129**:210.

Söll, D., Jones, D. S., Ohtsuka, E., Faulkner, R. D., Lohrmann, R., Hayatsu, H., Khorana, H. G., Cherayil, J. D., Hampel, A., and Bock, R. M., 1966a, *J. Mol. Biol.*, **19**:556.

Söll, D., Cherayil, J., Jones, D. S., Faulkner, R. D., Hampel, A., Bock, R. M., and Khorana, H. G., 1966b, *Cold Spring Harbor Symp. Quant. Biol.*, **31**:51.

Spirin, A. S., and Gavrilova, L. D., 1968, in: *The Ribosome*, Nauka, Moscow, p. 53 (in Russian).

Squires, C., and Carbon, J., 1971, *Nature, New Biol.*, **233**:274.

Srinivasan, P. R., and Borek, E., 1963, *Proc. Nat. Acad. Sci. (Washington)*, **49**:529.

Srinivasan, P. R., and Borek, E., 1966, *Progr. Nucleic Acid Res. Mol. Biol.*, **5**:157.

Staehelin, M., 1961, *Biochim. Biophys. Acta*, **49**:20.
Staehelin, M., 1963, *Progr. Nucleic Acid Res. Mol. Biol.*, **2**:169.
Staehelin, M., 1964a, *J. Mol. Biol.*, **8**:470.
Staehelin, M., 1964b, *Helv. Physiol. Acta*, **22**:C43.
Staehelin, M., 1971, *Experientia*, **27**:1.
Staehelin, M., Zachau, H. G., and Schweiger, M., 1962, *Biochem. J.*, **84**:107P.
Staehelin, M., Schweiger, M., and Zachau, H. G., 1963, *Biochim. Biophys. Acta*, **68**:129.
Staehelin, M., Rogg, H., Baguley, B. C., Ginsberg, T., and Wehrli, W., 1968, *Nature*, **219**: 1363.
Staehelin, M., Rouge, M., and Wehrli, W., 1969, *Abstr. 6th Meeting FEBS*, Madrid, p.. 60
Stanley, W. M., and Bock, R. M., 1965, *Anal. Chem.*, **13**:43.
Starr, J. L., 1963, *Biochem. Biophys. Res. Commun.*, **10**:428.
Steinschneider, A., and Fraenkel-Conrat, H., 1966, *Biochemistry*, **5**:2735.
Steitz, J. A., 1969, *Nature*, **224**:957.
Stent, G. S., 1964, *Science*, **144**:816.
Stephenson, M. L., and Zamecnik, P. G., 1961, *Proc. Nat. Acad. Sci. (Washington)*, **47**: 1627.
Stewart, T. S., Roberts, R. J., and Strominger, J. L., 1971, *Nature*, **230**:36.
Sueoka, N., and Yamane, T., 1962, *Proc. Nat. Acad. Sci. (Washington)*, **48**:1454.
Sugiura, M., and Takanami, M., 1967, *Proc. Nat. Acad. Sci. (Washington)*, **58**:1595.
Sugiyama, T., 1965, *J. Mol. Biol.*, **11**:856.
Sulkowski, E., and Laskowski, M., 1962, *J. Biol. Chem.*, **237**:2620.
Sulkowski, E., and Laskowski, M., 1966, *J. Biol. Chem.*, **241**:4386.
Sundharadas, G., Katze, J. R., Söll, D., Konigsberg, W., and Lengyel, P., 1968, *Proc. Nat. Acad. Sci. (Washington)*, **61**:693.
Suyama, Y., and Eyer, J., 1967, *Biochem. Biophys. Res. Commun.*, **28**:746.
Suzuki, T., and Ito, E., 1958, *J. Biochem. (Tokyo)*, **45**:403.
Svensson, I., Boman, H. G., Eriksson, K. G., and Kjellin, K., 1963, *J. Mol. Biol.*, **7**:254.
Szekely, M., and Sanger, F., 1969, *J. Mol. Biol.*, **43**:607.
Szer, D., 1965, *Biochem. Biophys. Res. Commun.*, **20**:182.
Szer, W., 1966, in: *Genetic Elements*, Academic Press, New York and London, p. 330.
Szer, W., and Shugar, D., 1961, *Acta Biochem. Polon.*, **8**:235.
Tada, M., Schweiger, M., and Zachau, H. G., 1962, *Z. Physiol. Chem.*, **328**:85.
Takanami, M., 1967a, *J. Mol.. Biol.*, **29**:323.
Takanami, M., 1967b, *J. Mol. Biol.*, **23**:135.
Takanami, M., Yan, Y., and Jukes, T. H., 1965, *J. Mol. Biol.*, **12**:761.
Takemura, S., and Miyazaki, M., 1969, *J. Biochem. (Tokyo)*, **65**:159.
Takemura, S., Miyazaki, M., Kawata, M., Mizutani, T., Hashimoto, S., and Murakami, M., 1967, *Abstr. 7th Internat. Congr. Biochem.*, Tokyo, p. 49.
Takemura, S., Mizutani, T., and Miyazaki, M., 1968a, *J. Biochem. (Tokyo)*, **63**:277.
Takemura, S., Mizutani, T., and Miyazaki, M., 1968b, *J. Biochem. (Tokyo)*, **64**:827.
Takemura, S., Murakami, M., and Miyazaki, M., 1969a, *J. Biochem. (Tokyo)*, **65**:489.
Takemura, S., Murakami, M., and Miyazaki, M., 1969b, *J. Biochem. (Tokyo)*, **65**:553.
Tamaoki, T., and Lane, B. G., 1967, *Biochemistry*, **6**:3583.
Tanaka, K., and Cantoni, G. L., 1963, *Biochim. Biophys. Acta*, **72**:641.
Tanaka, K., Richards, H. H., and Cantoni, G. L., 1962, *Biochim. Biophys. Acta*, **61**:846.
Tatarskaya, R. I., Abrosimova-Amel'yanchik, N. M., Aksel'rod, V. D., Korenyako, A. I., Venkstern, T. V., Mirzabekov, A. D., and Baev, A. A., 1964, *Dokl. Akad. Nauk SSSR*, **157**:725.
Tatarskaya, R. I., Abrosimova-Amel'yanchik, N. M., Aksel'rod, V. D., Korenyako, A. I., Niedra, N. Ya., and Baev, A. A., 1966a, *Biokhimiya*, **31**:1017.
Tatarskaya, R. I., Korenyako, A. I., and Abrosimova-Amel'yanchik, N. M., 1966b, *Priklad. Biokhim. i Mikrobiol.*, **2**:151.
Tatarskaya, R. I., Lvova, T. N., Abrosimova-Amel'yanchik, N. M., Korenyako, A. I., and Baev (Bayev), A. A., 1970, *Europ. J. Biochem.*, **15**:442.

Thiebe, R., and Zachau, H. G., 1965, *Biochim. Biophys. Acta*, **103**:368,
Thiebe, R., and Zachau, H. G., 1968a, *Europ. J. Biochem.*, **5**:546.
Thiebe, R., and Zachau, H. G., 1968b, *Biochem. Biophys. Res. Commun.*, **33**:260.
Thiebe, R., and Zachau, H. G., 1969a, *Biochem. Biophys. Res. Commun.*, **36**:1024.
Thiebe, R., and Zachau, H. G., 1969b, *FEBS Letters*, **5**:15.
Thirion, J.-P., and Kaesberg, P., 1968a, *Biochim. Biophys. Acta*, **161**:247.
Thirion, J.-P., and Kaesberg, P., 1968b, *J. Mol. Biol.*, **33**:379.
Tomasz, M., and Chambers, R. W., 1964, *J. Amer. Chem. Soc.*, **86**:4216.
Tomasz, M., and Chambers, R. W., 1965, *Biochim. Biophys. Acta*, **108**:510.
Tomasz, M., and Chambers, R. W., 1966, *Biochemistry*, **5**:773.
Tomasz, M., Sanno, Y., and Chambers, R. W., 1965, *Biochemistry*, **4**:1710.
Tomlinson, R. V., and Tener, G. M., 1962, *J. Amer. Chem. Soc.*, **84**:2644.
Tomlinson, R. V., and Tener, G. M., 1963a, *Biochemistry*, **2**:697.
Tomlinson, R. V., and Tener, G. M., 1963b, *Biochemistry*, **2**:703.
Townsend, L. B., and Robins, R. K., 1963, *J. Amer. Chem. Soc.*, **85**:242.
Tsutsui, E. A., Srinivasan, P. R., and Borek, E., 1966, *Proc. Nat. Acad. Sci. (Washington)*, **56**:1003.
Tsutsui, E. A., Srinivasan, P. R., and Borek, E., 1967, *Abstr. 7th Internat. Congr. Biochem.*, Tokyo, p. 642.
Tumaitis, T. D., and Lane, B. G., 1970, *Biochim. Biophys. Acta*, **224**:391.
Tumanyan, V. G., and Kiselev, L. L., 1963, *Biofizika*, **8**:147.
Tyndall, R. L., Jacobson, K. B., and Teeter, E., 1964, *Biochim. Biophys. Acta*, **87**:335.
Uchida, T., Arima, T., and Egami, F., 1970, *J. Biochem. (Tokyo)*, **67**:91.
Udenfriend, S., and Zaltzman, P., 1962, *Analyt. Biochem.*, **4**:349.
Udenfriend, S., Zaltzman-Nirenberg, P., and Cantoni, G. L., 1963, *Analyt. Biochem.*, **5**:258.
Uhlenbeck, O. C., Baller, J., and Doty, P., 1970, *Nature*, **225**:508.
Ukita, T., 1967, *Abstr. 7th Intern. Congr. Biochem.*, Tokyo, p. 49.
Ukita, T., Kikugawa, K. K., and Hayatsu, H., 1964, *Abstr. 6th Internat. Congr. Biochem.*, New York, p. 91.
Ulanov, B. P., Serebryanyi, A. M., and Kostyanovskii, R. G., 1967, *Dokl. Akad. Nauk SSSR*, **176**:474.
Uziel, M., and Cohn, W. E., 1965, *Biochim. Biophys. Acta*, **103**:539.
Uziel, M., and Gassen, W. E., 1969, *Federat. Proc.*, **28**:419.
Uziel, M., and Khym, J. X., 1969, *Biochemistry*, **8**:3254.
Vandenbusche, P., and Fiers, W., 1966, *Biochim. Biophys. Acta*, **114**:182.
Van Holde, K. E., Brahms, J., and Michelson, A. M., 1965, *J. Mol. Biol.*, **12**:762.
Vanyushin, B. F., Belozersky, A. N., Kokurina, N. A., and Kadirova, D. X., 1968, *Nature*, **218**:1066.
Vasilenko, S. K., 1964, *Biokhimiya*, **29**:1190.
Vasilenko, S. K., 1965, Dissertation for the degree of Candidate, Novosibirsk.
Vasilenko, S. K., 1968, in: *Third Conference on Nucleotide Sequence in the Transfer RNAs*, Kiev (in Russian).
Vasilenko, S. K., Demushkin, V. P., Budovskii, E. I., and Knorre, D. G., 1965, *Dokl. Akad. Nauk SSSR*, **162**:694.
Vasilenko, S. K., Dimitrova, F. F., and Chernaenko, V. M., 1969, *Molekul. Biol.*, **3**:228.
Vasilenko, S. K., Dimitrova, F. F., Obukhova, L. V., Podgornyi, V. F., and Serbo, N. A., 1970, *Molekul. Biol.*, **4**:205.
Venkstern, T. V., 1964, *Uspekhi Biol. Khimii*, **6**:3.
Venkstern, T. V., 1966, *Dokl. Akad. Nauk SSSR*, **170**:718.
Venkstern, T. V., 1967, *Uspekhi Biol. Khimii*, **8**:3.
Venkstern, T. V., and Baev, A. A., 1968, *Spectra of Nucleic Acid Compounds*, Plenum Press, New York.
Venkstern, T. V., Baev, A.A., Mirzabekov, A. D., and Gorshkova, V. I., 1963a, *Dokl. Akad. Nauk SSSR*, **151**:220.
Venkstern, T. V., Mirzabekov, A. D., Gorshkova, V. I., and Baev, A. A., 1963b, *Biokhimiya*, **28**:712.

Venkstern, T. V., Krutilina, A. I., Li, L., Aksel'rod, V. D., Mirzabekov, A. D., and Baev, A. A., 1968a, *Molekul. Biol.*, 2:394.

Venkstern, T. V., Li, L., Krutilina, A. I., Mirzabekov, A. D., Aksel'rod, V. D., and Baev, A. A., 1968b, *Molekul. Biol.*, 2:597.

Verwoerd, D. W., Kohlhage, H., and Zillig, W., 1961, *Nature*, 192:1038.

Verwoerd, D. W., and Zillig, W., 1963, *Biochim. Biophys. Acta*, 68:484.

Viale, G. L., Restelli, A. F., and Viale, E., 1967, *Tumori*, 53:533.

Vickers, J. D., and Logan, D. M., 1970, *Biochem. Biophys. Res. Commun.*, 41:741.

Vournakis, J. N., and Scheraga, H. G., 1966, *Biochemistry*, 5:2997.

Wainfan, E., Srinivasan, P. R., and Borek, E., 1966, *Cold Spring Harbor Symp. Quant. Biol.*, 31:525.

Walker, R. T., and RajBhandary, U. L., 1970, *Biochem. Biophys. Res. Commun.*, 38:907.

Warner, R. C., and Waimberg, P., 1958, *Federat. Proc.*, 17:331.

Warshaw, M. W., and Tinoco, J., Jr., 1965, *J. Mol.. Biol.*, 13:54.

Watanabe, K. A., and Fox, J. J., 1966, *Angew. Chem.*, 78:589.

Weber, K., 1967, *Biochemistry*, 6:3144.

Webster, R. E., Engelhardt, D. L., and Zinder, N. D., 1966, *Proc. Nat. Acad. Sci. (Washington)*, 55:155.

Weil, J. H., Befort, N., Rether B., and Ebel, J. P., 1964, *Biochem. Biophys. Res. Commun.*, 15:447.

Weisberger, A. S., Suhrland, L. G., and Griggs, R. C., 1954, *Blood*, 9:1095.

Weisblum, B., Benzer, S., and Holley, R. W., 1962, *Proc. Nat. Acad. Sci. (Washington)*, 48:1448.

Weiss, J. F., and Kelmers, A. D., 1967, *Biochemistry*, 6:2507.

Weiss, J. F., Pearson, R. L., and Kelmers, A. D., 1968, *Biochemistry*, 7:3479.

Weiss, S. B., and Legault-Demare, J., 1965, *Science*, 149:429.

Weith, H. L., and Gilham, P. T., 1967, *J. Amer. Chem. Soc.* 89:5473.

Weith, H. L., and Gilham, P. T, 1969, *Science*, 166:1004.

Wheeler, H. L., and Johnson, T. B., 1907, *J. Biol. Chem.*, 3:183.

Whitfeld, P. R., 1965, *Biochim. Biophys. Acta*, 108:202.

Whitfeld, P. R., and Markham, R., 1953, *Nature*, 171:1151.

Whitfeld, P. R., and Witzel, H., 1963, *Biochim. Biophys. Acta*, 72:338.

Wintermeyer, W., and Zachau, H. G., 1970, *FEBS Letters*, 11:160.

Witzel, H., 1963, *Progr. Nucleic Acid Res. Mol. Biol.*, 2:221.

Witzel, H., and Barnard, E. A., 1962, *Biochem. Biophys. Res. Commun.*, 7:289.

Wong, T., Weiss, S. B., Eliceiri, G. L., and Bryant, J., 1970, *Biochemistry*, 9:2376.

Yamada, Y., Murao, K., Saneyoshi, M., and Nishimura, S., 1969, *Biochem. Biophys. Res. Commun.*, 37:990.

Yamada, Y., Saneyoshi, M., Nishimura, S., and Ishikura, H., 1970, *FEBS Letters*, 7:207.

Yamane, T., and Sueoka, N., 1964, *Proc. Nat. Acad. Sci. (Washington)*, 51:1178.

Yaniv, M., and Barrell, B. G., 1969, *Nature*, 222:278.

Yaniv, M., and Barrell, B. G., 1971, *Nature, New Biol.*, 233:113.

Yaniv, M., Favre, A., and Barrell, B. G., 1969, *Nature*, 223:1331.

Yoshida, M., and Ukita, T., 1965a, *J. Biochem.* 57:818.

Yoshida, M., and Ukita, T., 1965b, *J. Biochem.*, 58:191.

Yoshida, M., and Ukita, T., 1966, *Biochim. Biophys. Acta*, 123:214.

Yoshida, M., and Ukita, T., 1968, *Biochim. Biophys. Acta*, 157:466.

Yoshida, M., Furuichi, Y., Ukita, T., and Kaziro, Y., 1967a, *Biochim. Biophys. Acta*, 149:308.

Yoshida, M., Ukita, T., Kaziro, Y., Hayashi, Y., and Miura, K., 1967b, *Abstr. 7th Internat. Congr. Biochem., Tokyo*, p. 670.

Yoshida, M., Furuichi, Y., Kaziro, Y., and Ukita, T., 1968a, *Biochim. Biophys. Acta*, 166:636.

Yoshida, M., Kaziro, Y., and Ukita, T., 1968b, *Biochim. Biophys. Acta*, 166:646.

Yoshida, M., Takeishi, K., and Ukita, T., 1970, *Biochem. Biophys. Res. Commun.*, 39:852.

Yoshida, M., Takeishi, K., and Ukita, T., 1971, *Biochim. Biophys. Acta*, 228:153.

Yoshikami, D., and Keller, E.B., 1969, *Federat. Proc.*, **28**:409.
Yoshikami, D., Katz, G., Keller, E. B., and Dudock, B. S., 1968, *Biochim. Biophys. Acta*, **166**:714.
Young, R. J., and Fraenkel-Conrat, H., 1971, *Biochim. Biophys. Acta*, **228**:446.
Yu, C.-T., and Allen, F. W., 1959, *Biochim. Biophys. Acta*, **32**:393.
Yu, C.-T., and Zamecnik, P. C., 1960, *Biochim. Biophys. Acta*, **45**:148.
Zachau, H. G., 1964, *Z. Physiol. Chem.*, **336**:176.
Zachau, H. G., 1965, *Z. Physiol. Chem.*, **342**:98.
Zachau, H. G., 1968a, *Structure and Function of tRNA and 5S RNA*, Universitätsforlaget, Oslo.
Zachau, H. G., 1968b, *Europ. J. Biochem.*, **5**:559.
Zachau, H. G., 1969, *Angew. Chem.*, **81**:645.
Zachau, H. G., Tada, M., Lawson, W. B., Schweiger, M., 1961, *Biochim. Biophys. Acta*, **53**:221.
Zachau, H. G., Schweiger, M., and Melchers, F., 1962, *Angew. Chem.*, **74**:515.
Zachau, H. G., Dütting, D., and Feldmann, H., 1966a, *Angew. Chem.*, **78**:392.
Zachau, H. G., Dütting, D., and Feldmann, H., 1966b, *Z. Physiol. Chem.*, **347**:212.
Zachau, H. G., Dütting, D., and Feldmann, H., 1966c, *Abstr. 3rd Meeting FEBS*, Warsaw, p. 7.
Zachau, H. G., Dütting, D., Feldmann, H., Melchers, F., and Karau, W., 1966d, *Cold Spring Harbor Symp. Quant. Biol.*, **31**:417.
Zachau, H. G., Riesner, D., Römer, R., and Maass, G., 1969, *FEBS Letters*, **5**:23.
Zaitseva, G. N., Dmitrieva, T. M., Hsü Ch'ang-fa, and Belozerskii, A. N., 1962, *Dokl. Akad. Nauk SSSR*, **147**:1211.
Zaitseva, G. N., Pang T'ing-chao, Kalyuzhnaya, A. P., and Belozerskii, A. N., 1964, *Biokhimiya*, **29**:1150.
Zaitseva, G. N., Glebov, R. N., Matveeva, L. I., Khvoika, L. A., and Belozerskii, A. N., 1966, *Biokhimiya*, **31**:740.
Zakharyan, R. A., Venkstern, T. V., and Baev, A. A., 1967, *Biokhimiya*, **32**:1068.
Zakharyan, R. A., Venkstern, T. V., and Baev, A. A., 1968, *Biokhimiya*, **33**:111.
Zamecnik, P. G., Stephenson, M. J., and Scott, J. F., 1960, *Proc. Nat. Acad. Sci. (Washington)*, **46**:811.
Zamir, A., Holley, R. W., and Marquisee, M., 1965, *J. Biol. Chem.*, **240**:1267.
Zavil'gel'skii, G. B., and Li, L., 1967, *Molekul. Biol.*, **1**:323.
Zavil'gel'skii, G. B., Venkstern, T. V., and Baev, A. A., 1966, *Dokl. Akad. Nauk SSSR*, **166**:978.
Zhilyaeva (Jilyaeva), T. I., and Kiselev, L. L., 1970, *FEBS Letters*, **10**:224.
Zhilyaeva, T. I., Tatarskaya, R. I., and Kiselev, L. L., 1970, *Molekul. Biol.*, **5**:139.
Ziff, E. B., and Fresco, J. R., 1969, *Biochemistry*, **8**:3242.
Zillig, W., Verwoerd, D. W., and Kholkage, H., 1962, *Acides ribonucléiques et polyphosphates. Structure, synthèse et fonction*, Paris, p. 229.
Zubay, G., 1962, *J. Mol. Biol.*, **4**:347.
Zubay, G., and Takanami, M., 1964, *Biochem. Biophys. Res. Commun.*, **15**:207.

INDEX